Communications in Computer and Information Science 473

T0215137

Leon Shyue-Liang Wang Jason J. June
Chung-Hong Lee Koji Okuhara
Hsin-Chang Yang (Eds.)

Multidisciplinary
Social Networks Research

International Conference, MISNC 2014
Kaohsiung, Taiwan, September 13-14, 2014
Proceedings

 Springer

Volume Editors

Leon Shyue-Liang Wang
National University of Kaohsiung, Taiwan
E-mail: slwang@nuk.edu.tw

Jason J. June
Chung-Ang University, Seoul, Korea
E-mail: j2jung@gmail.com

Chung-Hong Lee
National Kaohsiung University of Applied Sciences, Taiwan
E-mail: leechung@mail.ee.kuas.edu.tw

Koji Okuhara
Osaka University, Japan
E-mail: okuhara@ist.osaka-u.ac.jp

Hsin-Chang Yang
National University of Kaohsiung, Taiwan
E-mail: yanghc@nuk.edu.tw

ISSN 1865-0929   e-ISSN 1865-0937
ISBN 978-3-662-45070-3   e-ISBN 978-3-662-45071-0
DOI 10.1007/978-3-662-45071-0
Springer Heidelberg New York Dordrecht London

Library of Congress Control Number: 2014950067

Typesetting: Camera-ready by author, data conversion by Scientific Publishing Services, Chennai, India

Printed on acid-free paper

Springer is part of Springer Science+Business Media (www.springer.com)

Preface

Welcome to the proceedings of MISNC 2014, the 2014 Multidisciplinary International Social Networks Conference. The conference, at its inauguration, was established to provide an international platform to promote the collaboration and exchange of ideas and practices among researchers, professionals, experts, and students from all fields related to the study and application of social networks.

The conference was founded by a group of international scholars and professionals as well as members of Taiwanese Association for Social Networks (TASN) and their associates. The initial topics of interest for the 2014 conference included: business, management and social networks, social network issues in sociology, politics and statistics, information technology for social networks analysis and mining, social networks for global e-health and biomedics, and other topics of general interest.

This year, there were participants from 11 countries/regions around the world, and we invited three keynote speakers from the USA and Taiwan. The venue of the conference was the new waterfront exhibition landmark in the ocean city of Taiwan – the Kaohsiung Exhibition Center.

The success of MISNC 2014 depended on the people and organizations that support it. We must thank all the volunteers who helped organize this conference. In particular, we thank the honorary chairs Prof. Jiawei Han of the University of Illinois at Urbana-Champaign, USA, and President Cheng-Hong Yang of the National Kaohsiung University of Applied Sciences, Taiwan, for their kind support. We thank the program chairs — Prof. Jason J. June of Yeungnam University, Korea, Prof. Chung-Hong Lee of the National Kaohsiung University of Applied Sciences, Taiwan, and Prof. Koji Okuhara of Osaka University, Japan — who together with the Program Committee created a great technical program. In addition, we would like to thank the publicity chairs (Piotr Bródka, Han-Fen Hu, Luca Rossi), financial chair (Shu-Chen Yang), publications chairs (Hsin-Chang Yang), Steering Committee chairs (Christine Largeon, Johann Stan, I-Hsien Ting, Kai Wang), track chairs (Hsin-Lu Chang, Kai Wang, Chyi-In Wu, Christine Largeon, Jerry Chun-Wei Lin), and last but not least the local organization team.

The conference would not have been possible without partners and sponsors. The partners were Multigraphia and Intellinium and the sponsors included the IEEE Technical Committee of Granular Computing, IEEE Computational Intelligence Society, European Research Centre of Network Intelligence for Innovation

Enhancement (ENGINE), National University of Kaohsiung and its Social Network Innovation Center (SNIC), Ministry of Science and Technology of ROC, Taiwanese Association for Social Networks, and Harbin Institute of Technology, China.

August 2014   Leon Shyue-Liang Wang
                                                             Jason J. June
                                                           Chung-Hong Lee
                                                             Koji Okuhara
                                                          Hsin-Chang Yang

Organization

Honorary Chairs

Jiawei Han   University of Illinois at Urbana-Champaign, USA

Cheng-Hong Yang   National Kaohsiung University of Applied Sciences, Taiwan

General Chair

Leon Shyue-Liang Wang   National University of Kaohsiung, Taiwan

Program Chairs

Jason J. June   Chung-Ang University, Korea

Chung-Hong Lee   National Kaohsiung University of Applied Sciences, Taiwan

Koji Okuhara   Osaka University, Japan

Publicity Chairs

Piotr Bródka   Wroclaw University of Technology, Poland

Han-Fen Hu   University of Nevada, Las Vegas, USA

Luca Rossi   IT University of Copenhagen, Denmark

Financial Chair

Shu-Chen Yang   National University of Kaohsiung, Taiwan

Publication Chair

Hsin-Chang Yang   National University of Kaohsiung, Taiwan

Steering Committee Chairs

Christine Largeon   University of Jean-Monnet at St. Etienne, France

Johann Stan   National Institutes of Health, Washington D.C., USA

I-Hsien Ting   National University of Kaohsiung, Taiwan
Kai Wang   National University of Kaohsiung, Taiwan

Local Chairs

Hsin-Chang Yang   National University of Kaohsiung, Taiwan
I-Hsien Ting   National University of Kaohsiung, Taiwan

Track Chairs

Electronic Commerce, e-Business Management, and Social Networks Track

Hsin-Lu Chang   National Chengchi University, Taiwan
Kai Wang   National University of Kaohsiung, Taiwan
Kai-Yu Wang   Brock University, Canada

Social Networks Issues on Sociology, Politics and Statistics Track

Chyi-In Wu   Academia Sinica, Taiwan

Information Technology for Social Networks Analysis and Mining Track

Christine Largeon   University of Jean-Monnet at St. Etienne,
                                        France

Special Session Intelligent Data Analysis and its Applications

Jerry Chun-Wei Lin   Harbin Institute of Technology, China

Track 4: Social Networks for Global e-Health and Biomedics

Johann Stan   NIH, USA

Program Committee

Ajith Abraham   Machine Intelligence Research Labs
                                        (MIR Labs), USA
Rayner Alfred   Universiti Malaysia Sabah, Malaysia
Marenglen Biba   University of New York in Tirana, USA
Maria Bielikova   Slovak University of Technology in Bratislava,
                                        Slovakia
Guido Barbian Leuphana   University of Lüneburg, Germany

Table of Contents

Electronic Commerce, e-Business Management, and Social Networks

Social Networks Issues on Sociology, Politics and Statistics

Information Technology for Social Networks Analysis and Mining

Social Networks for Global eHealth and Bio-medics

Security, Open Data, E-Learning and Other Related Topics

Intelligent Data Analysis and Its Applications

Social Network Advertising: An Investigation of Its Impact on Consumer Behaviour

Hokyin Lai[*], Hiufung Cheng, and Hiuping Fong

Department of Computer Science,
Hong Kong Baptist University, Hong Kong

jeanlai@comp.hkbu.edu.hk, hfcheng5@gmail.com,
fpinky2006@hotmail.com

Abstract. The rise of Social Networking Services (SNSs) has not only transformed people as well as consumer behavior on the Internet, but also transformed the means by which various enterprises globally conduct their promotional and marketing campaigns. There are a variety of means by which enterprises have launched their marketing campaigns on Social Networking Services, and one of the most common techniques adopted is through extensive advertising campaigns on SNSs. This study seeks to examine consumer behaviors towards advertisements on Social Networking Services. Key factors affecting consumer behaviors include usage pattern, the credibility of a particular Social Networking Service as well as electronic word-of-mouth. This clearly illustrates that in today's virtual electronic world, social media have progressed from being merely a place to meet people, to being a virtual sales floor. It is unexpected that consumer behaviors are influenced by the electronic word-of-mouth of friends rather than that of strangers.

Keywords: Social Network Advertising, Social Network, Consumer Behaviour, Electronic word-of-mouth (e-WOM), Community of Interest.

1 Introduction

Over the last decade, Social Networking Services (SNSs) have become progressively better received and accepted for they allow users to engage in data sharing, communication, in addition to keeping in touch with community (Kim, Shim and Ahn, 2011). People tend to spend more time on SNSs rather than on traditional media outlets, such as print media, and broadcast news.

Given the extraordinary increase in popularity of Social Networking Sites, organizations globally have become aware of the vast potential commercial uses and business opportunities they provide (Clemons, Barnett and Appadurai, 2007). For instance, SNSs provide a secure environment for business organizations to sustain operations, by keeping track of user behavior patterns through methods such as

---

[*] Corresponding author.

L.S.-L. Wang et al. (Eds.): MISNC 2014, CCIS 473, pp. 1–13, 2014.
© Springer-Verlag Berlin Heidelberg 2014

discussion posts on SNSs (McClure, 2009). With the help of web 2.0, companies are currently qualified to create applications that allow users to retrieve their business information by forming partnerships with SNSs (Lytras and Pablos, 2009). Armed with the ability to share information, SNSs provide advantages for businesses without geographical constraints (Cao, Knotts, Xu and Chau, 2009). The Social Networking Site acts as a middleman to assist companies in targeting users through accessing and utilizing their personal information from users' individual profiles on Social Networking Sites (Gregurec, Vranesevic and Dobrinic, 2011). Information obtained from user profiles pertaining to demographic information, lifestyles, behaviors and habits that are key indicators of consumer behavior, can be turned into advantageous business opportunities. This definitely portrays SNSs as the new, innovative medium for advertising. Advertisers can very effortlessly target their potential market based on the provided information on SNSs (Hoy and Milne, 2010). Increasingly, as SNSs enable users to expand their social networks, social advertising appears to be the latest ingenious tool for reaching and engaging consumers (Nielsen, 2009).

It is common practice for companies to advertise on websites. The same is true for Social Networking Sites. In fact, a major source of income for SNSs is from advertising (Learmonth and Klaassen, 2009; Sledgianowski and Kulviwat, 2009). For instance, Facebook generated 85% of its revenue from online advertising, and is expected to make over $3.8 billion this year from advertising to its 800 million members (The Economist, 2012). Revenue from advertising on Social Networking Sites is ever-increasing and has attracted the attention of scholars as well as astute businesses, not only in terms of how the advertising works, but also how it works best.

This study aims to develop a new conceptual information communication model to investigate factors such as SNSs Usage Pattern, Credibility of SNSs and electronic Word-of-Mouth, affecting the effectiveness of social network advertising. The study also seeks to examine how these factors may prompt audiences to take further action, for example actual purchase of the product, or the action of investigating, with the purpose of finding more information on a product they like, etc. 97 participants were randomly picked up to respond on a questionnaire survey. This study endeavors to find out the key factors that determine why users have intention to take further action when they view an advertisement. The study would aim to benefit Information Systems and marketing scholars as well as business organizations. The study would make it easier for them to understand the types of users they should target in order to have an exceptionally effective marketing campaign.

2   Online Advertising on Social Networking Services

There are varied forms of online advertising, which aim to convey marketing messages to prospective customers. One very distinctive advantage of online advertising is that consumers possess the ability to determine whether they wish to check out the product being advertised or completely ignore the advertisement.

Advertisers also endeavor to enhance advertising campaigns by changing advertising formats and their representation. For example, advertisers have now merged couponing with social networks for marketing purposes, in order to target external consumers (Hayes, 2011). One of the mechanisms by which Groupon (a listed couponing company) operates is to attract a large group of consumers to purchase their vouchers through Social Networking Sites, as SNSs provide a huge data base of potential consumers, who have extensive links within their social network communities. Facebook recently announced that it would launch a new feature called "Buy With Friends", which has a coupon-like characteristic that enables users to get special discounts by sharing special offers from internet companies with their social network community friends. However, it would appear that this "sharing" would be in the virtual world of Facebook and not sharing in the real sense of the word "share." This experiment by Facebook seeks to emulate Groupon and yet it is not exactly like Groupon (Purewal, 2011).

3   Hypothesis Development

3.1   Intention for Further Action

The attempt to evaluate the marketing effectiveness of Social Networking Services is profoundly affected by the criteria prevalent in traditional marketing, as there are still no commonly agreed criteria and standards by which to appraise and evaluate marketing effectiveness on SNSs. In traditional advertising, various evaluation criteria such as expenditure (Chung and Kaiser, 1999), advertising content, organization of content, and the strategy of transmitting marketing messages, like messages pertaining to health or messages relating to entertainment (Ducoffe, 1996; Brackett and Carr, 2001), are studied. However, these types of traditional criteria fail to take into consideration the unique features of SNSs, such as peer-to-peer interactions, dialog participation, etc (Thorne, 2008; Owyang, 2010).

To limit the scope of the study, this study endeavors to observe how different constructs are related to advertising effectiveness. Advertising effectiveness relates to whether a consumer likes, and then chooses or purchases the advertised product due to the advertising campaign (Thomas, 2007; Strategic Marketing and Research, 2007). However, as this definition is not unequivocally specific on the extent of effectiveness and may generate some ambiguity on how to measure effectiveness, we therefore, present another definition for advertising effectiveness. We define advertising effectiveness as that juncture or point in time, where there is intention on the part of consumers for further action, no matter what that action may be. Further action on the part of consumers may include actual purchase or acquisition of the product, further exploration about product information, forwarding the product advertisement verbally or on SNSs to friends, joining a web-site or forum as an admirer of the product's Company, and numerous similar actions. In other words, intention for further action is the only dependent variable in this study.

3.2   Usage Patterns on Social Networking Services

Social Networking Sites do more than just connect people all over the world. They have totally changed the style and approach of people's interactions with their friends and family members. Personality traits, demographic variables and the frequency of SNSs usage form various discernible patterns. Shin (2007) pointed out that users who have more online experience and usage experience are more likely to explore different online functions available on SNSs (Joinson, 2008; Steinfield, Ellison and Lampe, 2008).

With the availability of diverse functions on SNSs such as comment, share and "like," the social network community would be better equipped to express their opinions with regard to advertisements on SNSs. Consumer behaviour is becoming progressively more dominated by peer opinions and collective intelligence. Users are increasingly more likely to voice their opinions about advertisements posted on SNSs (Gillin, 2007). With the emergence of online social advertisements, online behavior is considerably diversified as users are allowed to participate and assess content and share their opinions with their social network community of friends and relatives (Kozinets, 1999; Hoegg, Martignoni, Meckel and Stanoevska-Slabeva, 2006).

When viewing an advertisement on a Social Networking Service, there is a distinct possibility that users with a greater degree of involvement, will desire to take further action as compared to those with a lesser degree of involvement. Such users are more likely to evaluate the advertisement to appease their curiosity (Peter and Olson, 1987). Users who frequently read the contents on SNSs are also more involved in appraisal of advertisements on SNSs, because if they find the subject matter and content of the text significant and important enough to hold their interest, the users are engrossed and spend longer hours on the SNS. It follows as a result of their increased activity on Social Networking Sites that such users are more likely to assess, share or post comments on advertisements they find appealing and interesting.

Hypothesis 1: Users with a greater involvement on Social Networking Services (such as those who customize the profile page design and post or share their comments or "like" the advertisement) are more likely to take further action toward social advertisements.

3.3   Credibility of Social Networking Services

Enders, Hungenberg, Denker, and Mauch (2008) highlighted that the connectedness facilitated by Social Networking Services has given business organizations more opportunities to reach a greater number of consumers. With the application of Web 2.0, business organizations are enabled to deliver their messages on SNS (McClure, 2009). Besides, people are used to exchanging messages with friends and family and online social networks permit people to stay connected to each other. This perception arises from a very fundamental and essential need in human beings to live peacefully together in a community (Das and Sahoo, 2011; Gross and Acquisti, 2005). As the majority of people are willing to accept the credibility of the SNSs they use most frequently, these users are more enthusiastic and eager to accept and trust information that is shared from their Social Networking Service community.

A survey has disclosed that people are now more experienced in information sharing and generally tend to accept information from SNSs without difficulty. Therefore, advertising disseminated through Social Networking Services is increasingly being considered as more trustworthy and acceptable (Karimzadehgan, Agrawal and Zhai, 2009). However, people are more likely to trust and accept information and suggestions from their friends and neighbors in a social network environment, compared to information delivered through an advertising company (Senecal and Nantel, 2004). Sledgianowski and Kulviwat (2009) reported that trust is a crucial element on Social Networking Services that has a propensity to affect people's online behavior, which means that factors that affect credibility such as technology, application and the trust people have for their preferred SNS, would also have the ability to influence people's intentions on information exchange.

Hypothesis 2: A trustworthy environment on Social Networking Services has a positive influence towards users' intentions to take further action on social advertisements.

3.4   Electronic Word-of-Mouth: Comments from Friends

Hennig-Thurau, Gwinner, Walsh and Gremler (2004) defined electronic word-of-mouth (e-WOM) as a type of communication with a marketing aim in an online environment. Electronic word-of-mouth (e-WOM) transfers a variety of comments or feedback between people/parties to other people/parties via electronic media. These comments or feedback can be either positive or negative in nature. Trusov, Bucklin and Pauwels (2009) stated that e-WOM is an unparalleled method to transfer information in a social network community due to its superior accessibility and advanced reach. Strutton, Taylor and Thompson (2011) pointed out that e-WOM can be a format of delivering advertisements through social networking sites, blogs, forums, web-based opinion platforms and so on.

An incomparable advertising strategy is to sell a product and services through friends of consumers, who recommend the product, instead of the salespeople who sell the product (Dichter, 1966). Comments and views about an advertised product through e-WOM, are powerful and influential, and tend to create a ripple effect, either in a positive or negative way (Keller, 2007). Moreover, information shared by friends is a more effective and efficient method of boosting the popularity of a particular product, than if that same information is communicated by means of an advertisement.

Receivers of such messages and information are more inclined to pay greater attention and take further action on messages that are forwarded from their peers, due to the fact that they implicitly trust their friends (Dobele, Lindgreen, Beverland,

Vanhamme and Wijk, 2007). In other words, Keller (2007) concluded that positive e-WOM can have an influence on product sales, as it drives people to take further positive action, such as buying that particular product and even better, recommending that product to their social network community.

Hypothesis 3: Positive comments from friends via e-WOM gain users' attention and are the cause of their intention to take further action toward social advertisements.

3.5   Electronic Word-of-Mouth: Comments from Members of Community of Interest

In most studies conducted on Social Networking Services, it is found that the major incentive for using SNSs is to consolidate collective or economic benefits from interactions therein, which in turn, encourage users of SNSs to gain self-esteem, confidence and in some instances even attain happiness (Ellison, Steinfield and Lampe, 2007; Ji, Hwangbo, Yi, Rau, Fang, and Ling, 2010; Cortimiglia, Ghezzi, and Renga, 2011). The benefits derived and information gained induces people to spend more resources, such as time and money, to sustain online connections with individuals via SNSs (Westland, 2010; Lu and Hsiao, 2010). The very nature of this type of connectedness makes users group together, to create and form a new group, which is called Community of Interest (COI).

As defined by Henri and Pudelko (2003), COI is a community or collection of people sharing a common interest or passion. Social Networking Services present members of a Community of Interest with an uncomplicated means of interacting with each other (Shin and Kim, 2008). Communicating on SNSs through e-WOM among COI members allows them to stay connected as a more cohesive and mutually dependent community. A greater degree of involvement and connection exists among COI members, who stay in touch with each other frequently.

Han and Windsor (2011) established that a user's trust in other community members could have significant yet positive effects on a user's consumption intentions. The comments left by other community members have a tendency to exert influence on an advertisement viewer, especially if a large majority of COI members post similar comments. For example, beneficial comments that an advertised product is better than any that they have ever experienced by a bulk of COI members, would induce users who have viewed these constructive comments to gain a positive and affirmative impression about the advertised product, irrespective of the fact that they themselves have or have not, sampled the product.

Hypothesis 4: Positive comments from COI members via e-WOM gain users' attention and are the cause of their intention to take further action toward social advertisements.

4   Research Methodology

4.1   Questionnaire Survey

A questionnaire was designed based on the characteristics of Social Networking Services, with questions associated to SNSs Usage Pattern; Credibility of SNSs; e-WOM: Comments from Friends and Comments from COI Members; and Intention for Further Action. The study made use of a 5-point Likert scale to measure quantitative responses. As regards the SNSs Usage Pattern, the frequency of sharing, commenting and viewing content were measured. The measure of the credibility of SNSs was dependent on the trust users experienced for the content on SNSs, irrespective of whether the content was generated through friends or a business organization. With reference to questions on Intention for Further Action, whenever users viewed and were attracted by an advertisement on the SNS, their responses were measured by their intentions to: (1) investigate further and get additional information about the advertised product; (2) join a web-site or forum as an admirer of the product's Company; (3) forward the product advertisement verbally or on SNSs to friends; (4) actual purchase or acquisition of the product. For e-WOM, our measures were dependent on the manner in which comments of friends and members belonging to a particular Community of Interest, served to influence users' perception towards a product.

4.2   Sampling Method

Questionnaires were distributed to target audiences who satisfied one criterion. They were expected to satisfy the standard of possessing sufficient experience in the use of Social Networking Services, such as Facebook, Renren, Twitter and Google.

Overall, two tests were administered to participants. In the pilot test, 30 participants were randomly selected and a face-to-face interview conducted, for improving the design of the questionnaire before ultimately finalizing the questionnaire. The final survey consisted of a total of 97 participants out of which 53 were female participants and 44 were male participants. All of them have not involved in the pilot test. Amongst the participants, 68 were well qualified educationally, as they possessed a Bachelor degree or even a higher education qualification; 86 of the members in the survey belonged to the 18-25 age group; and 95 participants claimed to possess more than 4 years of internet experience. Of the 97 participants in the survey, over 80% admitted having more than a hundred friends on their most frequently used SNSs account. Besides, for more than half of the participants, 30% of their friends were close or confidential friends on their most frequently used SNS account. Figure 1 demonstrates the details of the demographical data.

		Count	Percentage
Gender			
	Male	53	54.6%
	Female	44	45.4%
Age			
	18-28	86	88.7%
	29-35	9	9.3%
	35-40	1	1.0%
	>40	1	1.0%
Education Level			
	High School/ Secondary School	9	9.3%
	Higher Diploma/ Associate Degree	20	20.6%
	Bachelor's Degree	64	66.0%
	Master's Degree	3	3.1%
	Doctoral Degree	1	1.0%
Internet Experience			
	1-2 years	1	1.0%
	3-4 years	1	1.0%
	>4 years	95	97.9%
No. of Friends in the Participants' Most Frequently Used SNS Account			
	0-50	5	5.2%
	51-100	11	11.3%
	101-200	26	26.8%
	201-400	27	27.8%
	>400	28	28.9%
Percentage of Close Friends in the Participants' Most Frequently Used SNS Account			
	0-10%	26	26.8%
	11-30%	31	32.0%
	21-30%	11	11.3%
	31-50%	12	12.4%
	>50%	17	17.5%

Fig. 1. Demographic Information of Participants

5   Data Analyses and Results

Our findings proved that around 60% of all our participants were unquestionably attracted by advertisements on Social Networking Services. It was found that it was easier to attract and engross females, 67.9% or 36 females, through advertisements, than it was to acquire the attention of males. A mere 54.5% or only 24 male participants paid attention to advertisements on Social Networking Sites.

Of the 60% unquestionably attracted by advertisements on Social Networking Services, it was established that 75% of them, i.e., 45 of those 60 participants, had intention for further action when they were attracted by those advertisements. Their further actions involved: (1) investigate further and get additional information about the advertised product; (2) join a web-site or forum as an admirer of the product's Company; (3) forward the product advertisement verbally or on SNSs to friends; (4) actual purchase or acquisition of the product. In this case, it was affirmed that the proportion of males and females having intentions for further action was similar viz.

People always attracted by Advertisements?		
Male	Yes	24 (54.5%)
	No	20 (45.5%)
Female	Yes	36 (67.9%)
	No	17 (32.1%)

Fig. 2. People always attracted by Advertisements and Gender Relation

Intention for Further Action?		
Male	Yes	19 (79.2%)
	No	5 (20.8%)
Female	Yes	26 (72.2%)
	No	10 (27.8%)

Fig. 3. Intention for Further Action and Gender Relation

A confirmatory factor analysis was completed, and from the rotated component matrix, we ascertained that there existed 5 different constructs. These 5 constructs are identical to what was discussed in our literature review: (1) SNSs usage pattern (composite reliability = 0.89); (2) Credibility of SNSs (composite reliability = 0.92); (3) Electronic Word-of-Mouth (e-WOM): Comments from Friends (composite reliability = 0.81); 4) Electronic Word-of-Mouth (e-WOM): Comments from Members of Community of Interest (composite reliability = 0.87); and (5) Intention for Further Action whenever people read or glance at a SNSs advertisement (composite reliability = 0.95).

Multiple regression analysis was employed to ascertain the summary of the relation between the dependent variable (Intention for further action) and the independent variables (SNSs Usage Pattern; Credibility of SNSs; Electronic Word-of-Mouth: Comments from Friends; Electronic Word-of-Mouth: Comments from Members of Community of Interest). Although all the independent variables and dependent variable are positively related, only the constructs of SNSs Usage Pattern and e-WOM: Comment from Friends are significant. And the model thus derived possesses considerably strong significance, with p smaller than 1% significant level.

Construct	Item	Loading	Mean	S.D.
SNS usage pattern	SUP1	0.775	1.94	0.977
CR=0.89; AVE=0.72	SUP2	0.722	2.19	1.064
	SUP3	0.808	1.86	1.041
	SUP4	0.798	1.86	0.957
	SUP5	0.726	1.90	0.895
	SUP6	0.510	2.48	1.042
Credibility on SNSs	COS1	0.733	2.56	1.020
CR=0.92; AVE=0.78	COS2	0.824	2.77	1.095
	COS3	0.840	2.72	1.106
	COS4	0.789	2.89	1.059
	COS5	0.710	2.33	0.965
	COS6	0.767	2.33	0.932
e-WOM: Comments from Friends	FRD1	0.637	3.44	1.000
CR=0.81; AVE=0.66	FRD2	0.785	2.88	0.916
	FRD3	0.553	2.75	1.233
e-WOM: Comments from COI Members	COI1	0.764	2.52	1.081
CR=0.87; AVE=0.80	COI2	0.785	2.41	1.018
	COI3	0.858	2.36	1.082
Intention for Further Action	IFA1	0.899	1.49	1.803
CR=0.95; AVE=0.91	IFA2	0.903	1.16	1.512
	IFA3	0.908	1.10	1.388
	IFA4	0.919	1.09	1.362

Fig. 4. Psychometric properties of measures (Note: CR-Composite Reliability, AVE-Average Variance Extracted)

The adjusted R^2 for the model is 0.687. This implies that when dependent variable (Intention for Further Action) varies, there is 68.7% explanatory power for the two independent variables (SNSs Usage Pattern and e-WOM: Comment from Friends).

The multicollinearity of the two independent variables (SNSs Usage Pattern and e-WOM: Comment from Friends) is checked. The VIF of SNSs Usage Pattern is 1.013 and that of e-WOM: Comment from Friends is 1.098, which means that they are not correlated and can co-exist in the research model.

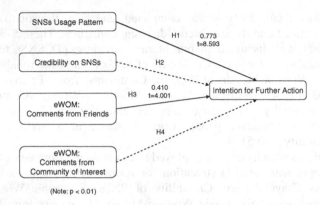

(Note: $p < 0.01$)

Fig. 5. Research Model

6   Conclusion

This study is based upon four constructs that determine our dependent variable, i.e., intention for further action, each time people view advertisements on Social Networking Sites. The four constructs are: (1) SNSs Usage Pattern; (2) Credibility of SNSs; (3) Electronic Word-of-Mouth: Comments from Friends; (4) Electronic Word-of-Mouth: Comments from Community of Interest Members. The results conclusively prove that two constructs viz. SNSs Usage Pattern and Electronic Word-of-Mouth (e-WOM): Comments from Friends are significantly and positively related to our dependent variable, i.e., Intention to take further action. This vindicates the fact that the more frequently a user utilizes SNSs and the greater the number of comments and views received by a user from his friends, the higher is that user's intention for further action.

Conversely, however, there exists a lack of statistical significance to prove the reliability of the remaining two constructs, i.e., Credibility of SNSs; and Electronic Word-of-Mouth (e-WOM): Comments from Community of Interest (COI) Members. This signifies that the degree of trust users have with regard to the environment prevalent on SNSs and Comments from COI Members, is not of any importance.

With the rapidly expanding popularity of SNSs and web advertising becoming the very foundation of a successful business enterprise, there appears to be very little by way of a similar academic study with regard to the effectiveness of advertisements on SNSs. This study hopes to make a noteworthy contribution to both researchers and marketing managers, to enable them to learn how best and most effectively to maximize the effectiveness of advertisement on SNSs.

References

1. Brackett, L., Carr, B.: Cyberspace advertising vs. other media: consumer vs mature student attitudes. Journal of Advertising Research 41(5), 23–29 (2001)
2. Cao, J., Knotts, T., Xu, J., Chau, M.: Word of mouth marketing through online social networks. The HKU Scholars Hub (2009),
   http://hub.hku.hk/handle/10722/127325 (retrieved March 12, 2012)
3. Chung, C., Kaiser, H.M.: Measurement of advertising effectiveness using alternative measures of advertising exposure [dagger]. Agribusiness 15(4), 525–537 (1999),
   http://search.proquest.com/docview/197424976?accountid=14548
   (retrieved May 25, 2012)
4. Clemons, E.K., Barnett, S., Appadurai, A.: The future of advertising and the value of social network websites: some preliminary examinations. In: ICEC, pp. 267–276 (2007)
5. Cortimiglia, M.N., Ghezzi, A., Renga, F.: Social applications: revenue models, delivery channels, and critical success factors - an exploratory study and evidence from the spanish-speaking market. Journal of Theoretical and Applied Electronic Commerce Research 6(2), 108–122 (2011)
6. Das, B., Sahoo, J.: Social networking sites: a critical analysis of its impact on personal and social life. International Journal of Business & Social Science 2(14), 222–228 (2011)
7. Dichter, E.: How word-of-mouth advertising works. Harvard Business Review 16, 147–166 (1966)
8. Dobele, A., Lindgreen, A., Beverland, M., Vanhamme, J., Wijk, R.V.: Why pass on viral messages. Business Horizons 50(4), 291–304 (2007)
9. Ducoffe, R.: Advertising value and advertising on the web. Journal of Advertising Research 36(5), 21–35 (1996)
10. Enders, A., Hungenberg, H., Denker, H.P., Mauch, S.: The long tail of social networking: revenue models of social networking sites. European Management Journal 26, 199–211 (2008)
11. Ellison, N.B., Steinfield, C., Lampe, C.: The benefits of Facebook friends: social capital and college students' use of online social network sites. Journal of Computer-Mediated Communication 12(4), 1143–1168 (2007)
12. Gillin, P.: The New Influencers: a marketer's guide to the new social media. Quill Driver Books, Sanger (2007)
13. Gross, R., Acquisti, A.: Information revelation and privacy in online social networks (The Facebook case). In: Proceeding from ACM Workshop on Privacy in the Electronic Society (WPES), pp. 1–11 (2005)
14. Gregurec, I., Vranesevic, T., Dobrinic, D.: The importance of database marketing in social network advertising. International Journal of Management Cases 13(4), 165–172 (2011)
15. Han, B., Windsor, J.: Users willingness to pay on social network sites. The Journal of Computer Information Systems 51(4), 31–40 (2011)
16. Hayes, H.M.: Why online coupons pay off. Dance Retailer News 10(4), 32–36 (2011)
17. Hennig-Thurau, F.T., Gwinner, K.P., Walsh, G., Gremler, D.D.: Electronic word-of-mouth via consumer-opinion platforms: what motivates consumers to articulate themselves on the internet? Journal of Interactive Marketing 18(1), 38–52 (2004)
18. Henri, F., Pudelko, B.: Understanding and analysing activity and learning in virtual communities. Journal of Computer Assisted Learning 19(4), 474–487 (2003)

19. Hoegg, R., Martignoni, R., Meckel, M., Stanoevska-Slabeva, K.: Proceedings from overview of business models for web 2.0 communities (2006), http://www.alexandria.unisg.ch/export/DL/31412.pdf (retrieved March 15, 2012)
20. Hoy, M.G., Milne, G.: Gender differences in privacy-related measures for young adult Facebook users. Journal of Interactive Advertising 10(2), 28–45 (2010)
21. Ji, Y.G., Hwangbo, H., Yi, J.S., Rau, P.L.P., Fang, X., Ling, C.: The influence of cultural differences on the use of social network services and the formation of social capital. International Journal of Human-Computer Interaction 26(11-12), 1100–1121 (2010)
22. Joinson, A.N.: Looking at, looking up or keeping up with people? motives and use of Facebook. In: Proceeding from the Twenty-sixth Annual Conference on Human Factors in Computing Systems, pp. 1027–1036 (2008)
23. Karimzadehgan, M., Agrawal, M., Zhai, C.X.: Towards advertising on social networks. In: Proceeding from the ACM SIGIR Workshop on Information Retrieval and Advertising, pp. 28–31 (2009)
24. Keller, E.: Unleashing the power of word of mouth: creating brand advocacy to drive growth. Journal of Advertising Research 47(4), 448–452 (2007)
25. Kim, J.Y., Shim, J.P., Ahn, K.M.: Social networking service: motivation, pleasure, and behavioral intention to use. Journal of Computer Information Systems 51(4), 92–101 (2011)
26. Kozinets, R.V.: E-Tribalized marketing?: the strategic implications of virtual communities of consumption. European Management Journal 17, 252–264 (1999)
27. Learmonth, M., Klaassen, A.: App revenue is poised to surpass Facebook revenue. Advertising Age 80(18), 3–40 (2009)
28. Lu, H., Hsiao, K.: The Influence of extro/introversion on the intention to pay for social networking sites. Information & Management 47(3), 150–157 (2010)
29. Lytras, M.D., Pablos, P.O.: Social web evolution: integrating semantic applications and web 2.0 technologies. Information Science Reference, Hershey (2009)
30. McClure, M.: Creating safe, collaborative cultures in a web 2.0 world. Econtent 32(5), 22–26 (2009)
31. Nielsen: Nielsen reports 17 percent of time spent on the internet in August devoted to social networking and blog sites, up from 6 percent a year ago (2009), http://socialnetworking.procon.org/sourcefiles/NielsenAug200 9.pdf (retrieved March 12, 2012)
32. Owyang, J.: The 8 success criteria for Facebook page marketing: analysis reveals brands lack maturity by not leveraging social features (2010), http://www.web-strategist.com/blog/2010/07/27/altimeter-report-the-8-success-criteria-for-facebook-page-marketing/ (retrieved March 3, 2012)
33. Peter, J.P., Olson, J.C.: Consumer behavior: marketing strategy perspectives. Homewood, Irwin (1987)
34. Purewal, S.J.: Facebook's groupon-style discounts. PC World 29(4), 24 (2011)
35. Senecal, S., Nantel, J.: The influence of online product recommendations on consumer online choice. Journal of Retailing 80(3), 159–169 (2004)
36. Shin, D.: User acceptance of mobile internet: implication for convergence technologies. Interacting with Computers 19(4), 472–483 (2007)
37. Shin, D., Kim, W.: Applying the technology acceptance model and flow theory to Cyworld user behavior. CyberPsychology and Behavior 11(3), 378–382 (2008)

38. Sledgianowski, D., Kulviwat, S.: Using social network sites: the effects of playfulness, critical mass and trust in a hedonic context. The Journal of Computer Information Systems 49(4), 74–83 (2009)
39. Steinfield, C., Ellison, N.B., Lampe, C.: Social capital, self-esteem, and use of online social network sites: a longitudinal analysis. Journal of Applied Developmental Psychology 29(6), 434–445 (2008)
40. Strutton, D., Taylor, D.G., Thompson, K.: Investigating generational differences in e-WOM behaviours: for advertising purposes, does X=Y? International Journal of Advertising 30(4), 559–586 (2011)
41. The Economist: The value of friendship. Economist 402(8770), 23–26 (2012)
42. Thomas, J.W.: Advertising effectiveness (2007),
    http://www.decisionanalyst.com/Downloads/
    AdvertisingEffectiveness.pdf (retrieved March 3, 2012)
43. Thorne, J.: Online community interaction - revolution or revulsion? Journal of Systemics, Cybernetics, and Informatics 6(1), 67–72 (2008)
44. Trusov, M., Bucklin, R., Pauwels, K.: Effects of word-of-mouth versus traditional marketing: findings from an internet social networking site. Journal of Marketing 73(5), 90–102 (2009)
45. Westland, J.C.: Critial mass and willingness to pay for social networks. Electronic Commerce Research and Applications 9(1), 6–19 (2010)

Effects of Knowledge Creation-Technology Fit on Creation Performance: Moderating Impact of Cognition Styles

Chien-Hsing Wu*, Jen-Yu Peng, and Cheng-Hua Chen

National University of Kaohsiung
700, Kaohsiung University Rd., Nanzih District, 811. Kaohsiung, Taiwan, R.O.C.
chwu@nuk.edu.tw, ponysocute@gmail.com, ron79913@yahoo.com.tw

Abstract. Knowledge creation has been developing its own characteristics with respect to its antecedents. A research model is proposed and empirically examined that describes knowledge creation by considering creation task-technology fit (CTTF), creation task mode (goal-driven, goal-free, and goal-frame) and information and communication technology (ICT). Based on the data analysis from 258 valid subjects from research institutes, manufacturing industry, and service industry, research findings suggest that (1) subjects from research institutes, manufacturing industry, and service industry are likely to be significantly concerned with the effect of CTTF on creation merits, (2) the relationships between independent variables and dependent ones are almost not moderated by cognitive style, except ICT showing that subjects with the analytical style regard ICT not a significant predicator of creation task-technology fit while those with intuitive style regard so, and (3) result of moderating effect on the creation task mode shows that goal-driven mode does not reveal significant for the analysis-styled subjects. Implications and discussions are also addressed.

Keywords: Knowledge creation, task-technology fit, cognitive style.

1 Introduction

The advanced information and communication technology (ICT) has increasingly enlarged the spectrum of knowledge creation covering creation of product and service, manufacturing, process, management, and even strategy [1, 2, 3]. To appropriately address the term of creation, Ettlie [4] defined that creation includes two stages. One is creativity that involves the idea generation and the second is idea implementation [5]. The term of creation in the current study follows Ettli's definition that contains idea generation and idea implementation [5, 6].

From the perspectives of idea generation and implementation, there are four major trends shown in the updated literature. The first deals with the deep understanding of behavioral intention to explore new creation sources. For example, extended from the SECI mode by Nonaka and Takeuchi [7], Schulze & Hoegl [1] reported that the main

L.S.-L. Wang et al. (Eds.): MISNC 2014, CCIS 473, pp. 14–27, 2014.
© Springer-Verlag Berlin Heidelberg 2014

creation sources for product design is socialization (S) and internalization (I). The main point for this is that both S and I within a free thinking environment will be likely to foster the idea generation and implementation while externalization (E) and combination (C) will likely limit their development due to organizational regulations. They finally suggested that when knowledge creation spiral apply on creative ideas of products, it is necessary to have a broader space of idea. The second one focuses on how ICT helps on the knowledge creation; that is based on a technical support aspect to analyze how to effectively use of ICT to help creation [8, 9, 10]. For example, Toikka [8] developed a called ICT supported inter-organizational knowledge creation system to assist inter-organizational transformation and sharing of knowledge innovation. This system is the application of ICT in four functions: analysis and presentation, storage and management, connection and communication, interaction and cooperation [11] to enhance its usefulness and ease of use. However, the application of the performance of this system needs further analysis, especially in the practical application of the views or opinions of users, has yet to be further discussed.

The third is how creation management process enhances the creative concepts, that is, from process analysis to construct an effective model of creation, in order to enhance the quality and quantity of creative ideas 7, 12, 13, 14]. For example, Martin-de-Castro et al. [12] analyzed the differences of SECI process of innovation between United States samples and France samples. Research found that there is no certain knowledge learning process in both of them. Also, it is found that knowledge innovation process needs to be considered the differences in field in order to improve the effectiveness of innovation. This proposal can further adjust aforementioned research found from Schulze & Hoegl [1]. In other words, if knowledge innovation model can consider the suitability, then it will be able to enhance its effectiveness. On the other hand, Esterhuizen et al. [3] also emphasized that knowledge innovation process plays a critical role in innovation performance. They analyzed and develop the process architecture of knowledge innovation, and this architecture can provide business reference. However, this architecture does not determine its acceptance through empirical investigation. Another example is that Kao et al. [6] investigate high-tech industry which has a certain number of patent cases. Research findings indicate that there is a better performance in goal-free mode. However, this study is based on the view from a macro perspective; the question about how individual cognitive moderates the knowledge performance affected by its antecedents needs to be disclosed. Rohrbeck [14] also analyzed through case studies, made a number of methods and processes to meet the needs of business innovation. However, this study did not prove to be the usefulness of these methods and the impact on organizational innovation performance, although analysis of the innovation process does provide a number of solutions for knowledge creation.

The final one is how human cognitive behavior affects knowledge creation; that is the analysis of individual differences in cognitive traits to look for more appropriate innovative practices in order to enhance the effectiveness of creative concepts [13, 15, 16, 17]. For example, Kickul et al. [16] analyzed Intuitive and analytical characteristics of student diversity on innovation self-efficacy, and their study found that intuitive characteristics students have more sensitivity to the opportunities for innovation than analytical characteristics students. However, the analytical characteristics students are more capable of planning and evaluation of resources needed for innovation, but they are less likely to find opportunities for innovation. However, the disadvantage of these findings are only limited to respondents' cognitive stage, as to how the actual situation is yet to be verified. The paper also pointed out that when further analyzes the impact of individual cognitive to knowledge innovation, it is necessary to focus on a specific topic. Even if, as Schulze & Hoegl [1] suggested, have a broader space of thinking, it is still necessary to defined within a certain range (for example, a particular industry or commodity). In fact, there are studies about individual cognitive for decades. To associate it with the knowledge innovation is a topic worthy of further in-depth research, and it also can find another solution for knowledge creation.

From the four directions of knowledge innovation-related point of view, it can be summarized in three major dimensions, including technology, personal cognitive behavior, and innovation management processes. In fact, Nonaka & Takuichi [7] initially mentioned organizational knowledge creation comes from the mutual agitation and conversion of tacit knowledge and explicit knowledge. Through the process of exchange, it can at the same time expand the quality and quantity of tacit and explicit knowledge. It is the product of a multi-dimensional interaction, so it will be affected by the environment, the use of tools, personal inner qualities and external regulation (e.g., management practices). Therefore, in order to meet this demand, there are a number of researches on the influence of tasks, technology and (or) individual fit to the individual task performance.

However, knowledge creation has its own characteristics that are different from other tasks. For example, in terms of technology, it is not simply a function of a certain system; it needs more functions such as connection and communication, interaction and cooperation [11]. Another example is that in terms of personal characteristics, computer self-efficacy can be made up through education and training, but in terms of knowledge creation, perhaps the impact of cognitive behavioral characters of using technology on innovation is more worthy of further analysis. However, current literature paid limit attention to this multidimensional phenomenon of knowledge creation. From the perspective of the integration of the associated view, what's the impact of the role of each dimension on Innovation performance? If we see from the aspect of Dimension Fit, whether there will be able to find a better combination? These problems have become the motivation of this research. To do so, this study integrates technology-performance fit proposed by Delone & McLean [18] and task-technology fit proposed by Goodhue & Thompson [19], with the consideration of relevant literature and the features of creation task, to propose and examine an research model called CTTF model (Creation Task-Technology Fit Model).

2   Related Concepts

The Knowledge creation performance is a multi-dimensional issue. Schulze & Hoegl [1] evaluated knowledge creation performance by considering the salience difference between before and after product functions while adopting SECI to analyze new product creation. While some studies consider such factors as creation quality, customer relationship, management efficacy, technology value, brand value and human resources, others paid attention to the efficiency, growth, and profits of organizations [20]. Importantly, literature aimed mostly at the perception degree of influential factors that include creation of product [1], process [21], management [21, 22], organization, and strategy. The current study excluded the factors of organization and strategy due to the reason that it was comparatively difficult to collect the degree of perception of organization and strategy. To ease the implementation of the current study, we took on the first three factors as the measuring factors of knowledge creation. These are product creation, process creation, and management creation.

The concept of fit argued that performance comes mainly from the outcome of a suitable systemization that involves several parties. The fit models have been extensively introduced and examined successful to describe performance in varied contexts, such as individual-environment fit (IEF) [23], individual-team fit (ITeF)[24], individual-task fit (ITaF) [25], task-technology fit (TTF) [18], and task-individual-technology fit [19]. Accordingly, the current study regarded knowledge creation performance is also a systematic issue that needs to consider such influential factors as individual (or cognition) characteristics, technology characteristics, and creation characteristics. For example, Smith & Mentzer [22] examined and reported that fit of a predictive decision support system (DSS) and decisional tasks significantly described performance. Moreover, to examine the performance (or intention) of adoption of social communication technology (SCT), Liu et al. [26] reported that the proposed TTF was significantly useful. This implies that fit is one of the most important factors that influence performance. Therefore, based on the TTF model, the current study argued and proposed a fit model, namely Creation Task-Technology Fit model (CTTF) to describe knowledge creation performance. We also included the adoption of technology as one of the factors influencing knowledge creation performance. Therefore, three hypotheses are defined as follows.

Hypothesis: CTTF is significantly related to knowledge creation performance

Hypothesis: CTTF is significantly related to the adoption of technology

Hypothesis: Use of technology is significantly related to knowledge creation performance

2.1   Information and Communications Technology

Please Literature indicates that that Information and communications technology (ICT) apparently plays an important supported role in knowledge development. In other word, ICT is the accelerator to promote knowledge development. For example, studies have shown that effect of knowledge sharing and dissemination depends on the appropriate use of ICT [13, 27, 29, 30]. Basically, these are four functions that

ICT helps knowledge creation performance: (1)Analysis and presentation which support knowledge objects or components, support user explanation, support cognitive development of receiver and enforce creative knowledge immersion status [11, 28]; (2)Storage and management used to support knowledge objects or components, support user cognitive development and support user knowledge externalization [10]; (3)Networking and communication is to support delivery [29, 39]; (4) Interaction and collaboration used to support interaction and collaboration between users .

During knowledge development process, the ICT will be affecting the knowledge externalization, knowledge objects development, and knowledge explanation [13]. Importantly, Wu et al. [13] reported that to analyze the features of knowledge structure effectively to develop new knowledge, knowledge creator should use data storage of database technology which is more suitable. However, knowledge externalization should use group system to construct knowledge components during the interaction. Moreover, knowledge objects should use data storage of database technology and knowledge explanation should use artificial intelligence software which is high media richness. The study also found another point is that the transferring context constructed by ICT is helpful to knowledge creator paying more attention and getting involved the results in stimulating more creation attention and enlarge the thinking space. We argue that ICT features should contribute to the fit of creation task toward creation performance. The next hypothesis is defined as follow. Hypothesis: ICT is significantly related to the creation task-technology fit

2.2   Creation Task Mode

Originally, Tsai & Li [20, p2] and Schulze & Hoegl [1, p1] indicated that SECI creation spiral of knowledge [7] is formed by four different creation modes which are socialization, externalization, combination and internalization. They analyzed every modes related to creation performance for corporation to make creation strategy. However, Kao et al [6] summarized two types of creation modes from literature: goal-free and goal-driven. Although there's no verified version, their study finally reported the creation mode was divided into three mode based on the perspective of goal definition: goal-free, goal-driven, and goal-frame after doing factor analysis and definite goals are a series of action and commitment. They argued that SECI includes four different stages of knowledge creation, which is in fact of creation process by which goal can be reached by SECI. Therefore, the current study regards that creation task can be divided into three creation mode by defining the ways to reach a goal.

Popper [31] argued that "if we know the goal, we can achieve" in the process of creation, implying that reaching a goal is a never-ending process and the goal is usually uncertain and complicated [32]. Kao et al. [6] examined and reported that there is a significant relationship between creation mode and creation performance in the context of hi- tech industry. Importantly, their research finding showed that the creation mode with no limitation will be having a better creation performance. In fact, prior to their study, Martins & Terblanche [33] emphasized that freedom is the core value to stimulate creation. This means that individual tends to freely achieve their goal, defined or not defined. For company, Judge et al. [34] called it a spontaneous chaos. In other words, employees work freely and find some ways and thoughts to improve their works under the environment constructed by the company. However, Hellström et al. [32] indicated that goal-free is lack of restriction, so the need of sub-group and individuals is contrary to the formal goals of organization. On the other hand, Scriven [35] showed that goal-driven is related to relevant knowledge and understanding of goal. This argument only pays attention to a goal, instead of the process which will be affected by outside environment such as intelligence, the stress, how much can be received, supporting tools, and the support of environment, etc. However, when doing knowledge creation with goal-driven within a certain area, set a goal will make creation more efficiently and has a clear way to follow. As a result, the research discuses that whether goal-driven creation mode is relevant to organizational creation performance; or organization creation performance will be changed by other factors. In the study by Kao et al. [6], goal-frame is another factor based on the exploratory factor analysis, in addition to the goal-driven and goal-free modes. This mode is somehow between goal-driven mode and goal-free mode, due to the reality that companies perform knowledge creation by balancing efficiency from effectiveness.

Although literature shows that there is a relationship between creation and organizational performance, addressing that creation mode is related to organization performance. Our research believes that creation mode should also consider other factors with a fit to increase creation performance. Therefore, another hypothesis is formed as follow.

Hypothesis: Creation task mode (CTM) is significantly related to creation task-technology fit

2.3   Cognitive Styles

The Cognitive style concerns the ways individual prefers to perform information processing according to their cognitive process. It is also related to individual preference in knowledge development For example, employees may prefer ICT with media-enriched tools to help sharing their knowledge, such that their idea to innovate products can be generated. Among typical styles introduced by Allinson & Hayes [15] are analytical and intuitive. They were developed from the theory of psychological type developed by Jung [36]. Moreover, a review of literature indicated that the role that cognitive style plays on the knowledge creation varies [16, 29, 30, 37, 38]. For example, Kolfschoten et al. [38] reported that effect of cognitive style is significant in the research of learning performance. They also placed an emphasis on that learning performance can be significantly improved by a pre-designed learning environment, implying that cognitive style may be better under some particular circumstances. Literature paid limitation to how cognitive style moderates the relationships between antecedents (e.g., creation task mode) and their dependent variable (e.g., creation task-technology fit and creation performance). Accordingly, the current study defined the next hypotheses as follows.

Hypothesis: Cognitive style is significantly moderate the relationship between ICT and CTTF
Hypothesis: Cognitive style is significantly moderate the relationship between CTM and CTTF
Hypothesis: Cognitive style is significantly moderate the relationship between CTTF and creation performance
Hypothesis: Cognitive style is significantly moderate the relationship between CTTF and technology use
Hypothesis: Cognitive style is significantly moderate the relationship between technology use and creation performance

3   Method

3.1   Research Model

The research model is illustrated in Figure 1. It contains five components and one moderator, namely ICT, knowledge creation task, creation-technology fit, technology use, creation performance, and cognition style (moderator). Five hypotheses are defined for independent variable and dependent ones while another five for the moderation effects on H1 to H5.

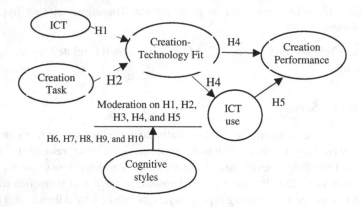

Fig. 1. Research Model

3.2   Sampling and Measure

The study targeted on the population of top-100 research institutes and universities, top-200 companies in the manufacturing industry, and top-200 companies in the services industry. The major concern for doing so is that comparatively these subjects perform more tasks on knowledge creation. The questionnaire was developed based on the aforementioned cognitive styles and variables and was modified according to the comments by 10 subjects at the stage of pretest. It contained three parts. The first part is the basic information of the subject, including gender, age, highest education, and company type (research institute, manufacturing industry, or service industry).

The second part was the questions used to differentiate cognitive styles, which included intuitive cognitive style and analytical one. The subjects were asked to answer the questions by choosing "agree", "disagree" or "not sure". The scale consisted of 38 questions, in which 21 questions were described on the "analytical type", and the rest of the 17 were described on the "intuitive type" [15]. For example: "When I am making a decision, I usually think calmly, then carry out and complete all relevant factors that may affect the outcomes" and "I usually perform better in generating idea, rather than handling data". The third part was the questions for variables based on the research model, including 33 questions in total in the 7-digit Likert scale. For example, one of the questions used in the ICT variable is that "When performing knowledge creation, I perceive that ICT is useful to interact and cooperate with others". One of the questions used in the knowledge creation task is "When setting knowledge creation goal, the supervisor does not limit the way we discuss". The data analysis was divided into three parts: (1) the descriptive statistics, (2) reliability and validity analysis, and (3) structural equation model for hypothesis test.

4   Data Analysis Results and Discussion

Of the 500 questionnaires sent to the targeted population, there were valid 258 returned, presenting 51.60% valid returned rate. The data analysis contains four parts. The first is cognitive style differentiation by using CSI questions. The results show that there are 132 samples that obtained more than or equal to 46 points. They are grouped into the analytical style. The rest samples, 126, obtained less than or equal to 45 points, who are regarded as the intuitive style group. The second is descriptive statistics that presents the sample characteristics. It is shown in Table 1. Importantly, it was found that although not quite equally distributed, there is no extreme proposition occurred.

Table 1. The descriptive statistics

Valid sample: 258			
	Features	Quantity	Percentage
Gender	Male	137	53.10%
	Female	121	46.90%
Age	Less than or equal to 30	85	32.95%
	Between 30 to 49	139	53.86%
	More than or equal to50	34	13.19%
Highest education	Below Bachelor	54	20.93%
	Bachelor	120	46.51%
	Master	84	32.56%
Business type	Manufacturing	84	32.56%
	Service	111	43.02%
	Research institutes	63	24.42%

The third is data reliability analysis. The Cronbach's α was used to ensure the internal consistency for each item and composite. It was set to be greater than 0.3;

implying that item less than 0.3 is deleted to increase the overall reliability, if necessary. It is realized that the critical value set to be 0.3 is comparatively low and would possibly influence the result of exploratory factor analysis. For example, item CTM1 (creation task mode) and CTM3 were 0.323 and 0.304, which were greater than, but close to the critical value 0.3. Yet, our study considered to include every collected information that contributed to the proposed model and was not unacceptable. Details are shown in Table 2. It was found that there was no item unacceptable. The Cronbach's α of the overall model was 0.856.

Moreover, the construct validity was also checked using the exploratory factor analysis (EFA) to measure the concept correctly because the proposed model contained several factors and sub-factors. First, the Kaiser-Meyer-Olkin (KMO), Bartlett's test of sphericity, and principle component analysis (PCA) were used to provide a lower-dimensional construct with the most informative viewpoint. We then utilized the PCA and varimax of orthogonal rotation to derive the final factors. Factor loading was between 0.568 and 0.949, which is greater than 0.5. Details are listed in Table 3. It was found that variable of creation task style was explored to have three sub-factors, namely goal-driven, goal-free, and goal-frame, which was exactly according to expectation. However, variable of knowledge creation performance was found grouped in a single factor. Moreover, the correlation coefficient of factors with average variance extracted (AVE) validity is shown in Table 4. For example, the AVE of creation task technology fit (0.706) in Table 4 is greater than 0.5, and the square root of AVE (0.840) is greater than the correlation coefficient of other factors. However, it was found that the AVE of creation task style is a little small (0.294), but the root of it is 0.541, which is not unacceptable. Although not with a very high reliability, overall the validity of the proposed model is acceptable.

Table 2. Reliability Analysis

Composite	Items	Mean	S.d.	Facet Mean S.d. Cronbach's α	Item to Total	Cronbach's α if item deleted
ICT	ITC1	5.99	1.000	5.985; 1.021; 0.943	0.858	0.926
	ITC2	6.04	1.015		0.858	0.926
	ITC3	5.97	1.041		0.907	0.911
	ITC4	5.95	1.027		0.828	0.936
Creation Task Mode	CTM1	5.06	1.464	5.013; 1.371; 0.763	0.323	0.738
	CTM2	3.62	1.813		0.396	0.731
	CTM3	4.05	1.643		0.304	0.743
	CTM4	5.64	1.051		0.404	0.728
	CTM5	5.43	1.083		0.418	0.727
	CTM6	5.37	1.157		0.446	0.723
	CTM7	5.47	1.133		0.412	0.727
	CTM8	5.43	1.304		0.433	0.723
	CTM9	5.62	1.049		0.408	0.728
	CTM10	4.40	1.653		0.443	0.721
	CTM11	5.06	1.460		0.455	0.719

Table 2. (*continued*)

Creation Task-Technology Fit	CTTF1	4.73	1.415	4.839; 1.419; 0.917	0.776	0.900
	CTTF2	5.06	1.373		0.789	0.898
	CTTF3	4.68	1.434		0.775	0.900
	CTTF4	4.81	1.484		0.720	0.908
	CTTF5	4.76	1.433		0.812	0.895
	CTTF6	5.00	1.374		0.715	0.908
Technology Use	TU1	5.40	1.200	5.543; 1.150; 0.868	0.776	0.792
	TU2	5.48	1.217		0.836	0.732
	TU3	5.74	1.023		0.657	0.897
Knowledge Creation Performance	KCP1	5.02	1.228	5.162; 1.200; 0.922	0.705	0.914
	KCP2	5.02	1.227		0.717	0.913
	KCP3	5.12	1.172		0.727	0.913
	KCP4	5.26	1.126		0.728	0.913
	KCP5	5.26	1.149		0.686	0.915
	KCP6	5.37	1.177		0.725	0.913
	KCP7	5.16	1.245		0.752	0.911
	KCP8	5.16	1.221		0.736	0.912
	KCP9	5.09	1.255		0.706	0.914

Table 3. Factor analysis and results

Item	Factor loadings						
	ICT	CGF	CGD	CGR	CTFF	TU	KCP
ICT1	**0.922**	0.121	0.322	0.189	-	-	-
ICT2	**0.922**	0.098	0.158	0.204	-	-	-
ICT3	**0.949**	0.102	-0.069	0.116	-	-	-
ICT4	**0.902**	0.214	0.114	0.069	-	-	-
CTM7	0.048	**0.840**	-0071	0.110	-	-	-
CTM6	-0.108	**0.778**	0.026	0.078	-	-	-
CTM4	0.241	0.154	0.134	0.032	-	-	-
CTM3	0.264	-0.113	**0.894**	0.075	-	-	-
CTM2	0.142	-0.028	**0.874**	0.182	-	-	-
CTM1	0.089	0.225	**0.568**	0.012	-	-	-
CTM10	0.067	0.050	0.217	**0.861**	-	-	-
CTM11	0.103	0.221	0.008	**0.842**	-	-	-
CTTF5	-	-	-	-	**0.876**	-	-
CTTF3	-	-	-	-	**0.861**	-	-
CTTF2	-	-	-	-	**0.852**	-	-
CTTF3	-	-	-	-	**0.847**	-	-

Table 3. (*continued*)

CTTF4	-	-	-	-	0.804	-	-
CTTF6	-	-	-	-	0.802	-	-
TU1	-	-	-	-	-	0.903	-
TU2	-	-	-	-	-	0.933	-
TU3	-	-	-	-	-	0.833	-
KCP7	-	-	-	-	-	-	0.810
KCP8	-	-	-	-	-	-	0.795
KCP4	-	-	-	-	-	-	0.793
KCP6	-	-	-	-	-	-	0.791
KCP3	-	-	-	-	-	-	0.791
KCP2	-	-	-	-	-	-	0.781
KCP9	-	-	-	-	-	-	0.772
KCP1	-	-	-	-	-	-	0.771
KCP5	-	-	-	-	-	-	0.758

1. KMO= 0.898; Bartlett's test of sphericity: ***p<0.01;
2. ICT: Information and communication technology; CTM: Creation task mode; CTTF: Creation task technology fit; TU: Technology use; KCP: Knowledge creation performance

Table 4. Correlation coefficient of factors

	CR	AVE	ICT	CTS	CTTF	TU	KCP
ICT	5.99	0.852	**0.923**				
CTM	4.83	0.294	0.246	**0.541**			
CTTF	4.84	0.706	0.167	0.442	**0.840**		
TU	5.54	0.792	0.377	0.338	0.552	**0.890**	
KCP	5.16	0.616	0.254	0.383	0.698	0.612	**0.785**

1. The number in the bold diagonal course is the square root of AVE.
2. CR: Composite reliability
3. ICT: Information and communication technology; CTM: Creation task mode; CTTF: Creation task technology fit; TU: Technology use; KCP: Knowledge creation performance

Finally, for the significance of the hypothesis tests, the structure equation model is employed to derive the results (Table 5). It was found that the test results support all hypotheses, which are exactly according to our expectation. However, R-square of creation task-technology fit (CTTF) shows a little low, which is 0.205. Moreover, hypothesis 1 representing the relationship between ICT and creation task-technology fit shows weak significance in comparison with other hypotheses. In particular, in the variable of creation task mode, its sub-factors, goal-driven mode, goal-free mode, and goal-frame mode also reveal significant. However, the goal-driven mode does not show a strong significance. For the moderating effect tests, table 6 shows that H6 is not supported, implying that ICT does not show a significant effect on creation performance for the analysis-styled subjects. Moreover, result of moderating effect of the creation task mode on CTTF does not totally accord to our research expectation, indicating that goal-driven mode does not reveal significant for the analysis-styled subjects.

Table 5. Path test results

Hypothesis	Path	Std. Coe. (β)	R-Square	t- value	p-value	S/NS
H1	ICT → CTTF	0.167	0.205	2.716*	0.07	S
H2	CTM → CTTF	0.446		7.969***	0.00	S
H3	CTTF → KCP	0.698	0.561	15.585***	0.00	S
H4	CTTF → TU	0.552	0.305	10.602***	0.00	S
H5	TU → KCP	0.612	0.561	12.383***	0.00	S
H2-a	CGD → CTTF	0.106		1.702*	0.09	S
H2-b	CGF → CTTF	0.426	0.351	7.542***	0.00	S
H2-c	CGR → CTTF	0.511		9.502***	0.00	S

*p<0.1; **p<0.05; ***p<0.01

ICT: Information and communication technology; CTM: Creation task mode; CTTF: Creation task technology fit; TU: Technology use; KCP: Knowledge creation performance; CGD: Goal-driven creation mode; CGF: Goal-free creation mode; CGR: Goal-frame creation mode; S/NS: Support/not support

Table 6. Moderation effects

Path		Path moderated	Cognitive styles						S/NS
			Analytical (n=132)			Intuitive (n=126)			
			R-square	t- value	p-value	R-square	t- value	p-value	
H6	ICT → CTTF	0.163		1.384	0.17	0.244	2.711*	0.08	S
H7	CTM → CTTF			5.054***	0.00		5.951***	0.00	N
H8	CTTF → KCP	0.507		9.368***	0.00	0.615	13.356***	0.00	N
H9	CTTF → TU	0.300		7.462***	0.00	0.335	8.025***	0.00	N
H10	TU → KCP	0.507		8.938***	0.00	0.615	8.635***	0.00	N
H7-a	CGD → CTTF	0.397		-1.322	0.19	0.302	3.293***	0.00	S
H7-b	CGF → CTTF			5.812***	0.00		4.756***	0.00	N
H7-c	CGR → CTTF			6.830***	0.00		6.380***	0.00	N

*p<0.1; **p<0.05; ***p<0.01

5   Concluding Remarks

We draw attention to knowledge creation performance by considering the creation task- technology fit. A model that defined creation task mode and ICT as two important antecedents of creation merits is proposed and empirically examine. The use of technology is also examined for both whether it is determined by creation and task technology fit and whether it affects creation performance. A salient test is that human cognitive style moderates the independent variable and dependents. Our research findings show that subjects from research institutes, manufacturing industry, and service industry are likely to be significantly concerned with the effect of independent variables on dependent ones. Special emphasis is placed that knowledge creation has been developing its own characteristics. One of our findings confirms that its merits are greatly influenced by whether the antecedents fit well or not. Moreover, the relationships between independent variables and dependent ones are almost not moderated by cognitive style, except ICT. Importantly, subjects with the intuitive style regard ICT a significant predicator of creation task-technology fit while those with analytical one do not think so. This finding will be useful for ICT-mediated creation providers and consultants to develop strategy and policy creation.

References

[1] Schulze, A., Hoegl, M.: Organizational knowledge creation and the generation of new product ideas: A behavioral approach. Research Policy 37, 1742–1750 (2008)

[2] Camison, C., Fores, B.: Knowledge creation and absorptive capacity: The effect of intra-district shared competences. Scandinavian Journal of Management 27, 66–86 (2011)

[3] Esterhuizen, D., Schutte, C.S.L., du Toit, A.S.A.: Knowledge creation processes as critical enablers for innovation. International Journal of Information Management 32(4), 354–364 (2012)

[4] Ettlie, J.E.: Managing Technological Innovation. Wiley & Sons, Inc., New York (2000)

[5] Popadiuk, S., Choo, C.W.: Innovation and knowledge creation: How are these concepts related? International Journal of Information Management 26, 302–312 (2006)

[6] Kao, S.C., Wu, C.H., Su, P.J.: Which mode is better for knowledge creation? Management Decision 49(7), 1037–1060 (2011)

[7] Nonaka, I., Takeuchi, H.: The Knowledge-Creating Company. Oxford University Press, INC., New York (1995)

[8] Toikka, S.: ICT Supported inter-organizational knowledge creation: application of change laboratory. In: Duval, E., Klamma, R., Wolpers, M. (eds.) EC-TEL 2007. LNCS, vol. 4753, pp. 337–348. Springer, Heidelberg (2007)

[9] Vezzetti, E., Moos, S., Kretli, S.: A product lifecycle management methodology for supporting knowledge reuse in the consumer packaged goods domain. Computer-Aided Design 43, 1902–1911 (2011)

[10] McAfee, A.: Mastering the three worlds of information technology. Harvard Business Review, 132–144 (November 2006)

[11] Martin-de-Castro, G., Lopez-Saez, P., Navas-Lopez, J.E.: Processes of knowledge creation in knowledge-intensive firms: empirical evidence from Boston's Route 128 and Spain empirical. Technovation 28, 222–230 (2008)

[12] Wu, C.H., Kao, S.C., Shih, L.H.: Assessing the suitability of process and information technology in supporting tacit knowledge transfer. Behavior and Information Technology 29(5), 513–525 (2010)

[13] Rohrbeck, R.: Exploring value creation from corporate-foresight activities. Futures 44(5), 440–452 (2012)

[14] Allinson, C.W., Hayes, J.: The Cognitive Style Index: A measure of intuition-analysis for organizational research. Journal of Management Studies 33(1), 119–136 (1996)

[15] Kickul, J., Gundry, L.K., Barbosa, S.D., Whitcanack, L.: Intuition versus analysis? Testing differential models of cognitive style on entrepreneurial self-efficacy and the new venture creation process. Entrepreneurship Theory and Practice, 439–450 (2009)

[16] Bloodgood, J.M., Chilton, M.A.: Performance implications of matching adaption and innovation cognitive style with explicit and tacit knowledge resources. Knowledge Management Research and Practice 10, 106–117 (2012)

[17] DeLone, W.H., McLean, E.R.: The DeLone and McLean model of information systems success: A ten-year update. Journal of Management Information Systems 19(4), 9–30 (2003)

[18] Goodhue, D.L., Thompson, R.L.: Task-technology fit and individual performance. MIS Quarterly 19(2), 213–236 (1995)

[19] Tsai, M.-T., Li, Y.-H.: Knowledge creation process in new venture strategy and performance. Journal of Business Research 60, 371–381 (2007)

[20] Junglas, I., Abraham, C., Watson, R.T.: Task-technology fit for mobile locatable information systems. Decision Support Systems 45, 1046–1057 (2008)

[21] Smith, C.D., Mentzer, J.T.: Forecasting task-technology fit: The influence of individuals, systems and procedures on forecast performance. International Journal of Forecasting 26, 144–161 (2010)

[22] Posner, B.Z.: Person-organization values congruence: No support for individual differences as a moderating influence. Human Relations 45, 351–361 (1992)

[23] Werbel, J.D., Gilliland, S.W.: Person-environment fit in the selection process. Research in Personnel and Human Resource Management 17, 209–243 (1999)

[24] Edwards, J.R.: Person-job fit: A conceptual integration, literature review, and methodological critique. In: Cooper, C.L., Robertson, I.T. (eds.) International Review of Industrial/Organizational Psychology, pp. 283–357. Wiley, New York (1991)

[25] Liu, S., Lee, Y., Chen, A.N.K.: Evaluating the effects of task–individual–technology fit in multi-DSS models context: A two-phase view. Decision Support Systems 51, 688–700 (2011)

[26] Garavelli, C., Gorgoglione, M., Scozzi, B.: Manage knowledge transfer by knowledge technologies. Technovation 22, 269–279 (2002)

[27] Chen, Y.J.: Development of a method for ontology-based empirical knowledge representation and reasoning. Decision Support Systems 50, 1–20 (2010a)

[28] Chen, Y.J.: Knowledge integration and sharing for collaborative molding product design and process development. Computers in Industry 61, 659–675 (2010b)

[29] Mampadi, F., Chen, Y.S., Ghinea, G., Chen, M.P.: Design of adaptive hypermedia learning systems: A cognitive style approach. Computers & Education 56, 1003–1011 (2011)

[30] Popper, K.R.: Unended Quest: An Intellectual Autobiography. Fontana, Glasgow (1976)

[31] Hellstrom, T., Jacob, M.: Knowledge without goals? Evaluation of knowledge management programs. Evaluation 9(1), 55–72 (2003)

[32] Martins, E.C., Terblanche, F.: Building organizational culture that stimulates creativity and innovation. European Journal of Innovation Management 6(1), 64–74 (2003)

[33] Judge, W.Q., Fryxell, G.E., Dooley, R.S.: The new task of R&D management: creating goal directed communities for innovation. California Management Review 39(3), 72–85 (1997)

[34] Scriven, M.: Goal-free evaluation. In: Hamilton, D., MacDonald, B., King, C., Jenkins, D., Parlett, M. (eds.) Beyond the Numbers Game, pp. 134–138. McCutcheon, Berkeley (1977)

[35] Jung, C.G.: Analytical Psychological: It's Theory and Practice. Vintage Book, New York (1968)

[36] Austin, K.A.: Multimedia learning: Cognitive individual differences and display design techniques predict transfer learning with multimedia learning modules. Computers & Education 53, 1339–1354 (2009)

[37] Kolfschoten, G., Lukosch, S., Verbraeck, A., Valentin, E., deVreede, G.J.: Cognitive learning efficiency through the use of design patterns in teaching. Computers & Education 54, 652–660 (2010)

[38] Zhou, S., Chin, K.-S., Yarlagadda, P.K.D.V.: Internet-based intensive product design platform for product design. Knowledge-Based Systems 16, 7–15 (2003)

Online Knowledge Community Evaluation Model: A Balanced Scorecard Based Approach

Shu-Chen Kao[1], Chieh-Lin Huang[1], and Chien-Hsing Wu[2]

[1] Dept. of Information Management, Kun Shan University, Tainan, Taiwan
kaosc@mail.ksu.edu.tw
[2] Dept. of Information Management, National University of Kaohsiung, Taiwan
chwu@nuk.edu.tw

Abstract. For the last decade, knowledge community has become popular resources of knowledge and their measurement also become more complex and critical for users and community managers. In this research, an integrated framework based on BSC (Balanced Score Card) - OKC_{BSC} is proposed to evaluate the performance of online knowledge community in four dimensions: customer, internal process, learning and growth, and performance. After conducting Delphi approach and AHP methods, the measurements and their weights are derived. Two typical cases (Wikipedia and Yahoo!Kimo Knowledge$^+$) that conduct empirical surveys with 822 samples each are used to demonstrate the use of OKC_{BSC}. Results and discussion are also addressed.

Keywords: Online knowledge community, Balanced score card, Performance evaluation.

1 Introduction

As the rapid development of information and communication technology, virtual communities are viewed as chief social media attracting those users with similar interests or hobbies via Internet. For individual member, online knowledge communities (OKCs) can help he(she) sharing their own new ideas or experiences to others, meanwhile he (she) also can enhance his (her) capability by getting more problem solving knowledge and absorbing them from others [1]. Thus, OKCs have become one kind of powerful platform to enhance intrinsic knowledge transition and innovation among the group members by evolving from bulletin board systems and web forums [2]. As more and more OKCs are available, how to identify the effectiveness of an OKC in a more complete viewpoint is necessary for its future development.

Because evaluating the performance for an OKC is a complex and multidiscipline issue, varieties of measurement methods for OKCs are still under developing and trying to find the solutions [3,4,5,6,7,8,9,10]. With respect with the quantitative measurements for OKCs, most criteria are concentrate on the number of members, the time of staying in the communities, the frequency of discussing with others,

L.S.-L. Wang et al. (Eds.): MISNC 2014, CCIS 473, pp. 28–35, 2014.

difference between articles supply and article demand, the number of knowledge categories and contents, etc. [11]. On the other hand, the qualitative dimensions for OKCs extremely considered the operation process of community, the usefulness of knowledge acquired, and the process of knowledge management [12,13]. Furthermore, Chen & Hsiang [14] argued that the strategy, technology, process, and personnel are critical factors which can foster the knowledge community. From the prior research, it can be found that most of evaluation models are proposed focusing on some specific measuring dimensions instead of an integrated consideration for evaluating OKCs.

Balanced score card (BSC), which is an internal performance analysis method proposed by Kaplan and Norton [15,16], can be used to evaluate the performance of a system with a combination of measure in four perspectives: financial performance, customer knowledge, internal business process, and learning and growth. Owing to integrate financial measurement and other key performance indicators, BSC is proven a strategic approach and performance management system which organizations can use for vision and strategic implementation [17,18]. According, recent studies have conducted on BSC in various fields, including organizational and IS performance evaluation [18,19,20,21,22,23].

Hence, this research would like to propose a framework (OKC$_{BSC}$) which is based on balanced score card approach that can support online knowledge community evaluation and assist community managers in enhancing the overall performance. Accordingly, Delphi approach and AHP method are used to derive evaluation measure indexes and their weights. And then two cases are illustrated the evaluation results.

2   Framework of OKC$_{BSC}$

In this research, an evaluation model based on balanced scorecard is proposed for evaluating the performance of online knowledge communities. The four dimensions: customer, internal process, learning and growth, and financial are the main foundation to be assessed. However, the non-profit characteristics of knowledge communities will lead to hard evaluate financial performance, so that the financial measurements are replaced with the measurements referring to strategic operation in the past period in this research. The four dimensions of OKC$_{BSC}$ are illustrated as Fig. 1.

After figuring out the four different dimensions and corresponding measure indices, this research inducts Delphi method by inviting twenty experts including knowledge experts, knowledge community managers, and IT engineers to obtain their opinions and confirms the appropriation refer to the framework of OKC$_{BSC}$. Through two rounds of Delphi meeting, the consensus on the framework is possessed by all interviewers and the measurements are listed in Table 1.

Once the measurements of OKC$_{BSC}$ are obtained, the corresponding importance of the indices with respective to the measurement will subsequently be decided. To realize the importance of the indices or the sub-measurements towards the main measurement, the AHP method (Analytical hierarchy process) is conducted in this research. The twenty experts as in previous Delphi meeting are invited. The AHP evaluation questionnaire was available which was according to the evaluation index in a hierarchical structure. Every expert was asked to identify the relative importance between two criteria which were belonging to the same dimension. The research adopted the Expert choice software derive the results of collected AHP questionnaires (in Table 1). In Table 1, "Customer" is the most important dimension (0.28), second ones are "Learning and growth" and "Strategic operation" dimensions (0.26), and "Internal process" dimension is the least important dimension (0.2). If looking at "Customer" dimension in details, it can be found that users regard "Member satisfaction" as a key point. As to the "learning and growth" dimension, "Relationship growth" is viewed as the most vital criteria and "Goal achievement" is highly crucial to the whole strategic operation of knowledge communities. And the "administration process" is thought as a critical consideration from internal process perspective.

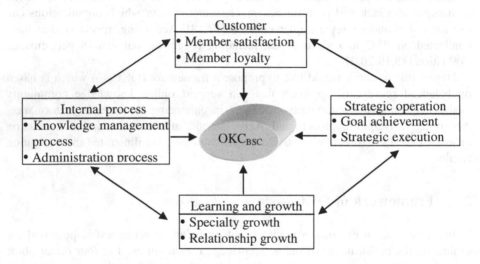

Fig. 1. Architecture of OKC$_{BSC}$

Finally, based on the aforementioned evaluation dimensions and the corresponding weights, the evaluation model for OKC can be expressed as follows:

$$KC\_Score = \sum_{i=1}^{m} A_i \times W_i$$

    A_i : the calculated score of main dimension
    W_i : the weight of main dimension
    m : the number of main dimensions, $m = 4$

$$A_i = \sum_{i=1}^{m} \sum_{j=1}^{n} A_{ij} \times W_{ij}$$

    A_i : the calculated score of main dimension

Table 1. Measurements and their corresponding evaluating weights in OKC$_{BSC}$

Dimensions	Sub-dimensions	Measurement	Measure indices	References
Customer perspective (CP) (0.28)	Member satisfaction (0.58)	Quality of service (0.46)	CP-1. Satisfaction with functions offered by knowledge virtual communities (0.41)	[3,6,9,24,25]
			CP-2. Satisfaction with services offered by knowledge virtual communities (0.59)	
		Quality of knowledge (0.26)	CP-3. My problems always can be resolved by the knowledge from other members (0.18)	
			CP-4. The knowledge contents are usually be accurate (0.36)	
			CP-5. The represented format of knowledge is clear and easy to identify on knowledge virtual communities (0. 20)	
			CP-6. I am satisfied with knowledge contents offered by knowledge virtual communities (0.26)	
		Quality of system (0.28)	CP-7. Satisfaction with the system because it is easy to use (0.51)	
			CP-8. Satisfaction with the system because it offers flexible and customized usage (0.49)	
	Member loyalty (0.42)	Relationship satisfaction (0.48)	CP-9. Satisfaction with the interaction with other members (0.23)	[3,5,26]
			CP-10. Satisfaction with the assistant from other members (0.29)	
			CP-11. Individual identified in the virtual group (0.17)	
			CP-12. individual reputation in this virtual group (0.31)	
		Sociality satisfaction (0.52)	CP-13. Common values exists among the members attracting members to involve (0.3)	[3, 25,26,27]
			CP-14. Individual involvement in this virtual group (0.22)	
			CP-15. I am respected and supported by other members (0.29)	
			CP-16. I have strong sense of belonging to this virtual group (0.19)	
Internal process (IP) (0.2)	Knowledge management process (0.45)	Knowledge acquisition (0.18)	IP-1. Knowledge categories can be recognized and searched easily (0.23)	[18]
			IP-2. Virtual community offers adequate search functions (0.32)	
			IP-3. Users can obtain extra knowledge by hyper linking to other web sites (0.19)	
			IP-4. Users can customize his own knowledge list easily (0.26)	
		Knowledge dissemination (0.27)	IP-5. Virtual community offers multimedia to share knowledge (0.44)	[18]
			IP-6. Virtual community offers multiple channels to share knowledge (0.31)	
			IP-7. Virtual community offers reward mechanisms to improve knowledge sharing (0.25)	
		Knowledge utilization (0.23)	IP-8. Virtual community offers billboard of knowledge utilization (0.33)	[18]
			IP-9. Virtual community can automatically recommend related knowledge to users (0.67)	
		Knowledge creation (0.32)	IP-10. Virtual community offers forums to proceed synchronous or asynchronous discussion (0.55)	[18]
			IP-11. The knowledge from other members can usually inspire development of new ideas or new knowledge (0.45)	

Table 1. (*continued*)

Administration process (0.55)	Member Security control (0.5)	IP-12. Virtual community offers account management to control valid users (0.3)		[28]
		IP-13. Virtual community offers account management to ensure statement or discuss (0.3)		
		IP-14. Virtual community offers mechanisms to guarantee users' privacy (0.4)		
	Intellectual protection (0.5)	IP-15. Knowledge sources or references have been properly cited (0.4)		[28]
		IP-16. Virtual community emphasize the intellectual rights (0.32)		
		IP-17. Virtual community maintains regularly and filters out unsuitable or unrelated knowledge (0.28)		
Learning and growth (LG) 0.26	Specialty growth (0.43)	Professional skills (0.36)	LG-1. Virtual community can improve my knowledge or capability of specific field (1)	[18,29]
		Problem solving skills (0.64)	LG-2. Virtual community can improve my capability of problem solving (0.51)	[18,29]
			LG-3. Virtual community can help me resolve some problems more fast (0.49)	
	Relationship growth (0.57)	Cooperation skills (0.39)	LG-4. Virtual community can improve my capability of cooperating with others	[30,31]
		Communication skills (0.32)	LG-5. Virtual community can improve my capability of coordinating with others (0.52)	
			LG-6. Virtual community can improve my capability of forming group consensus (0.48)	
		Expression skills (0.29)	LG-7. Virtual community can improve my opinions expression capability (0.48)	
			LG-8. Virtual community can improve my opinion comprehension capability (0.52)	
Strategic operation (SO) 0.26	Goal achievement (0.72)	Popularity (0.5)	SO-1. Most of people heart about the brand of virtual knowledge community (0.31)	[19]
			SO-2. The brand of this virtual knowledge community is popular (0.69)	
		Reputation (0.5)	SO-3. This virtual community is worth to visit frequently (0.28)	[19]
			SO-4. This virtual community is worth to recommend to others (0.12)	
			SO-5. This virtual community has great reputation (0.31)	
			SO-6. This virtual community is useful and valuable (0.29)	
	Strategies execution (0.28)	Promotion (1)	SO-7. This virtual community usually conduct promote activities to improve brand awareness (0.71)	[19]
			SO-8. This virtual community usually conduct promote activities to improve brand preference (0.29)	

A_{ij} : the score of sub-dimension towards main dimension

W_{ij} : the weight of sub-dimension towards main dimension

m : the number of main dimensions

n : the number of sub-dimension towards main dimension

$$A_{ij} = \sum_{i=1}^{s}\sum_{j=1}^{t}\sum_{k=1}^{u} A_{ijk} \times W_{ijk}$$

A_{ij} : the calculated score of sub-dimension

A_{ijk} : the score of measurements (or indices) towards sub-dimension

W_{ijk} : the weight of measurements (or indices) towards sub-dimension

s : the number of main dimensions
t : the number of sub-dimensions
u : the number of measurements (or indices) towards sub-dimension

3   Illustrative Cases and Evaluation Results

To demonstrate the proposed evaluation model, we conducted two case surveys to collect their opinions for knowledge community "Wikipedia" (Wikipedia, 2014) and "Yahoo!Kimo knowledge+" (Yahoo!Kimo knowledge+, 2014) separately. By the evaluation questionnaire with Likert five-point scale, the ultimate evaluation score of the communities was obtained by using the OKC$_{BSC}$. From 2012/2/1 to 2013/4/30, 847 questionnaires were returned in which 822 questionnaires were valid with 25 invalid questionnaires.

From the evaluation results of "Wikipedia" and of "Yahoo!Kimo Knowledge+", the finding are as follow.

(1) Wikipedia:

The result shows that the dimension of "customer" performs the best (1.0544), and the next one is "Strategic operation" (1.0163). "Learning and growth" is the third place (1.0038) and "Internal processes" needs to be improved further (0.745). If look at the details, it is found that Wikipedia has excellent performance of strategic operation (2.8), which including popularity and reputation, are the competitive advantage of this online community. Besides, members' satisfaction with the qualities of knowledge community (2.2475) is also the key factors to attract users to involve in. In Wikipedia, relationship growth (2.1616) is thought as one main benefit after involving this knowledge community because the large knowledge pool can attract more and more people to collaborate and cooperation. As a well-known online knowledge community, Wikipedia is evaluated to possess more complete administration process (2.016), including member security control and intellectual protection. On the other hand, Wikipedia is thought relatively insufficient on "Knowledge acquisition" (0.7122) and "knowledge utilization" (0.8742).

(2) Yahoo!Kimo Knowledge+:

The result shows this community performs well on "Customer" dimension (1.0369), "Strategic operation" (1.0234) is the second one, the third one "Learning and growth" (0.9787), and the least one is "Internal process" (0.7584). The evaluation results illustrate this community is thought as a valuable one in improving individual skills for specific problem solving (2.4385) because most of the knowledge relies on the solution experience exchange. In additions, popularity (2.0255) is another competitive advantage of this community. Alternatively, the knowledge acquisition (0.6957) and knowledge utilization (0.877) are evaluated inadequate in the whole process of knowledge management. As to the quality of knowledge in this online community (0.9428), it is considered need to be further improved in the future.

4   Conclusion

For the last decade, knowledge community has become popular resources of knowledge because it covers various types of knowledge over Internet. As more and more knowledge communities are developed, their evaluation become more complex and critical for users and community managers. In this research, an internal performance approach is adapted and an integrated evaluated framework based on Balanced Scorecard (BSC) is proposed to evaluate the performance of online knowledge community by four dimensions: customer, internal process, learning and growth, and strategic operation. The evaluation result of OKC_{BSC} can not only help users distinguish better knowledge communities from others, but also can give managers a valuable feedback used to precede self-improvement for online community in the future.

References

1. Chen, C.J., Hung, S.W.: To give or to receive? Factors influencing members' knowledge sharing and community promotion in professional virtual communities. Information & Management 47(4), 226–236 (2010)
2. Wang, J.F.: E-commerce communities as knowledge bases for firms. Electronic Commerce Research and Applications 9(4), 335–345 (2010)
3. Lin, H.: Determinants of successful virtual communities: Contributions from system characteristics and social factors. Information & Management 45(8), 522–527 (2008)
4. Lin, M.J.J., Hung, S.W., Chen, C.J.: Fostering the determinants of knowledge sharing in professional virtual communities. Computers in Human Behavior 25(4), 929–939 (2009)
5. Casalo, L.V., Flavian, C., Guinaliu, M.: Relationship quality, community promotion and brand loyalty in virtual communities: Evidence from free software communities. International Journal of Information Management 30(4), 357–367 (2010)
6. Chen, Y.J., Chen, Y.M., Wu, M.S.: An empirical knowledge management framework for professional virtual community in knowledge-intensive service industries. Expert Systems with Applications 39, 13135–13147 (2012)
7. Elliot, S., Li, G., Choi, C.: Understanding service quality in a virtual travel community environment. Journal of Business Research 66(8), 1153–1160 (2013)
8. Alali, H., Salim, J.: Virtual Communities of Practice Success Model to Support Knowledge Sharing Behaviour in Healthcare Sector. Procedia Technology 11, 176–183 (2013)
9. Hsu, F.M., Fan, C.T., Lin, C.M., Chiu, C.M.: Factors Affecting the Satisfaction of Participants in Community. Procedia - Social and Behavioral Sciences 73(27), 418–423 (2013)
10. Hong, H.Y., Scardamalia, M.: Community knowledge assessment in a knowledge building Environment. Computers & Education 71, 279–288 (2014)
11. Thomson, A.J.: Indicator-based knowledge management for participatory decision-making. Computers and Electronics in Agriculture 49(1), 206–218 (2005)
12. Ruuska, I., Vartiainen, M.: Characteristics of knowledge sharing communities in project organizations. International Journal of Project Management 23(5), 374–379 (2005)
13. Lin, F.R., Lin, S.C., Huang, T.P.: Knowledge sharing and creation in a teachers' professional virtual community. Computers & Education 50(3), 742–756 (2008)
14. Chen, R.S., Hsiang, C.H.: A study on the critical success factors for corporations embarking on knowledge community-based e-learning. Information Sciences 177(2), 570–586 (2007)
15. Kaplan, R.S., Norton, D.P.: The balance scorecard: measures that drive performance. Harvard Business Review 70(1), 71–79 (1992)

16. Kaplan, R.S., Norton, D.P.: Using the balance scorecard as a strategic management system. Harvard Business Review 74(1), 75–85 (1996)
17. Huang, H.C.: Designing a knowledge-based system for strategic planning: A balanced scorecard perspective. Expert Systems with Applications 36, 209–218 (2009)
18. Chen, M.Y., Huang, M.J., Cheng, Y.C.: Measuring knowledge management performance using a competitive perspective: An empirical study. Expert Systems with Applications 36, 8449–8459 (2009)
19. Wu, H.Y., Lin, Y.K., Chang, C.H.: Performance evaluation of extension education centers in universities based on the balanced scorecard. Evaluation and Program Planning 34, 37–50 (2011)
20. Kunz, H., Schaaf, T.: General and specific formalization approach for a Balanced Scorecard: An expert system with application in health care. Expert Systems with Applications 38, 1947–1955 (2011)
21. Sordo, C.D., Orelli, R.L., Padovani, E., Gardini, S.: Assessing global performance in universities: an application of balanced scorecard. Procedia - Social and Behavioral Sciences 46, 4793–4797 (2012)
22. Sainaghi, R., Phillips, P., Corti, V.: Measuring hotel performance: Using a balanced scorecard perspectives' approach. International Journal of Hospitality Management 34, 150–159 (2013)
23. Lee, S., Park, S.B., Lim, G.G.: Using balanced scorecards for the evaluation of "Software-as-a-service". Information & Management 50, 553–561 (2013)
24. Tsai, H.T., Pai, P.: Explaining members' proactive participation in virtual communities. Int. J. Human-Computer Studies 71(4), 475–491 (2013)
25. Zheng, Y.M., Zhao, K., Stylianou, A.: The impacts of information quality and system quality on users' continuance intention in information-exchange virtual communities: An empirical investigation. Decision Support Systems 56, 513–524 (2013)
26. Tsai, H.T., Pai, P.: Why do newcomers participate in virtual communities? An integration of self-determination and relationship management theories. Decision Support Systems 57, 178–187 (2014)
27. Zhao, L., Lu, Y., Wang, B., Chau, P.Y.K., Zhang, L.: Cultivating the sense of belonging and motivating user participation in virtual communities: A social capital perspective. International Journal of Information Management 32(6), 574–588 (2012)
28. Tamjidyamcholo, A., Baba, M.S.B.: Evaluation model for knowledge sharing in information security professional virtual community. Computers & Security 43, 19–34 (2014)
29. Papalexandris, A., Ioannou, G., Prastacos, G., Soderquist, K.E.: An integrated methodology for putting the balanced scorecard into action. European Management Journal 23(2), 214–227 (2005)
30. Henttonen, K.: Exploring social networks on the team level—A review of the empirical literature. Journal of Engineering and Technology Management 27(1-2), 74–109 (2010)
31. Luo, J.D.: Social network structure and performance of improvement teams. International Journal of Business Performance Management 7(2), 208–223 (2005)

Opinion Leadership and Negative Word-of-Mouth Communication

Chih-Chien Wang[1], Pei-Hua Wang[1], and Yolande Y.H. Yang[2]

[1] Graduate Institute of Information Management, National Taipei University, Taiwan
wangson@mail.ntpu.edu.tw, gnb10731@gmail.com
[2] Department of Business Administration, National Taipei University, Taiwan
yolande@mail.ntpu.edu.tw

Abstract. Customers may feel negative emotion when they experience service failure. The negative emotion may induce unsatisfactory customers to spread negative word-of-mouths (WOM). However, not all unsatisfactory will spread negative WOM. The current study conducted an experimental design to explore the influence of opinion leadership tendency to negative word-of-mouth communication intention. The results revealed that customers will spread negative WOMs when the service failure is serious. However, when the service failure is minor, customers with a higher opinion leadership tendency are with higher intention to spread negative WOMs. The findings of the current are useful in exploring the role of opinion leadership tendency in negative WOM communication.

Keywords: Negative word-of-mouth, Service failure, Negative emotion, Opinion leadership.

1 Introduction

With the growing popularity of the Internet, consumers can share their experiences of products and services and search for others' experiences through the Internet (Liu, 2006; Hung & Li, 2007). People usually search for information online, and gather information about products or brands through eWOM (Hennig-Thurau & Walsh, 2004). Literatures revealed that eWOMs are important message sources (Litvin, Goldsmith, & Pan, 2008; Simpson & Siguaw, 2008).

Past researches also revealed that sharing negative experiences is an effective way to vent negative emotions and to reduce anxiety (Nyer, 1997; Richinsn1984). After the consumption process, the customer will turn the feelings and thoughts into internal reference of psychological assessment. When consumers felt any dissatisfaction to a product or service, they may tell friends and relatives, and persuade them not to use the product of the service (Litvin et al., 2008; Parra-López, Bulchand-Gidumal, Gutiérrez-Taño, & Díaz-Armas, 2011). Singh (1990) pointed out that negative WOM is among the worst responses a business can get, so managing negative WOM has been an important task for business managers.

L.S.-L. Wang et al. (Eds.): MISNC 2014, CCIS 473, pp. 36–47, 2014.
© Springer-Verlag Berlin Heidelberg 2014

2   Literature Review and Hypotheses Development

2.1   Word of Mouth (WOM) and e-WOM Communication

Arndt [1] defined WOM as "face-to-face communication between people who were not commercial entities for products, services or companies." Harrison-Walker [2] defined WOM as "informal, person-to-person communication about brands, products, organizations or services between a noncommercial communicator and a receiver." WOM has received extensive attention from marketing managers and researchers. Literatures had shown that interpersonal conversations and information exchange influenced consumers' choices and pre-purchase decision-making [1, 3]. WOMs would shape consumer expectations [4], pre-usage attitudes [5], and even post-usage perceptions of a product or service [6, 7]. Previous research has demonstrated that WOMs are more persuasive than advertising and personal selling [8-10].

Litvin, Goldsmith and Pan [11] revealed that electronic words of mouth (e-WOMs) are informal communications about "the usage or characteristics of products and services in Internet-based context." Due to the advancement and popularity of network technology, people now can use the Internet to spread or receive WOM. Customers now can post their opinions, comments and reviews of products or services on discussion forums, review websites, bulletin board systems, and social networking sites [12].

Compared to traditional WOM, e-WOM communications are more persistent and more accessible for an indefinite period [13-18]. Online e-WOMs are also more abundant   than traditional WOM through contacts in the offline world [18]. Therefore, the importance of e-WOMs is greater than ever for product and service providers.

2.2   Negative WOM

Not all WOMs are positive ones. Negative WOMs usually exist when dissatisfaction or service failure happen. Consumers distribute negative WOM to communicate a dissatisfying consumption experience [19]. Consumers may narrate the cause of their dissatisfaction in order to get a solution through a negative WOM [20] or use WOM to vent negative feelings to reduce anxiety [21, 22]. Generally speaking, negative WOM is more influential than positive WOM, as the "negativity effect" caused by negative information has more impact on consumers' judgment and perceptions than the effect of positive information [5, 23, 24]. Consumers tend to pay more attention to negative WOM than positive WOM [25], and trust negative messages more than they trust positive messages. Literature revealed that the negative WOMs have a stronger influence on purchase decision than the positive ones [26] since they tend to weight negative WOMs more than they weight positive ones in the purchase decision process [27].

2.3   Service Failure

The most frequently reported action of dissatisfied consumers was telling friends about the negative consumption experience [28, 29]. Thus, service failure is highly connected with negative WOMs. According to Hess Jr, Ganesan and Klein [30], service failure can be defined as service performance that is below customers' expectations, where customers find the service to be flawed and irresponsible [31]. Literature indicated a negative correlation between service failure severity and customer relationships [32]. After customers experienced service failure, they may spread negative WOM about the service provider. They may also directly or indirectly complain to the service providers as well as third parties [33-37].

2.4   Service Failure, Negative Emotion and Negative WOM Intention

Wetzer, Zeelenberg and Pieters [38] indicated that individuals' engaging in negative e-WOM was positively influenced by negative emotion. Therefore, negative emotions often occur in a situation after service failure [21, 39, 40]. Negative emotions are likely to elicit individuals to post and engage in negative e-WOM [38]. It is obvious that the unfavorable consumption experience usually cause negative emotion which makes customers engage in a series of subsequent coping behavior [36, 41, 42]. Service failure may also trigger negative emotional responses. Some customers are more likely to post a negative e-WOM to vent their negative feelings. Previous researches have revealed the impact on negative WOM communication of negative emotions, such as anger [21, 43-45], disappointment [36], regret [36, 46], worry [43, 47] and guilt [48].

This study focuses on the influence of anger and disappointment on negative e-WOM. Based on the literature mentioned above, we proposed the following hypothesis:

H1:   *Consumers who have encountered serious service failure will generate stronger negative emotion, and have a higher intention to spread or share negative e-WOM than the ones who encounter less serious service failure.*

2.5   Opinion Leadership Tendency and Negative WOM Intention

Opinion leadership was defined as the individuals' disposition who were able to influence opinions, attitudes, beliefs, motivations, and behaviors of others [10, 49-53]. Opinion leaders are always capable of proposing new information, thoughts, and their opinion, and then spreading to others or the general public. They tend to offer the most representative opinions and turn into an influential node [54]. Overall, opinion leaders can be viewed as the central communicators of market, who determine the decisions of other consumers.

Opinion leaders always have higher levels of interest, knowledge, and opinion about social issues than other persons [55]. They viewed themselves as the forerunners of social trends, and early adopters of innovations about a new things or products [56, 57]. They often regard themselves as intelligent, independent, and possess personal judgments about public issues.

Opinion leaders are more talkative and have a stronger interpersonal interaction and social orientation [58, 59]. Besides, opinion leaders are regarded to be credible because of their expertise. They frequently share positive and negative WOMs with others. Hence, they tend to become a major source of WOM communication [59-63]. Thus, we presumed that the individual with a higher opinion leadership tendency will join the discussion of a negative WOM to share their opinion and thoughts with others. Thus, the study proposed the following hypothesis:

H2: *Consumers with a higher opinion leadership tendency have a higher intention to spread or share negative e-WOM.*

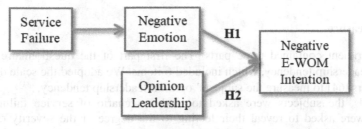

Fig. 1. Research model

3   Empirical Study

The current study conducted an experimental design to verify the relationship between opinion leadership and negative WOM intention. We also explore the influence of negative emotion and service failure on intention of spreading e-WOM.

3.1   Scenario Designs

This study created two scenarios: one was serious service failure, and the other was minor service failure. The two scenarios are based on the real negative WOMs.

Before conducting the experiment, we conducted a pilot test to check the difference of two scenarios. The degree of severity of the service failure was used to check the effectiveness of the manipulation, on the 7-point Likert-type scale with 1 representing minor and 7 representing serious service failure. A total of 30 subjects participated in the pilot questionnaire. Among them, 15 were tested with the scenario of serious service failure and 15 were tested with the scenario of minor service failure. We performed an independent sample t-test analysis, and the result revealed that the service failure serverity was a significant difference ($t = 7.419;$ $p < 0.001$) between the serious and minor sevice failure scenarios, which indicated a successful manipulation.

3.2   Samples

The study recruited subjects from a large online community. We posted call for voluntarily subject messages on a virtual community for a two weeks period. When volunteers clicked the hyperlink to join this study, they were randomly assigned to one of two different scenarios.

A total of 178 voluntary subjects participated in the study. After eliminating samples with incomplete data, 148 complete responses were accepted for analysis. Among participants, 77(52.03%) were male, and 71(49.97%) were female. The age of the subjects were 24.34 years on average (SD = 4.85), with an age range from 11 to 43. Among the participants, 93(62.84%) had a college degree, and 44(29.73%) had a master degree. 85(57.43%) of all participants were students, and 49(33.11%) were employed. 14(9.46%) participants were unemployed.

3.3   Measure

The experiment contained three parts. The first part of the questionnaire was the opinion leadership tendency, which included 6 items. We adopted the scale developed by Childers [64] to measure the subjects' opinion leadership tendency.

Afterwards, the subjects were asked to read a scenario of service failure. Then, subjects were asked to reveal their feeling to the degree of the severity of service failure reported in the scenario. In addition, they were asked to reveal their feelings of disappointment and anger to the scenario. We adopted the scale developed by Yi and Baumgartner [41] to measure the subjects' disappointment, which included three items. The study measure anger using three items used by Zeelenberg and Pieters [36] and Yi and Baumgartner [41].

In the third part, the subjects were asked to answer the intention to generate or spread negative e-WOM to friends and the public using a scale of five items. Among the five items, two items for private WOM communication was proposed by Zeelenberg and Pieters [36], while the other three items for public WOM communication were developed by the current study.

The final part was demographic variables which included gender, age, education, and occupation. The participants who completed the questionnaire were offered 200 virtual points of the virtual community as incentive. All the measurement items were on a seven-point Likert scale, with responses anchored by 1 ("strongly disagree") and 7 ("strongly agree").

3.4   Reliabilities and Validity

The Cronbach's alpha coefficients were 0.92, 0.74, and 0.96 for opinion leadership, disappointment, and anger, respectively. All Cronbach's alpha scored above 0.7. The composite reliability (CR) of opinion leadership was 0.94, disappointment was 0.85, anger was 0.97, spread negative WOMs to friends was 0.74, and spread negative WOMs to public was 0.87. All CR and Cronbach's alpha coefficients exceeded the recommended threshold of 0.7, which indicated acceptable reliability and stability for the measurement items.

The AVE value was 0.71 for opinion leadership, 0.90 for disappointment, 0.93 for angers, 0.59 for sprending negative WOMs to friends, and 0.69 for spreading negative WOMs to the public. All of the AVE values exceeded 0.5, which indicated a good convergent validity. The square root of AVE of each dimension exceeded the correlation coefficients, which also indicated a good discriminant validity, as shown in Table 1.

3.5   Data Analysis and Result

The study designed two different scenarios; one was severe service failure, and the other was minor service failure. Through the two scenarios, we could further explore the relationship between opinion leadership, service failure, emotion and negative e-WOM. The study compared the two scenarios through the independent sample t-test analysis. Table 2 showed the means, standard deviations, and t-test analysis result of the two scenarios.

Table 1. Correlations and Discriminant validity

	Variables	AVE	1	2	3	4	5
1.	Opinion leadership	.71	.84				
2.	Disappointment	.90	.174*	.95			
3.	Anger	.93	.197*	.792**	.97		
4.	Spread NWOM to friend	.59	.232**	.228**	.343**	.77	
5.	Spread NWOM to public	.69	.319**	.496**	.634**	.474**	.83

$*p < .05$; $**p < .01$; the diagonal line values are the square root of AVE while other values are correlation coefficients.

Table 2. The independent sample t-test of serious and minor service failure scenarios

		Serious service failure (N=74)	Minor service failure (N=74)	t value	Sig.
Failure degree	Mean (S.D.)	6.18 (0.85)	4.26 (1.48)	9.667	.000***
Disappointment	Mean (S.D.)	6.14 (0.90)	4.04 (0.43)	18.017	.000***
Anger	Mean (S.D.)	5.38 (1.06)	2.96 (1.16)	13.304	.000***
WOM to Friends	Mean (S.D.)	3.90 (1.21)	3.47 (1.05)	2.323	.022*
WOM to Public	Mean (S.D.)	4.32 (1.30)	3.27 (1.19)	5.112	.000***

$*p < .05$; $***p < .001$

The t-test analysis reported that there were significant different in failure degree ($t = 9.667; p < 0.001$), disappointment ($t = 18.017; p < 0.001$), anger ($t = 13.304; p < 0.001$), WOM to friends ($t = 2.323; p < 0.05$), and WOM to public ($t = 5.112; p < 0.001$). The Table 2 reveals that the degree of severity of service failure was significant different in the two scenarios. When subjects read severe service failure scenario, they were have higher scores of disappointment and anger. Besides, the result also revealed that when customer experienced serious service failure, they had greater intention to spread negative e-WOM to friends and public than minor service failure. Thus, the *H1* was supported.

We further explore the influence of opinion leadership in the two different scenarios. The negative emotion included two dimensions—disappointment and anger—which suited well in a second-order structure. For the second order structure, Partial Least Squares (PLS) could be used to perform the structure model [65]. Hence, we used the SmartPLS software to examine the relationship among opinion leadership, negative emotion, and intention of WOM.

In the serious service failure scenario, the PLS analysis result of Figure 2 revealed that the negative emotion has a positive influence on the intention of WOM. It means consumer will more likely spread negative WOM to friends (*path coefficient = 0.233; p < 0.001*) and the public (*path coefficient = 0.362; p < 0.001*) when they have negative emotion, while opinion leadership had only a significant influence on the intention to spread negative WOM to the public (*path coefficient = 0.286; p < 0.001*). The PLS analysis result as shown in Figure 3 revealed that negative emotion also positively affect the intention to spread negative WOM to friends (*path coefficient = 0.226; p < 0.001*) and the public (*path coefficient = 0.641; p < 0.001*) in the minor service failure scenario, while the opinion leadership has significant influence on the intention to spread negative WOM to friends (*path coefficient = 0.399; p < 0.001*) and the public (*path coefficient = 0.208; p < 0.001*). Thus, *H2* was supported.

We found that opinion leadership did not have significant influence on the intention to share their negative WOM to friends in the serious service failure scenario. A possible explanation was that people will spread negative WOMs to friends when they encounter serious service failure no matter they are with high or low opinion leadership tendency. Opinion leadership tendency will not impact the intention of e-WOM to friends in the serious service failure. However, in a minor service failure scenario, people with a less opinion leadership tendency will not spend negative WOMs to their friend. Thus, in a minor service failure scenario, opinion leadership was positively related with negative WOM intention.

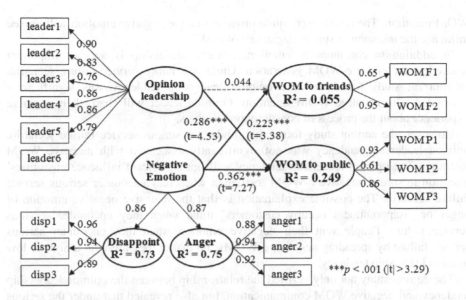

Fig. 2. The PLS analysis of the serious service failure scenario

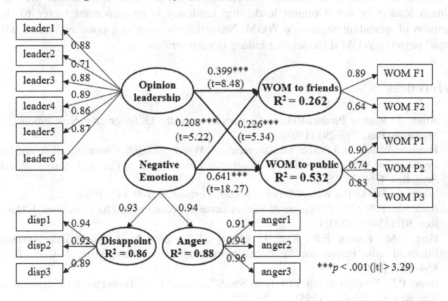

Fig. 3. The PLS analysis of the minor service failure scenario

4   Conclusions

Previous research indicated that the service failure will cause customers' negative emotions, which are likely to lead them to engage in negative e-WOM. We created two scenarios to examine the role of opinion leadership and negative emotion in

WOM intention. The results were quite obvious that the negative emotion will indeed influence the intention to spread negative e-WOM.

In addition to consumption emotion, opinion leadership is another important antecedent factor for e-WOM generation. Under the minor service failure scenario, the current study found that the opinion leadership tendency was significantly associated with negative WOM intention. Opinion leader will share their negative experience about the process of consumption to others.

However, the current study found that, under the serious service failure scenario, opinion leadership tendency was not significantly associated with negative WOM intention. The opinion leadership tendency will not significant influence customers' intention to spread negative e-WOM to friends when they encounter serious service failure scenario. The possible explaination is that the intensive negative emotion of anger or disappointment occupy customers' mind when they encounter servious service failure. People vent their negative emotion when they encounter serious service failure by spreading negative e-WOMs, no matter they are with a high or low opinion leadership tendency.

The current study not only verified the relationship between the opinion leadership tendency and negative WOM communication, but also revealed that under the serious service failure scenario, consumers will spread negative e-WOM, no matter they are opinion leaders or not. Opinion leadership tendency is an important factor to the intention of spreading negative e-WOM. Nevertheless, even non-opinion leaders will spread negative WOM if the service failure is very serious.

References

1. Arndt, J.: Role of Product-Related Conversations in the Diffusion of a New Product. J. Marketing Res., 291–295 (1967)
2. Harrison-Walker, L.J.: The Measurement of Word-of-Mouth Communication and an Investigation of Service Quality and Customer Commitment as Potential Antecedents. J. Serv. Res. 4, 60–75 (2001)
3. Whyte Jr., W.H.: The Web of Word of Mouth. Fortune 50, 140–143 (1954)
4. Anderson, E.W.: The Formation of Market-Level Expectations and Its Covariates. J. Cons. Res. 30, 115–124 (2003)
5. Herr, P.M., Kardes, F.R., Kim, J.: Effects of Word-of-Mouth and Product-Attribute Information on Persuasion: An Accessibility-Diagnosticity Perspective. J. Cons. Res., 454–462 (1991)
6. Bone, P.F.: Word-of-Mouth Effects on Short-Term and Long-Term Product Judgments. J. Bus. Res. 32, 213–223 (1995)
7. Burzynski, M.H., Bayer, D.J.: The Effect of Positive and Negative Prior Information on Motion Picture Appreciation. J. Soc. Psychol. 101, 215–218 (1977)
8. Engel, J.F., Kegerreis, R.J., Blackwell, R.D.: Word-of-Mouth Communication by the Innovator. J. Marketing, 15–19 (1969)
9. Feldman, S.P., Spencer, M.C.: The Effect of Personal Influence on the Selection of Consumer Services. Center for regional studies (1965)
10. Katz, E., Lazarsfeld, P.F.: Personal Influence, the Part Played by People in the Flow of Mass Communications. Free Press, New York (1955)

11. Litvin, S.W., Goldsmith, R.E., Pan, B.: Electronic Word-of-Mouth in Hospitality and Tourism Management. Tour. Manage. 29, 458–468 (2008)
12. Cheung, C.M., Lee, M.K.: What Drives Consumers to Spread Electronic Word of Mouth in Online Consumer-Opinion Platforms. Decis. Support Syst. 53, 218–225 (2012)
13. Hennig-Thurau, T., Gwinner, K.P., Walsh, G., Gremler, D.D.: Electronic Word-of-Mouth Via Consumer-Opinion Platforms: What Motivates Consumers to Articulate Themselves on the Internet? J. Interact. Mark. 18, 38–52 (2004)
14. Hung, K.H., Li, S.Y.: The Influence of Ewom on Virtual Consumer Communities: Social Capital, Consumer Learning, and Behavioral Outcomes. J. Advertising Res. 47, 485 (2007)
15. Lee, J., Park, D.-H., Han, I.: The Effect of Negative Online Consumer Reviews on Product Attitude: An Information Processing View. Electron. Commer. R. A. 7, 341–352 (2008)
16. Park, C., Lee, T.M.: Information Direction, Website Reputation and Ewom Effect: A Moderating Role of Product Type. J. Bus. Res. 62, 61–67 (2009)
17. Park, D.-H., Lee, J., Han, I.: The Effect of on-Line Consumer Reviews on Consumer Purchasing Intention: The Moderating Role of Involvement. Int. J. Electron. Comm 11, 125–148 (2007)
18. Sen, S.: Determinants of Consumer Trust of Virtual Word-of-Mouth: An Observation Study from a Retail Website. J. Am. Acad. Bus. Camb. 14, 30 (2008)
19. Anderson, E.W.: Customer Satisfaction and Word of Mouth. J. Serv. Res. 1, 5–17 (1998)
20. Thøgersen, J., Juhl, H.J., Poulsen, C.S.: Complaining: A Function of Attitude, Personality, and Situation. Psychology and Marketing 26, 760–777 (2009)
21. Nyer, P.U.: A Study of the Relationships between Cognitive Appraisals and Consumption Emotions. J. Acad. Mark. Sci. 25, 296–304 (1997)
22. Richins, M.L.: Word of Mouth Communication as Negative Information. Adv. Consum. Res. 11, 697–702 (1984)
23. Fiske, S.T.: Attention and Weight in Person Perception: The Impact of Negative and Extreme Behavior. JPSP 38, 889 (1980)
24. Mittal, V., Ross Jr., W.T., Baldasare, P.M.: The Asymmetric Impact of Negative and Positive Attribute-Level Performance on Overall Satisfaction and Repurchase Intentions. J. Marketing, 33–47 (1998)
25. Yang, J., Mai, E.S.: Experiential Goods with Network Externalities Effects: An Empirical Study of Online Rating System. J. Bus. Res. 63, 1050–1057 (2010)
26. Mizerski, R.W.: An Attribution Explanation of the Disproportionate Influence of Unfavorable Information. J. Cons. Res., 301–310 (1982)
27. Sen, S., Lerman, D.: Why Are You Telling Me This? An Examination into Negative Consumer Reviews on the Web. J. Interact. Mark. 21, 76–94 (2007)
28. Lau, G.T., Ng, S.: Individual and Situational Factors Influencing Negative Word-of-Mouth Behaviour. Canadian Journal of Administrative Sciences 18, 163–178 (2001)
29. Leonard-Barton, D.: Experts as Negative Opinion Leaders in the Diffusion of a Technological Innovation. J. Cons. Res., 914–926 (1985)
30. Hess Jr., R.L., Ganesan, S., Klein, N.M.: Interactional Service Failures in a Pseudorelationship: The Role of Organizational Attributions. J. Retailing 83, 79–95 (2007)
31. Palmer, A., Beggs, R., Keown-McMullan, C.: Equity and Repurchase Intention Following Service Failure. J. Serv. Mark. 14, 513–528 (2000)
32. Bejou, D., Palmer, A.: Service Failure and Loyalty: An Exploratory Empirical Study of Airline Customers. J. Serv. Mark. 12, 7–22 (1998)
33. Day, R.L., Landon, E.L.: Toward a Theory of Consumer Complaining Behavior. Consum. Ind. Buy. Behav. 95, 425–437 (1977)

34. Day, R.L., Grabicke, K., Schaetzle, T., Staubach, F.: The Hidden Agenda of Consumer Complaining. J. Retailing (1981)
35. Singh, J.: Consumer Complaint Intentions and Behavior: Definitional and Taxonomical Issues. J. Marketing, 93–107 (1988)
36. Zeelenberg, M., Pieters, R.: Beyond Valence in Customer Dissatisfaction: A Review and New Findings on Behavioral Responses to Regret and Disappointment in Failed Services. J. Bus. Res. 57, 445 (2004)
37. Kim, M.G., Wang, C., Mattila, A.S.: The Relationship between Consumer Complaining Behavior and Service Recovery: An Integrative Review. Int. J. Contemp. Hosp. Manage. 22, 975–991 (2010)
38. Wetzer, I.M., Zeelenberg, M., Pieters, R.: "Never Eat in That Restaurant, I Did!": Exploring Why People Engage in Negative Word-of-Mouth Communication. Psychology and Marketing 24, 661–680 (2007)
39. Richins, M.L.: Measuring Emotions in the Consumption Experience. J. Cons. Res. 24, 127–146 (1997)
40. Laros, F.J., Steenkamp, J.-B.E.: Emotions in Consumer Behavior: A Hierarchical Approach. J. Bus. Res. 58, 1437–1445 (2005)
41. Yi, S., Baumgartner, H.: Coping with Negative Emotions in Purchase-Related Situations. J. Consum. Psychol. 14, 303–317 (2004)
42. Mattila, A.S., Ro, H.: Discrete Negative Emotions and Customer Dissatisfaction Responses in a Casual Restaurant Setting. J. Hosp. Tour. Res. 32, 89–107 (2008)
43. Maute, M.F., Dubés, L.: Patterns of Emotional Responses and Behavioural Consequences of Dissatisfaction. Appl. Psychol. 48, 349–366 (1999)
44. Dubé, L., Maute, M.: The Antecedents of Brand Switching, Brand Loyalty and Verbal Responses to Service Failure. Adv. Serv. Mark. Manage. 5, 127–151 (1996)
45. Bougie, R., Pieters, R., Zeelenberg, M.: Angry Customers Don't Come Back, They Get Back: The Experience and Behavioral Implications of Anger and Dissatisfaction in Services. J. Acad. Mark. Sci. 31, 377–393 (2003)
46. Zeelenberg, M., Pieters, R.: Comparing Service Delivery to What Might Have Been Behavioral Responses to Regret and Disappointment. J. Serv. Res. 2, 86–97 (1999)
47. Derbaix, C., Vanhamme, J.: Inducing Word-of-Mouth by Eliciting Surprise–a Pilot Investigation. J. Econ. Psych. 24, 99–116 (2003)
48. Soscia, I.: Gratitude, Delight, or Guilt: The Role of Consumers' Emotions in Predicting Postconsumption Behaviors. Psychology and Marketing 24, 871–894 (2007)
49. Rogers, E.M.: Diffusion of Innovations. Free Press, New York (1962)
50. Flynn, L.R., Goldsmith, R.E., Eastman, J.K.: Opinion Leaders and Opinion Seekers: Two New Measurement Scales. J. Acad. Mark. Sci. 24, 137–147 (1996)
51. Hellevik, O., Bjørklund, T.: Opinion Leadership and Political Extremism. Int. J. Opin. Res. 3, 157–181 (1991)
52. Mowen, J.C.: Consumer Behavior. Macmillian, New York (1990)
53. Rogers, E.M.: Diffusion of Innovations. Free Press, New York (1983)
54. Song, X., Chi, Y., Hino, K., Tseng, B.: Identifying Opinion Leaders in the Blogosphere. In: Proceedings of the Sixteenth ACM Conference on Information and Knowledge Management, pp. 971–974. ACM (Year)
55. Weimann, G.: The Influentials: People Who Influence People. State University of New York Press, Albany (1994)
56. Rogers, E.M.: Diffusion of Innovations, 4th edn. Free Press, New York (1955)
57. Summers, J.O.: The Identity of Women's Clothing Fashion Opinion Leaders. J. Marketing Res. 7 (1970)

58. Weimann, G.: The Influentials: Back to the Concept of Opinion Leaders? Public Opin. Q. 55, 267–279 (1991)
59. Venkatraman, M.P.: Opinion Leaders, Adopters, and Communicative Adopters: A Role Analysis. Psychology and Marketing 6, 51–68 (1989)
60. Fieck, L.F., Price, L.L., Higie, R.A.: People Who Use People: The Other Side of Opinion Leadership. Adv. Consum. Res. 13 (1986)
61. Feick, L.F., Price, L.L.: The Market Maven: A Diffuser of Marketplace Information. J. Marketing, 83–97 (1987)
62. Bertrandias, L., Goldsmith, R.E.: Some Psychological Motivations for Fashion Opinion Leadership and Fashion Opinion Seeking. J. Fash. Mark. Manag. 10, 25–40 (2006)
63. Schiffman, L.G., Kanuk, L.L.: Consumer Behavior (1991)
64. Childers, T.L.: Assessment of the Psychometric Properties of an Opinion Leadership Scale. J. Marketing Res., 184–188 (1986)
65. Wetzels, M., Odekerken-Schröder, G., Van Oppen, C.: Using Pls Path Modeling for Assessing Hierarchical Construct Models: Guidelines and Empirical Illustration. MIS Quart. 33 (2009)

Recommendations of E-commerce Seller
Based on Buyer Feedbacks

Yin-Fu Huang and Yu-Chin Yang

Department of Computer Science and Information Engineering
National Yunlin University of Science and Technology
{huangyf,m10017016}@yuntech.edu.tw

Abstract. Online shopping will become increasingly popular as more and more people have been changing their shopping ways from traditional to online shopping. However, sometimes it is difficult for a buyer to decide which seller is good for his/her purchases. Therefore, it is necessary to design a system that can recommend sellers for a buyer to purchase his/her products from them. In this paper, a seller recommendation system using user feedback is proposed. The seller recommendation system is to rank the sellers selling the products specified by buyers. Buyers can specify products and budgets as they need. Finally, the experimental results show that an seller being able to provide excellent services should be experienced for more than 10 years and our system can actually reflect the real situation by contrasting the top rated sellers recognized by the eBay site.

Keywords: E-commerce, recommendation system, user feedback, XML document.

1   Introduction

Online shopping will become increasingly popular as more and more people have been changing their shopping ways from traditional to online shopping. However, sometimes it is difficult for a buyer to decide which seller is good for his/her purchases. For example, when the product with the same price can be purchased from different sellers, which one is a good choice for the purchase. Therefore, it is necessary to design a system that can recommend sellers for a buyer to purchase his/her products from them.

In recent years, many studies focused on auctions and e-commerce platforms, but most of them tended to explore the prevention of fraudulence [1, 7], security [3, 8] and transactional behaviors [9] and few studies investigated the quality of sellers' services. Morzy and Jezierski [4] proposed the topology of connections between sellers and buyers to derive knowledge about trustworthy sellers; they discovered that clusters of densely connected sellers can be used as the prediction of future performance for a seller. However, user feedback which can be as an indicator of a seller's reputation [2, 5, 6] was not considered there. In this paper, a seller

L.S.-L. Wang et al. (Eds.): MISNC 2014, CCIS 473, pp. 48–59, 2014.
© Springer-Verlag Berlin Heidelberg 2014

recommendation system using user feedback is proposed. The seller recommendation system is to rank the sellers selling the products specified by buyers. Buyers can specify products and budgets as they need. First, the system searches matched products, then finds out the corresponding sellers, and finally ranks them according to their scores. The seller scores can be calculated using feedback rating, seller popularity, and seller registration year.

The remainder of this paper is organized as follows. In Section 2, we review basic concepts used in this paper. Then, the framework of a seller recommendation system is proposed in Section 3. The system implementation and some experimental results are described in Section 4. Finally, we make conclusions in Section 5.

2   Basic Concepts

There are various tools and techniques closely related to our work. We would briefly review these tools and techniques.

2.1   XML Documents

An XML document can be represented as a tree structure. A document tree could contain element nodes, attribute nodes, and text nodes which are all mapped from elements, attributes, and texts specified in an XML document. For the XML document as shown in Fig. 1, it has five elements (i.e., book, title, prod, chapter, and para), two attributes (i.e., id and media), and their texts. And, these elements, attributes, and texts are denoted by circles, triangles, and rectangles in the tree representation as shown in Fig. 2, respectively.

```
<book>
        <title>My First XML</title>
        <prod id="33-657"media="paper"></prod>
        <chapter>Introduction to XML
                    <para>What is HTML</para>
        </chapter>
</book>
```

Fig. 1. XML document

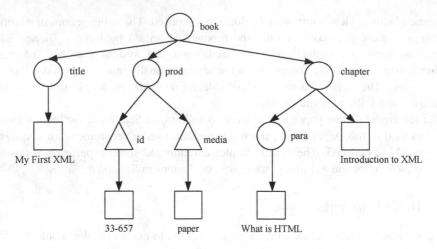

Fig. 2. Tree representation of the XML document

2.2   XML Path Language

The XML Path Language [13] (i.e., XPath) was defined by the W3C (World Wide Web Consortium), and it is a major element in XSLT (Extensible Stylesheet Language Transformations). XPath uses path expressions to select nodes or node-sets in an XML document. A path expression is the path from an XML node (or the current context node) to another node, or a group of nodes written as a sequence of steps. The character "/" in a path expression is used to separate each step, and three constituent elements can be employed to reach a target node; i.e., 1) directly approaching the target node, 2) specifying the element node for screening, and 3) filtering the node with the attribute having a specified value. Given an example as follows, we can use the path expression "// name[@kind='Tablet']" to find the node "name" with the attribute "kind" and the value "Tablet", and then call function text() to get the text "Nexus 7".

```
<html>
        <body>
                <name kind="Camera">Nikon D3000</name>
                <name kind="Tablet">Nexus 7</name>
        </body>
</html>
```

2.3   Jsoup

Jsoup [11] is a Java library for working with a real-world HTML, and it can directly parse a URL address or HTML contents. Jsoup also provides a very convenient API for extracting and manipulating data, using DOM (Document Object Model), CSS (Cascading Style Sheets), and jQuery-like methods. For example, we can use

Element.attr(key) and Element.text() to get the text of an element with an attribute value. With Jsoup, we can be easy to manipulate HTML contents.

2.4   JDOM

There are several methods commonly used to manipulate an XML file. The most well-known open source API is JDOM developed by Jason Hunter and Brett McLaughlin [10]. JDOM is based on a tree structure and designed to represent an XML document and its contents to the typical Java developer in an intuitive and straightforward way. As the name indicates, JDOM is Java-optimized and an establishment of a complete solution based on the Java platform for accessing, manipulating, and outputting XML data. Therefore, JDOM users can get their jobs done in XML without tremendous expertise.

3   System Framework

In this section, we propose the framework of a seller recommendation system as shown in Fig. 3, which consists of five components: 1) downloading product and seller feedback HTML pages, 2) extracting product and seller information, 3) building seller XML files, 4) seller ranking, and 5) user interface. In general, although each buyer has his/her own spending habit, they always hope sellers can provide good services. Therefore, a seller recommendation system is designed to provide relevant sellers ranked according to different buyer needs. First, we download product and seller feedback pages from the eBay website, and then extract product and seller feedback information from the downloaded pages. Finally, the product and seller feedback data are integrated into an XML file. Afterwards, buyers can specify products and budgets in a query, and get back seller ranking results.

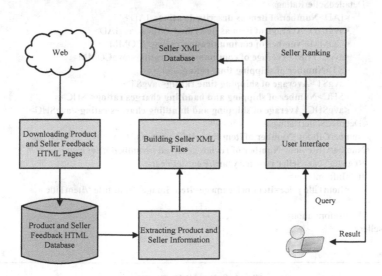

Fig. 3. System framework

3.1   Downloading Product and Seller Feedback HTML Pages

For the eBay home page, a registered seller can classify products according to the categories established by the eBay website; therefore, buyers can specify product categories and brands to download the specified product information. Buyers can download the product pages classified by a seller, and then download the seller feedback page when they consider purchasing products from this seller.

3.2   Extracting Product and Seller Information

From the product page, we can use Jsoup to extract the following information: 1) seller ID, 2) product title, 3) product price, and 4) product image. From the seller feedback page, we can also extract the following information: 1) recent feedback during last 12 months, 2) detailed seller ratings during last 12 months, 3) the feedback received, and 4) seller registration year.

3.3   Building Seller XML Files

Finally, the product and seller feedback data are integrated into an XML file by JDM. As illustrated in Fig. 4, a seller XML file stores all the information used to compare all the sellers registered in the eBay website.

```
<kind ProductKind=Product kind>
    <seller Id=Seller ID>
        <RecentFeedback>
            <positive>Number of positive rating</positive>
            <neutral>Number of neutral rating</neutral>
            <negative>Number of negative rating</negative>
        </RecentFeedback>
        <DetailedSellerRatings>
            <IAD>Number of item as described rating</IAD>
            <avgIAD>Average of item as described rating</avgIAD>
            <COMM>Number of communication rating</COMM>
            <avgCOMM>Average of communication rating</avgCOMM>
            <ST>Number of shipping time rating</ST>
            <avgST>Average of shipping time rating</avgST>
            <SHC>Number of shipping and handling charges rating</SHC>
            <avgSHC>Average of shipping and handling charges rating</avgSHC>
        </DetailedSellerRatings>
        <NumberOfItem>Number of item for sell</NumberOfItem>
        <NumberOfReview>Number of review received</NumberOfReview>
        <RegisterYear>Seller register year</RegisterYear>
        <ItemInformation>
            <ItemTitle price=Item price image=Item image>Item title</ItemTitle>
                                :
        </ItemInformation>
    </seller>
        :
</kind>
```

Fig. 4. Seller XML file

3.4 Seller Ranking

Our seller recommendation system is to rank the sellers selling the products specified by buyers. Buyers can specify products and budgets as they need. First, the system searches matched products, then finds out the corresponding sellers, and finally ranks them according to their scores. The seller scores could be calculated as follows.

$$Score_i = FR_i \times PW_i \times WSRY_i$$

where i is seller i, FR_i is the Feedback Rating, PW_i is the Popularity Weight, and $WSRY_i$ is the Weight of Seller Registration Year.

3.4.1 Feedback Rating (FR)

For a seller feedback page, feedback data can be divided into two parts: 1) recent feedback during last 12 months (i.e., PNR), and 2) detailed seller ratings during last 12 months (i.e., DSR). Therefore, feedback ratings (i.e., FR) could be calculated as follows.

$$FR_i = PNR_i \times DSR_i$$

$$PNR_i = \frac{Positive_i - Negative_i}{Positive_i + Neutral_i + Negative_i}$$

where $Positive_i$, $Neutral_i$, and $Negative_i$ are the numbers of recent feedback for seller i, including positive, neutral, and negative.

$$DSR_i = \frac{rating_{score}}{total_{score}}$$

$$rating_{score} = \sum_{j=1}^{4} C_j \times AVGR_j$$

$$total_{score} = \sum_{j=1}^{4} C_j \times 5$$

where C_j is the number of criterion j among detailed seller ratings (i.e., IAD, $COMM$, ST, and SHC), and $AVGR_j$ is the average rating of criterion j.

3.4.2 Popularity Weight (PW)

The more popular a seller is, the more discussion he/she has. Here, we use the number of feedback received to represent the seller popularity. PW could be calculated as follows.

$$PW_i = \frac{ln(FBR_i + 1)}{ln(max(FBR) + 1)}$$

where FBR_i is the number of feedback received for seller i, and $max(FBR)$ is the maximum number of feedback received among all sellers.

3.4.3 Weight of Seller Registration Year (WSRY)

In general, the earlier the seller registration year is, the more experiences he/she has in the e-commerce. Thus, it is believed that registration time has tight correlation with service quality. WSRY could be calculated as follows.

$$WSRY_i = exp\left(\frac{min(SRY_i)-SRY_i}{\beta}\right), \beta = t - min(SRY_i)$$

where SRY_i is the registration year of seller i, t is the current year, and β normalizes WSRY value in the range (0.36, 0.99).

3.5 User Interface

We have implemented a user interface through which buyers can specify products and their budgets. Through the seller recommendation system, buyers can get back the ranking results of all sellers selling these matched products. As shown in Fig. 5, two blocks are provided in the user interface: 1) top sellers – displaying the ranking results of all relevant sellers, and 2) seller information – displaying the feedback data and final score of a clicked seller.

4 Implementations

4.1 Developing Environment

The seller recommendation system is implemented using Java, and the experiments are conducted on an Intel Core i7 2.93GHz CPU with 4G main memory in Windows 7 ultimate.

4.2 Data Collection

In this system, the dataset used in the experiments is collected from the eBay home page, which consists of three kinds of products: 1) cellphone, 2) camera, and 3) tablet, as shown in Table 1. Here, to ensure that inexperienced sellers not included in the system, we filter out those sellers whom the amount of items sold by is less than 5. Finally, more than 200 sellers are collected for each kind of products, respectively.

Table 1. Dataset

Product Categories / Number of Collections	Cellphones	Cameras	Tablets
Number of Sellers	251	202	204
Number of Items	14,489	11,617	5,072

4.3   Ranking Results

While a buyer specifies a kind of products, an at-most price, and a query with terms, and then presses button "Execute", the system would display the ranking results of all matched sellers, as shown in Fig. 5. On the top seller block, sellers are ranked based on their scores from high to low. Buyer can also view the seller detailed information on the seller information block by clicking a specific seller.

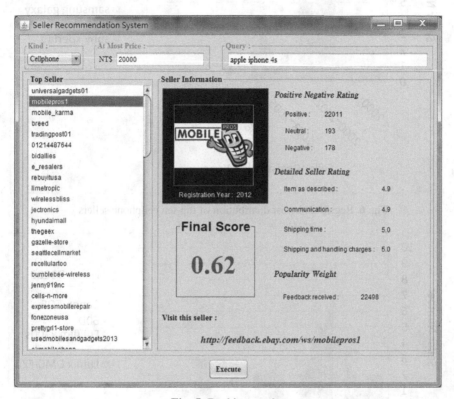

Fig. 5. Ranking results

4.4   Experimental Results

According to the formula used to calculate seller scores, the weight of seller registration year (i.e., *WSRY*) is favorable for early registered sellers. In this experiment, we would like to observe what ranking happens to beginners and veterans in these three kinds of products. Without specifying an at-most price, we test five brands for each kind of products. As shown in Fig. 6, Fig. 7, and Fig. 8, the registration year distribution of top-ten sellers with the specified brands is found, respectively for "cellphone", "camera", and "tablet". The results show that regardless of the products they sell, most top-ten sellers are with the registration year 2000~2005. Therefore, we conclude that an seller being able to provide excellent services should be experienced for more than 10 years.

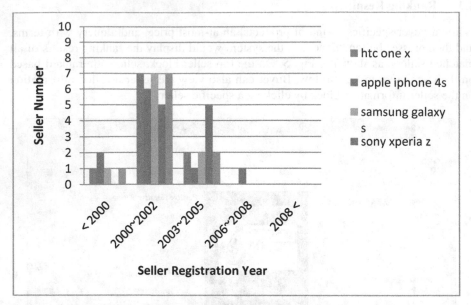

Fig. 6. Registration year distribution of top-ten cellphone sellers

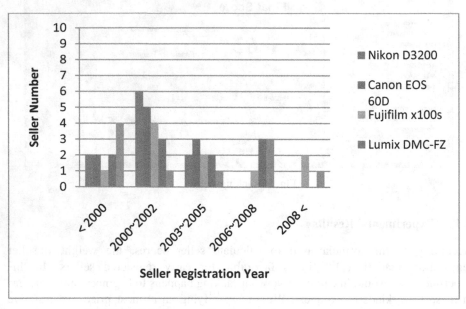

Fig. 7. Registration year distribution of top-ten camera sellers

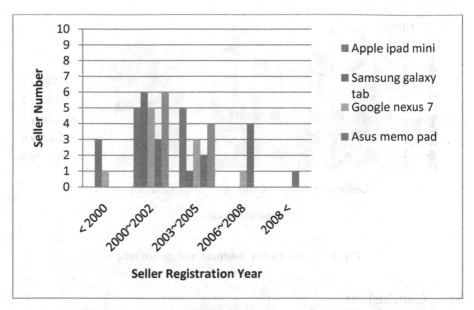

Fig. 8. Registration year distribution of top-ten tablet sellers

Besides, in order to observe the effectiveness of our recommendation system, we use the top rated sellers recognized by the eBay site [12] as a baseline to evaluate the recommendations. The top rated sellers are those who consistently deliver on outstanding services. To become a top rated seller, his/her eBay account must have been active for at least 90 days, and must be satisfied with the following conditions: 1) consistently receiving high ratings from buyers, 2) shipping items quickly, and 3) having earned a track record of excellent services. Table 2 shows the distribution of the top rated sellers collected in our dataset. As shown in Fig. 9, the top rated seller distribution of top-ten sellers is found, respectively for "cellphone", "camera", and "tablet". The results show that most top-ten sellers appearing in the recommendation list are also top rated sellers. This verifies that our system can actually reflect the real situation, although some top-ten sellers recommended by the system are not recognized as the top rated sellers by the eBay site. The reasons could be 1) the seller score calculated in our system is to combine three factors, but not to consider the threshold of each factor, or 2) the popularity weight and seller registration year are not yet considered by the eBay site.

Table 2. Distribution of the top rated sellers

	Cellphone	Camera	Tablet
Number of Top Rated Sellers	76	51	54
Ratio of Top Rated Sellers	30.3%	25.2%	26.5%

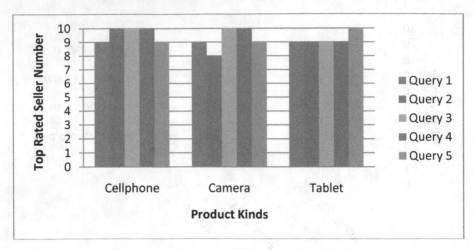

Fig. 9. Top rated seller distribution of top-ten sellers

5   Conclusions

For diverse shopping ways, online shopping will become increasingly popular as more and more buyers purchase their products using e-commerce platforms. Thus, finding excellent sellers among a large amount of sellers on the Internet is crucial to purchasing products. In this paper, we propose a seller recommendation system where buyers can specify product kinds, budgets, and queries to get back the ranking results of all sellers selling these matched products. The recommendation list is ranked according to the seller scores calculated using feedback rating, seller popularity, and seller registration year. The experimental results show that an seller being able to provide excellent services should be experienced for more than 10 years. Also, it is verified that our system can actually reflect the real situation by contrasting the top rated sellers recognized by the eBay site.

Acknowledgments. This work was supported by National Science Council of R.O.C. under grant MOST 103-2221-E-224-049.

References

1. Balingit, R., Trevathan, J., Read, W.: Analysing Bidding Trends in Online Auctions. In: Proc. the 6th International Conference on Information Technology: New Generations, pp. 928–933 (2009)
2. Huang, Y.F., Lin, H.: Web Product Ranking Using Opinion Mining. In: Proc. IEEE Symposium on Computational Intelligence and Data Mining, pp. 184–190 (2013)
3. Khatua, S., Iyengar, N.C.S.N.: Analyzing the Security System Applied in E-Shopping System Using Elliptic Curve Cryptography. International Journal of Latest Trends in Computing 2(2), 253–264 (2011)

4. Morzy, M., Jezierski, J.: Cluster-based Analysis and Recommendation of Sellers in Online Auctions. In: Proc. the 3rd International Conference on Trust, Privacy and Security in Digital Business, pp. 172–181 (2006)
5. Nariman, D.: Analyzing Text-based User Feedback in E-Government Services Using Topic Models. In: Proc. the 7th International Conference on Complex, Intelligent, and Software Intensive Systems, pp. 720–725 (2013)
6. Pagano, D., Maalej, W.: User Feedback in the AppStore: an Empirical Study. In: Proc. the IEEE 21st International Conference on Requirements Engineering Conference, pp. 125–134 (2013)
7. Xin, P., Ju, X.F.: A Research on Development of Chinese B2C E-commerce Trust. In: Proc. International Conference on E-Business and Information System Security, pp. 1–4 (2009)
8. Zhang, C.: Analyzing Encryption Technology Applied in Farm Product E-commerce System Security. In: Proc. the 8th International Conference on Electronic Measurement and Instruments, pp. 4-433–4-436 (2007)
9. Zhang, J., Zhang, Y.: Research on Duration and Bid Arrivals in eBay Online Auctions in the Internet. In: Proc. the 2nd International Conference on Artificial Intelligence, Management Science and Electronic Commerce, pp. 1379–1382 (2011)
10. JDOM, http://www.jdom.org/
11. Jsoup, http://jsoup.org/
12. Top Rated Plus Items, http://pages.ebay.com/topratedplus/index.html
13. XPath (XML Path Language), http://www.w3.org/TR/xpath-20/

A Study on Development of Local Culture Industry: The Case Study of Community Colleges and Community Development Associations in Taiwan

Tain-Fung Wu[1], Nien-Tsu Hou[2], Cheng-Feng Cheng[3], and Chi-Hsiang Ting[4,*]

[1] Department of Business Administration, Asia University, 500, Lioufeng Rd.,
Wufeng, Taichung 41354, Taiwan
thomaswu@asia.edu.tw
[2] Department of Social Work, Asia University, Taiwan
nthou@asia.edu.tw
[3] Department of International Business, Asia University, Taiwan
cheng-cf@asia.edu.tw
[4] Department of Business Administration, Asia University, Taiwan
djs55211@ms29.hinet.net

Abstract. The study is an exploratory study to investigate interactions between the strategic alliances of community colleges and community development associations on development of local culture industry. Their strategic alliances are identified through the methods such as fuzzy-set qualitative comparative analysis (fs/qca) and social network analysis, etc. According to the resource dependency theory, it was discovered that community colleges and community development associations are influenced by tripartite resources such as community empowerment, lifelong learning, and social citizens in terms of the importance of resources, and of these three, lifelong learning accounts for a majority. In terms of resource ownership control, community colleges can provide community development associations with administrative support and assistance. In terms of degree of resource substitution, community colleges and community development associations are influenced by factors such as principal attitudes, environmental change, appearance of alternative institutions, etc.

Keywords: Culture Industry, strategic alliances, social network analysis, fuzzy-set qualitative comparative analysis.

1 Introduction

Taiwan officially began the history of community colleges in 1998 and it is an adult education establishment and lifelong learning institution unique to this nation. At present, there are 83 community colleges. The lifelong education learning act, promulgated and implemented by the government's Ministry of Education in 2002,

---

* Corresponding author.

L.S.-L. Wang et al. (Eds.): MISNC 2014, CCIS 473, pp. 60–75, 2014.
© Springer-Verlag Berlin Heidelberg 2014

was for non-formal lifelong learning institutions. An analysis of this country's community college education shows that it is more akin to the European Volkshochschule and does not operate in the mode of U.S. community colleges [1]. Currently, the positioning of our country's community colleges is to only allow city and county governments to grant certificates of completion and not degrees [2].

Currently, modes of operation are focused on the academic, arts and crafts, and association categories. Arts and crafts courses currently make up the largest portion of the curriculum followed by academic courses. The initiation of association courses that tend to include higher levels of public affairs is the most unsatisfactory. In addition, public forums and cultural characteristics are a part of the community college curriculum. Furthermore, community colleges combine with the overall community to create job promotion through strategies such as developing local industrial culture, community participation, socializing courses, turning associations into courses, etc. [3]. The community colleges referred to in this study are educational institutions self-organized or commissioned by the competent authorities of municipalities and counties (cities) to provide community residents with lifelong education.

Taiwan's Community Development Association was established according to the "Community Development Outline" amended by the Ministry of the Interior in 1999. The goal was to promote community development organizations and activity areas. Regarding the work of the Community Development Association, it focuses on the characteristics and resident needs of each community, in coordination with designated government community development job items, annual items recommended by the government, and self-developed community projects, to formulate a community development plan and compile budgets and promotion. According to Ministry of the Interior statistics (2012), there are currently 6,443 community development associations in Taiwan.

Thus, community colleges play the role of lifelong learning institutions and also require development of strategic alliances or partnerships [4]. In research by Wu [5] on the partnership between the community colleges and community development associations in Changhua, it was discovered that the two can form partnerships, assist each other, and share resources.

In 2008, Taiwan government announced 6 new key industries for future economic development and cultural and creative industries were one of them. There were many community development associations of villages and towns in Taiwan. The association used the characteristics of the community, local and government resources for increasing residents' live and economy. The community colleges assist them to develop their local culture industry [4, 5].

The goals of this study are (1) establish a resource sharing strategic alliance model for community colleges and community development associations. (2) Construct key indicators for community college and community development association strategic alliances. (3) Provide community colleges and community development associations with partnership references during strategic alliance process.

2   Literature

2.1   Overall Community Development

The term "General Community Development" in Taiwan is derived from the book, "The Concept of Overall Community Development" written by the Director of the National Taiwan Craft Research and Development Institute, Weng Xu De and Chiba University professor Miyazaki Kiyoshi. It divides the content and process of community development into 5 major portions, full resident participation, further review of local culture, symbiosis between man and nature, mutual support and friendship, and value innovation and dissemination of community resources [6]. Its primary target is "establishing community culture, building community consensus, and constructing the concept of a living community entity as a class of new thinking and policy in cultural administration." Its main goal is to generate policy terms to integrate the 5 primary community development orientations of "people, culture, place, scenery, production." Japan's Professor Miyazaki Kiyoshi advocated the division of these issues into the 5 major classifications of "people," "culture," "place," "production," and "scenery."

2.2   Lifelong Learning

Adult education scholar Cropley [7], explained the implications of lifelong learning through 4 different orientations: (1) from the perspective of time, lifelong learning begins at birth and ends at death. (2) From the perspective of type, lifelong learning occurs in educational context that are formal, informal, and unofficial. (3) From the perspective of results, lifelong learning can cause one to update knowledge, skills, and attitudes. (4) From the perspective of goals, the ultimate goal of lifelong learning is to promote individual self-realization.

Huang [8] proposed to divide the community college curriculum into 3 major categories (1) academic courses, (2) association courses, (3) life courses. Hu [9] believed that lifelong learning refers to the individual as a learner, conducting planned or unplanned learning activities, according to personal interests and needs, between birth and death at every stage of their lives.

2.3   Civil Society

Civil society refers to a domain or system composed of intermediaries of spontaneous organizations in a system of government. These organizations are relatively independent of public power and families and private sector production or reproduction such as corporations [10]. Cooper [11] believed civil society refers to a public domain where citizens conduct public affairs interactions. In this domain, the public can participate in complicated political or public activities through dialogue. In the governance process, the public plays an active role as a participant and supervisor instead of the roles of malcontent or victim. The public can realize that the freedom to participate in public policy is not a compulsion to participate, but instead, stems from the personal expression of autonomy and the development of social care responsibility.

In addition, in public society, people learn self-respect, group identity, the ability to deal with public affairs, the value of cooperation, and civic virtue. Furthermore, everyone needs to comply with community norms and obligations in order to maintain the existence of a public domain. [12]. Chen [3] once explored the very important roles played by civil society and community colleges from the angle of the 3 perspectives of the function of community education (education perspective), the practice of citizen participation (participation perspective), and the shaping of civil society (construction perspective).

According to the aforementioned literary analysis, it was discovered that the strategic alliance model between community colleges and community development associations is divided into lifelong learning, community empowerment, and civic education. In addition, resources are mutually shared and mutual assistance provided in seeking common development cooperation models through methods such as course types, holding activities, etc.

2.4   Strategic Alliance

A strategic alliance is: "Two of more enterprises, in order to achieve common strategic objectives, possessing mutual strategic cooperative behavior, and competition. The enterprises still retain independent autonomy and maintain or enhance competitive advantages through mutually mastering each other's positive and negative resources, and mutually sharing responsibilities, risks, and rewards through members [13]."

Pfeffer and Salancik [14] proposed the concept of resource dependence. This concept believes that there are many resources in the environment and organizations must rely on obtaining environmental resources to maintain their existence. Viewing resource outsourcing from the perspective of the resource dependence theory, an organization's tasking environment (concentration, openness or connectivity) will determine this organization's resource perspectives (importance, prudent allocation, selectivity) and these resource perspectives and organization established strategies will mutually influence the strategy of organization outsourcing functions [15].

Therefore, it can be discovered through the resource dependence theory: environmental resources are limited, if organizations wish to strengthen or maintain competitive advantages, they should not be limited to internal resources and abilities, but should focus on their core competitive technologies through a professional division of labor model. Businesses should commission external execution of non-competitive or poor resource performance manufacturing activities, retrieve complementary external resources, and thus, derive interdependent partnerships between organizations [14,16,17].

The degree in which organizations rely on outside resources depends on 3 factors including: degree of resource importance, control capability of resource ownership, and degree of resource substitution [18]. This study utilizes the 3 aforementioned factors to measure the dependencies between each relationship in outsourcing scenarios.

2.5   Social Network Theory

The so-called social network refers to the specific relationships between every person in a group. Network relationships formed by different relationships are different. Thus, the individual's social behavior in the group can be seen through link relationships and constituent structure. [19] Therefore, social networks are composed of 3 factors: "actors," "relationship," and "ties." Scott [20] believed social network analysis primarily utilizes dotted line and graphical representations after rendering all kinds of relationships into values to show the directional and relationship distance of each member in the network. The tightness of these relationships can divide central indicators into 3 categories for measurement: (1) degree centrality; (2) betweeness centrality; (3) closeness centrality [21].

2.6   Fussy-set Qualitative Comparative Analysis

Fussy-set Qualitative Comparative Analysis (fs/qca) is a technique, originally developed by Ragin in 1987. It is used for analyzing data sets by listing and counting all the combinations of variables observed in the data set, and then applying the rules of logical inference to determine which descriptive inferences or implications the data supports. In the case of categorical variables, QCA begins by listing and counting all types of cases which occur, where each type of case is defined by its unique combination of values of its independent and dependent variables [22,23,24,25,26].

2.7   Cultural Industry

Since 1998, the UK Government recognized the creative industries with the distinct economic contribution. The Taiwanese authorities consider the concept of creative industries and combined cultural industries into the 'Cultural and Creative Industries' (CCI), it would be a new economy. The Taiwanese Government proposed the Challenge 2008: Six-Year National Development Plan, a six-year policy guideline aimed at generating a new set of competitive advantages. The one of sub-plan was aimed to construct new hometown communities through integrating cultural traditions and communities can develop local attractions, offer various employment opportunities at retaining local talent by refocusing on the roles of local community [27]. Wang [28] investigated nine community industries developed in northern Taiwan. His recommendations for Community Development Association (CDA) to promote community industry, is effectively link and integrate resources within and outside the community, helping communities to promote industry. In addition to, Wu [4,5] research the community college and community development association in terms of locate both playing roles in the processing of community whole construction. They can share each other's resource for developing cultural industry.

3   Research Methods

This study's first stage of expert interviews utilized the purposive sample method on research subjects to select 2 community colleges and 4 community development associations to conduct semi-structured interviews. The second stage of focused interviews obtained 1 community college and 5 community development associations for focused interviews [29]. According to the data accumulated through the interviews, key indicators were extracted using the grounded theory. Classification was then conducted through the resource dependence theory. The third stage involved conducting mail surveys of 83 community colleges according to national community college advancement associations and 500 community development associations. Lastly, according to the survey data, social network analysis was utilized to conduct an analysis of key indicators to establish a complete architectural model of strategic alliances between community colleges and community development associations.

4   Data Analysis

4.1   Examination of Community College and Community Development Association Strategic Alliance Framework

According to the content of stage 1 expert interviews and stage 2 focused interviews and after utilizing grounded theory analysis, it was discovered that the types of mutual strategic alliances between community colleges and community development associations are the four areas of cooperation of overall community development, lifelong learning, citizen social education, and administrative support. The cooperative process was influenced by the factors of degree of resource importance, control capability of resource ownership, and degree of resource substitution. Divided into 3 factor categories according to Grover, Cheon, & Teng [18] resource theory.

1) Degree of Resource Importance: in the area of resource importance, that which community colleges provide to community development associations can be divided into the following types: (1) lifelong learning, (2) community empowerment, (3) social citizen.

2) Control Capability of Resource Ownership: in the area of control capability of resources ownership, community colleges provide community development associations with administrative support.

3) Degree of Resource Substitution: the influence of community colleges providing community development associations with a degree of resource substitution can be divided into the following factors:
   a) The attitude of the prime mover affects the cooperative relationship: the attitude of the community college or community development association prime mover will affect the cooperative relationship.

b) Policy affects cooperative relationship: i.e. Ministry of Education supervision requests community colleges possess interactive and supportive relationships with community development associations.

c) Environmental change: after community college moves, is affected by additional distance, causing increasing alienation in bilateral relations.

d) Alternative groups: community development association has already found a suitable management consulting firm to replace the support status of community college.

4.2   Key Indicators of Community College and Community Development Association Strategic Alliances

The third stage mail surveys of 83 community colleges according to national community college advancement associations were subjected to the grounded theory to conduct key indicator classification to establish a complete architectural model of strategic alliances between community colleges and community development associations. This strategic alliance model and critical analysis diagram are shown on figure 1. Analysis explanations of key indicators are as follows:

1) The key indicators of community college and community development association strategic alliances in the area of degree of resource importance (RI) are as follows:

a) Lifelong learning (LL): beauty studies, catering studies, arts studies, financial seminars, ecological education, cultivating local culture, and information education.

b) Community empowerment (CE): community greening, promotion of local cuisine, development of unique artifacts, festivals, industrial development, environmental protection, and other indicators.

c) Social citizen (SC): low-carbon movement activities, community care, health education, cultural festivals and ceremonies, green citizen, civic literacy, community disaster prevention.

2) The key indicators of community college and community development association strategic alliance control capability of resource ownership (CO) are: (a) course learning (CL) (b) teacher support (TS) (c) venue rental (VR) (d) event organization (EO).

3) The key indicators of community college and community development association strategic alliance degree of resource substitution (RS) are: (a) Influence of prime mover attitude (IA) (b) Policy influence (PI) (c) Environmental change (EC) (d)Alternative groups(AG).

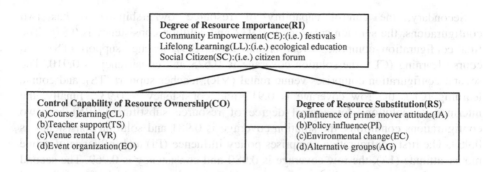

Fig. 1. Key Indicators of Community College and Community Development Association Strategic Alliances

4.3   Sampling and Data Collection

This study was carried out in Taiwan. The authors delivered a total of 583 copies of the questionnaire to participants, from which they received 250 completed questionnaires from 25 Community Colleges and 225 community development associations, producing a response rate of 42.8 percent. To check that the sample of responses obtained was representative of the population, non-response bias was examined through a comparison of early and late waves of returned surveys.

4.4   Fuzzy Set Qualitative Comparative Analysis (fs/QCA)

According to fuzzy set qualitative comparative analysis, the research constructs including causal conditions and outcomes in this study are transforming into fuzzy set scores (i.e., fuzzy set scores) by using the calibrating function of fs/QCA software and these scores are ranging from 0.00 to 1.00 based on Ragin [24]. Ragin [25] indicates that the threshold for full membership (i.e., fuzzy score equals to 0.95), cross-over point (i.e., fuzzy score equals to 0.50), and full non-membership (i.e., fuzzy score equals to 0.05) are three qualitative anchors of fuzzy set. Therefore, this study set the original values of 5.0, 3.0, and 1.0 from Likert-type five points scales to correspond to the full membership, cross-over point, and full non-membership anchors, respectively.

The intermediate solutions for degree of resource importance (RI), control capability of resource ownership (CO), and control degree of resource substitution (RS) are shown in Table 1. First of all, the intermediate solutions for degree of resource importance (RI) has three configurations, the solution coverage is 0.868 and solution consistency is 0.997. The first configuration comprises social citizen (SC) and lifelong learning (LL), the row coverage is 0.721 and consistency is 1.000; the second configuration comprises social citizen (SC) and community empowerment (CE), the row coverage is 0.635 and consistency is 1.000; and the third configuration comprises lifelong learning (LL) and community empowerment (CE), the row coverage is 0.770 and consistency is 0.996.

Secondary, the control capability of resource ownership (CO) has two configurations, the solution coverage is 0.932 and solution consistency is 0.878. The first configuration comprises event organization (EO), teacher support (TS), and course learning (CL), the solution coverage is 0.625 and consistency is 0.910. The second configuration comprises venue rental (VR), teacher support (TS), and course learning (CL), the row coverage is 0.912 and consistency is 0.932. Finally, the intermediate solutions for control degree of resource substitution (RS) has two configurations conditions, the solution coverage is 0.981 and solution consistency is 0.969. The first configuration comprises policy influence (PI) and influence of prime mover attitude (IA), the row coverage is 0.980 and consistency is 0.969. The second configuration comprises alternative groups (AG), environmental change (EC), and policy influence, the row coverage is 0.323 and consistency 1.000.

Table 1. QCA Output-intermediate solution

| Factors | Conditions | Coverage | | Consistency | Solution | |
		Raw	Unique		Coverage	Consistency
Resource importance(RI)	SC*LL	0.721	0.091	1.000	0.868	0.997
	SC*CE	0.635	0.006	1.000		
	LL*CE	0.770	0.141	0.996		
Control capability of resource ownership(CO)	EO*TS*CL	0.625	0.910	0.910	0.932	0.878
	VR*TS*CL	0.912	0.912	0.932		
Resource substitution(RS)	PI*IA	0.980	0.658	0.969	0.981	0.969
	AG*EC*PI	0.323	0.001	1.000		

The results of fs/QCA indicate that all configurations are sufficient conditions causing degree of resource importance (RI), control capability of resource ownership (CO), and control degree of resource substitution (RS) (i.e., consistency values exceed 0.865). The consistency indices, can see as significance metrics in statistical, which measures the degree to which configurations are subsets of the outcome [24, 26]. Table 1 represents that all of the solution consistency values is above 0.878. High consistency also indicates that a subset relation exists and supports an argument of sufficiency [25]. Furthermore, the coverage indices can see as the effects size [26] that raw coverage and solution coverage measure the extent to which the configurations account for the outcome [24]. Table 1 also represents that most of the raw and solution coverage values is above 60%, indicating that the configurations explain a large proportion of degree of resource importance (RI), control capability of resource ownership (CO), and control degree of resource substitution (DS).

4.5   Social Network Analysis of Strategic Alliances

Social network analysis was utilized to analyze the integration and use of resources by community college and community development association strategic alliances construct. We use the 2 mode networks method to find the cohesion and centrality for the community college and community development association strategic alliances.

1) SNA of degree of resource importance

Social network analysis was conducted on the degree of resource importance of community colleges and community development associations. The neutral lifelong learning indicator was highest at degree centrality 0.640, closeness centrality 0.633, and betweenness centrality 0.255. This shows that in the area of degree of resource importance, community colleges and community development associations most frequently cooperate in lifelong education. This is followed by social citizen education and lastly community empowerment. The three elements all have high degree centrality, closeness, shown on Table 2. They have high density of cohesion measure is 0.62 (Fig. 2).

Table 2. Network centrality of resource importance

Degree of resource importance	Degree	Closeness	Betweenness
Community empowerment	0.600	0.608	0.216
Lifelong learning	0.640	0.633	0.255
Social citizen	0.600	0.608	0.236

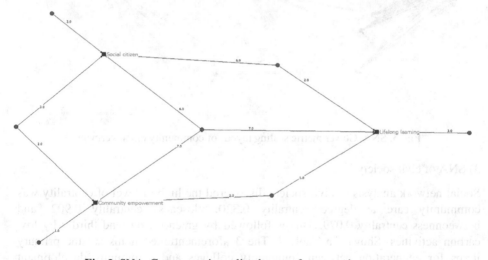

Fig. 2. SNA, Gower metric scaling layout of resource importance

2) SNA of community empowerment

Social network analysis of community empowerment discovered the highest level of centrality was development of unique artifacts at degree centrality 0.520, closeness centrality 1.207, and betweenness centrality 0.107. This is followed by festivals, and thirdly by industrial development, shown on Table 3. The 3 aforementioned items are the primary items for cooperation between community colleges and community development associations in community empowerment. The density of cohesion measure is 0.44 (Fig 3).

Table 3. Network centrality of community empowerment

Community empowerment	Degree	Closeness	Betweenness
Community greening	0.400	1.000	0.021
Promotion of local cuisine	0.400	1.000	0.026
Development of unique artifacts	0.520	1.207	0.107
Festivals	0.480	1.129	0.077
Industrial development	0.480	1.129	0.053
Environmental protection	0.360	0.946	0.027

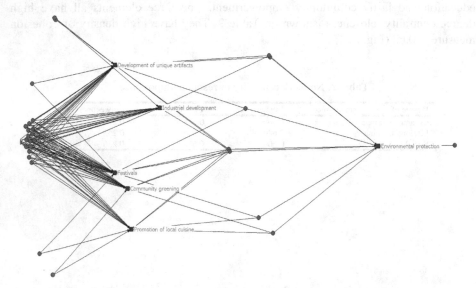

Fig. 3. SNA, Gower metric scaling layout of community empowerment

3) SNA of civil society

Social network analysis of civil society discovered the highest level of centrality was community care at degree centrality 0.320, closeness centrality 0.902, and betweenness centrality 0.079. This is followed by green citizen and thirdly by low carbon activities. Shown on Table 4. The 3 aforementioned items are the primary items for cooperation between community colleges and community development associations in civil society. The density of cohesion measure is 0.24(Fig 4).

Table 4. Network centrality of civil society

Civil society	Degree	Closeness	Betweenness
Low carbon movement activities	0.280	0.860	0.104
Community care	0.320	0.902	0.079
Health education	0.240	0.822	0.068
Cultural festivals and ceremonies	0.200	0.787	0.087
Green citizen	0.280	0.860	0.075
Civic literacy	0.240	0.822	0.044
Community disaster prevention	0.120	0.725	0.002

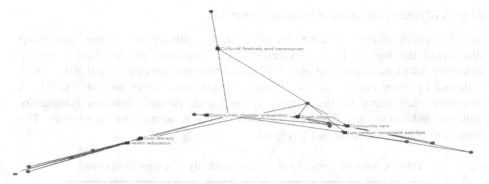

Fig. 4. SNA, Gower metric scaling layout of civil society

4) SNA of lifelong learning

Social network analysis of lifelong learning discovered the highest level of centrality was arts studies at degree centrality 0.480, closeness centrality 1.028, and betweenness centrality 0.194. This is followed by ecological education and thirdly by cultivating local culture. Shown on Table 5. The 3 aforementioned items are the primary items for cooperation between community colleges and community development associations in lifelong learning. The density of cohesion measure is 1.0 (Fig 5).

Table 5. Network centrality of lifelong learning

Lifelong learning	Degree	Closeness	Betweeness
Beauty studies	0.240	0.771	0.018
Catering studies	0.240	0.771	0.018
Arts studies	0.480	1.028	0.194
Financial seminars	0.280	0.804	0.062
Ecological education	0.360	0.881	0.094
Cultivation local culture	0.280	0.804	0.032
Information education	0.240	0.771	0.027

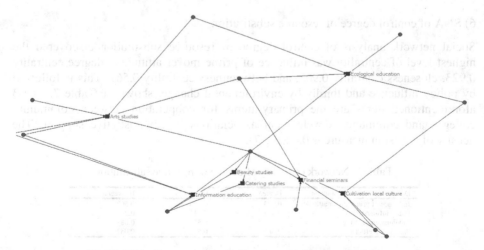

Fig. 5. SNA, Gower metric scaling layout of lifelong learning

5) SNA of control capability of resource ownership

Social network analysis of centrality of control capability of resource ownership discovered the highest level of centrality was seminars and courses at degree centrality 0.440, closeness centrality 0.972, and betweenness centrality 0.205. This is followed by event organization and thirdly by teachers, shown on Table 6. The 3 aforementioned items are the primary items for cooperation between community colleges and community development associations in administrative support. The density of cohesion measure is 1.0 (Fig 6).

Table 6. Network centrality of control capability of resource ownership

Control capability of resource ownership	Degree	Closeness	Betweenness
Teachers	0.320	0.833	0.104
Venues	0.200	0.729	0.006
Equipment	0.200	0.729	0.006
Event organization	0.400	0.921	0.138
Teaching materials	0.200	0.729	0.006
Seminars and courses	0.440	0.972	0.205

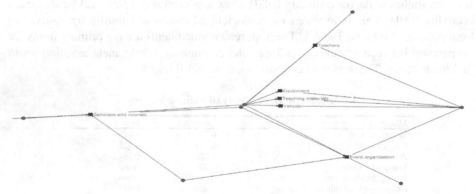

Fig. 6. SNA, Gower metric scaling layout of control capability of resource ownership

6) SNA of control degree of resource substitution

Social network analysis of control degree of resource substitution discovered the highest level of centrality was influence of prime mover attitude at degree centrality 0.923, closeness centrality 0.905, and betweenness centrality 0.562. This is followed by policy influence and thirdly by environmental change, shown on Table 7. The 3 aforementioned items are the primary items for cooperation between community colleges and community development associations in administrative support. The density of cohesion measure is 0.62(Fig.7).

Table 7. Network centrality of degree of resource substitution

Degree of resource substitution	Degree	Closeness	Betweenness
Influence of prime mover attitude	0.923	0.905	0.562
Policy influence	0.538	0.613	0.203
Environmental change	0.462	0.576	0.083
Alternative groups	0.385	0.543	0.057

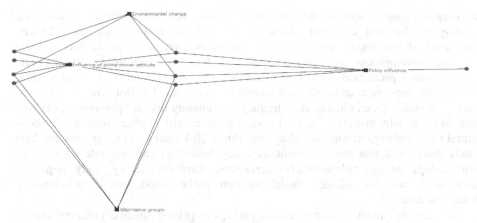

Fig. 7. SNA, Gower metric scaling layout of degree of resource substitution

According to the social network analysis of community college and community development association strategic alliances construct. We can find lifelong learning indicator was highest at degree centrality 0.640, the degree centrality of social citizen education was 0.60 and lastly the degree centrality of community empowerment was the same value. It conducted degree of resource importance. They are all high degree centrality, like as unique artifacts, arts studies, seminars and courses. These indexes influence the development of local culture industry.

5   Conclusion

5.1   Establishment of the Strategic Alliance Model

The strategic alliance model between community colleges and community development associations was discovered through expert interviews and focus group interviews. Cooperation is conducted in the areas of lifelong learning, community empowerment, civic education, administrative support, etc. The perspective of the resource dependence theory was used to investigate the cooperative relationship between the two. Investigating from the perspective of resource importance, lifelong learning, community empowerment, civic education possess correlations. In terms of control capability of resource ownership, the feasibility of community colleges providing community development associations with administrative resource abilities will influence the willingness of the two parties to cooperate. In terms of resources substitution, the influence of prime mover's attitude, policy influence, environmental change, and alternative group factors will all lower the willingness of the two parties to cooperate.

5.2   Key Indicators of the Strategic Alliance Model

The surveys taken of national community colleges discovered that the key indicators of mutual strategic alliances can be divided into 4 major areas (1) Key indicators of lifelong learning: beauty studies, catering studies, arts studies, financial seminars, ecological education, cultivating local culture, and information education. (2) Key indicators of community empowerment: community greening, promotion of local cuisine, development of unique artifacts, festivals, industrial development, environmental protection, and other indicators. (3) Key indicators of social citizen: low-carbon movement activities, community care, health education, cultural festivals and ceremonies, green citizen, civic literacy, community disaster prevention. (4) Key indicators of administration are: (a) course learning (b) teacher support (c) venue rental (d) event organization (e) equipment rental (f) lectures and study courses. This study discovered that the aforementioned key indicators can generate cooperative partnerships through administrative resources. This exploratory study hopes to establish a strategic alliance model through the establishment of accumulated interview data.

The findings identify several causal paths, comprising specific combinations of key indicators of degree of resource importance (RI), control capability of resource ownership (CO) and control degree of resource substitution (DS), to achieve successful strategic alliances for community colleges and community development associations by using FsQCA to analyze. Social network analysis helps increase understanding of the degree centrality, closeness and betweenness in configurations. The conditions that associate with strategic alliances for community colleges and community development associations cooperation relationship include the lifelong learning, social citizen education, and community empowerment. Accordingly, the three elements for community colleges and community development associations are critical for successful strategic alliances. Likewise, the facts of mover attitude, policy, environmental change, alternative groups, they will impact community colleges and community development associations toward to cooperate or not.

Acknowledgements. This paper was supported by Asia University Research Projects (100-asia51).

References

1. Wu, M.L., Li, H.C., Chung, W.C.: A model of strategic alliances in Community University with regional Universities and Colleges. Taiwan Ministry of Education commissioned research projects (2005)
2. Chang, H.Y.: To analyze for the importance and actual application of the professional competency of the program planner on Community University. Department of Adult & Continuing education, National Chung University, Taiwan (2002)
3. Chen, T.M.: A Study of the Community Universities in Taiwan: the practice of the structure of civil society and lifelong learning policy. Department of Public Administration, National Chengchi University, Taiwan (2002)
4. Wu, M.L.: Community college's role in the positioning and sustainable development strategy. Adult and Lifelong Education Review 21, 34–43 (2009)
5. Wu, C.C.: Analyzing cooperation relation between community college and community development association from social exchange point of views- Example of Chung-hwa County. Department of Human Resources Relations, Dayeh University, Taiwan (2006)
6. Weng, X.D., Kiyoshi, M.: Community Development Concept. Nation Taiwan Craft Research and Development institute (1997)

7. Cropley, A.J.: Towards a system of lifelong education: some practical considerations. UNESCO Institute for education, Hamburg (1980)
8. Huang, W.H.: The reconstruction of education in Taiwan. Yuan-Liou Publishing Co., Ltd. (1996)
9. Hu, M.C.: The development and practice for lifelong education model. Nation Taiwan Normal University publish (1997)
10. Schmitter, P.C.: On civil society and the consolidation of democracy. Paper present in the 3rd Wave Democracies Conference, Taipei (1996)
11. Cooper, T.: An ethics of citizenship for public administration. Prentice Hall, Englewood Cliffs (1991)
12. O'Connell, B.: Civil Society: Definitions and Descriptions. Nonprofit and Voluntary Sector Quarterly 29(3), 471–478 (2000)
13. Carpenter, M.A., Sanders, G.W.: Strategic management -a dynamic perspective concepts. Person Eduction, Inc., New Jersery (2009)
14. Pfeffer, J., Salancik, G.R.: The External Control of Organizations: A Resource Dependence Perspective. Harpr and Row, NY (1978)
15. Grover, V., Teng, J.T.: Theoretical perspectives on the outsourcing of information systems. Journal of Information Technology 26(1), 75–103 (1995)
16. Galaskiewicz, J.: Exchange networks and community politics. Academic Press, New York (1979)
17. Bourantas, D.: Avoiding Dependence on Suppliers and Distributors. Long Range Planning 22, 140–149 (1989)
18. Grover, V., Cheon, M.J., Teng, J.T.C.: Decisions to outsourcing information systems, testing a strategy-theoretic discrepancy Model. Decision Sciences 26(1), 75–103 (1994)
19. Mitchell, J.C.: The Concept and Use of Social Networks, Social network in Urban Situation Manchester. Manchester University Press, England (1969)
20. Scott, J.: Social Network Analysis: A Handbook, 2nd edn. Sage Publication, London (2000)
21. Freeman, L.C.: Centrality in social networks: conceptual clarification. Social Networks 1, 215–239 (1979)
22. Ragin, C.C.: Fuzzy-set social science. University of Chicago Press, Chicago (2000)
23. Ragin, C.C.: Redesigning social inquiry: Fuzzy sets and beyond. University of Chicago Press, Chicago (2008a)
24. Ragin, C.C.: User's guide to fuzzy-set/qualitative comparative analysis (2008b), http://www.fsqca.com
25. Ragin, C.C.: Qualitative comparative analysis using fuzzy sets (fsQCA). In: Rihoux, B., Ragin, C.C. (eds.) Configurational Comparative Methods: Qualitative Comparative Analysis (QCA) and Related Techniques (Applied Social Research Methods), pp. 87–121. Sage, Thousand Oaks (2009)
26. Woodside, A.G., Zhang, M.: Identifying x-consumers using causal recipes: "Whales" and "jumbo shrimps" casino gamblers. Journal of Gambling Studies 28, 13–26 (2011)
27. Chung, H.L.: Rebooting the dragon at the cross-roads? Divergence or convergence of cultural policy in Taiwan. International Journal of Cultural Policy 18(3), 340–355 (2012)
28. Wang, P.R.: A Study of Promoting Community Industry by Community Development Association: Examples of Community In Northern Taiwan. Shih Chin University, Taiwan (2013)
29. Wu, T.F., Hou, N.T., Ting, C.H.: A cooperation model in community university and community development association. Asia University Research Projects (2011)

PLM Usage Behavior and Technology Adaptation

Chuan-Chun Wu[1], Chin-Fu Ho[2], Wei-Hsi Hung[3], and Kao-Hui Kung[4]

[1] I-Shou University of Department of Information Management, Taiwan
miswucc@isu.edu.tw
[2] Takming University of Science and Technology of Department of Multimedia Design, Taiwan
cfho@takming.edu.tw
[3] National Chung Cheng University of Department of Information Management, Taiwan
fhung@mis.ccu.edu.tw
[4] I-Shou University of Department of Information Engineering, Taiwan
kh.kung@msa.hinet.net

Abstract. It is frequently faced unforeseen managerial problems when users start to use the IT. Once the problem is identified, it is necessary to ensure an adaptation existed among the technology, the organization, and groups. The appropriateness of such adaptation even plays a critical role in successful IT implementation projects. The original equipment manufacturer (OEM) industry has been the most representative industrial model in Taiwan. This research found that during the use of Product lifecycle management (PLM) software system in Taiwan for supporting research and development (R&D) projects, the degree of adaptation depends on the interaction condition and appropriation among the technology, the organization, and groups. In the process of adaptation, discrepant events are resulted from the appropriation of technology. The degree of solving these discrepant events is a critical factor in determining whether the enterprise is able to gain competitive advantages from using an IT.

Keywords: Adaptive Structuration Theory, Product Lifecycle Management, Collaborative Design.

1 Introduction

The history of product development research traces back to a stream of innovation research, focusing at a microlevel regarding how specific products are developed based on an organizations-oriented approach (Brown and Eisenhardt, 1995), i.e., decisions on product development projects within a single firm. The existing literature on product development is vast, covering perspectives from different research communities: marketing, organizations, engineering design, and operations management (Krishnan and Ulrich, 2001). However, the mainstream research still lies in structures and processes by which individuals create products, which emphasizes the organizational structures, roles, and processes that are related to enhanced product development.

In managing a development project, decisions are made about the relative priority of development objectives. These include process performance metrics like lead-time

L.S.-L. Wang et al. (Eds.): MISNC 2014, CCIS 473, pp. 76–91, 2014.
© Springer-Verlag Berlin Heidelberg 2014

and productivity, and financial performance metrics like profits, revenues, and market share. Product development process is often executed in the "black box" of the development team. This is because flow of information in R&D groups is versatile, subject to different organizational requirements in different stages of product design and development. The complexity of communication in the project team has triggered unpredictable impacts, leading to the need of managerial intervention to ensure the realization of organizational objectives. On the other hand, the main issue of product development is about creative ideas and knowledge contribution from team members. Autonomy of the project team is often granted by the organization to exert its best possible productivity.

The development of new information technologies appears to be revolutionizing the practices of product development to a considerable degree. The benefit of new tools like product life management (PLM) systems to manage product knowledge and support development decision making within the extended enterprise has been explored in greater detail (Liberatore and Stylianou 1995, Ruecker and Seering 1996).

Whether the adaptation of PLM system occurs in the adoption process of new products is a key judgment basis of the success or failure of CPC business model. Leonard-Barton's (1988) proposed the mutual adaptation model of the technology and organization to reduce the misalignment between techniques, delivery systems, and performance criteria in the technology adoption process. The model focuses on large or small adaptive cycles. In terms of the adaptation in technology development, the large adaptive cycle means that the developer needs to repeat the activities in the very beginning stage when the prototype is complete. In contrast, the small adaptive cycle denotes smaller change in the development process, such as moving back to test stage from product prototype stage. For the adaptation of delivery systems, the large adaptive cycle means going back the initial stage, and the small adaptive cycle means moving back the closest stage. In terms of the adaptation of performance criteria, the large adaptive cycle is the concern of turning back to strategic level considerations from technological mutualization level whereas the small adaptive cycle is the status of moving back to operational level from technological mutualization level. Concerning human resource and budget, the large adaptive cycle is more costly. Thus, Leonard-Barton's (1988) believed that the misalignment between technology and organization and the two types of adaptive cycles must be turned into alignment status, so that the alignment can provide benefit to the organization.

Orlikowski and Iacono (2001), "IT artifacts are always embedded in some time, place, discourse, and community. As such, their materiality is bound up with the historical and cultural aspects of their ongoing development and use, and these conditions, both material and cultural, cannot be ignored, abstracted, or assumed away (p. 131)." Hence, we conceptualize PLM systems as embedded in specific organization and group contexts and these contexts have fundamental influences on the way PLM systems are implemented and used as well as the outcomes of PLM implementation and use. Due to the complexity and high cost in developing enterprise information systems (EIS), package software becomes the prevailing approach for firms to implement EIS, and the PLM system is no exception. During the system at installation, the organization behavior, by plan, is a top-down approach that integrates system functions with business processes through business process engineering (BPR). To achieve organizational performance goals, the behavior of individual users is a bottom-up approach which means that the user poses system function needs based on the operation of business processes. Hence, during the system implementation, the manager focuses on achieving organizational performance whereas the individual user places more concern on simplifying the operation of tasks. When a confliction occurs between achieving performance and simplifying the operation of tasks, the manager and user must assess operation processes of the system and make proper adjustment in order to achieve technology adaptation.

The Research of This Study Is Set as Below
In consideration of achieving organizational objectives for the implementation of PLM system, how would the organization make technology adaptation based on the usage behavior?

The rest of the paper is structured as follows. First, provides a discussion of depict the relationships between NPD and PLM and the theoretical base of technology adaptation. Second, introduce the PLM system and description of PLM technology. Third, we design the implementation technology adaptation model. Fourth, According to implementation a PLM system as package software, we discuss the findings from case study and then use technology adaptation model to implement PLM system and verify the organization KPI. Finally, we discuss the findings from case results and provide valuable implications on the technology adaptation issue for academics and practitioners.

2   Literature Review

2.1   PLM in NPD

The field of new product development (NPD) has been defined as including the set of activities "beginning with the perception of a market opportunity and ending in the production, sales, and delivery of a product" (Ulrich and Eppinger, 2000, p. 2). Therefore, the dominant models in NPD emphasize an interdisciplinary mode of inquiry and call for contributions from most business functional areas.

In the new product development process the PLM systems link product design information with customer data, processing data, cost data and resource planning data and make them accessible to everyone inside and outside the firm that requires the information at the right time (customers, suppliers or other partners working together in the same project). (Serrano and Fischer, 2007) Product life-cycle management (PLM) is a powerful approach for breaking through organizational and technical barriers to collaborative New Product Process Development. PLM is a management process aimed at enabling collaboration activities spanning all product value chain elements. In NPD, PLM systems provide design engineers with customer and product features information, existing designs that might meet the design needs (eliminates re-inventing the wheel), tooling requirements and availability of existing tools (process capabilities), sources of parts and components (unique vs. off-the-shelf items). PLM is a means for ensuring that the knowledge generated at any point in the product value chain is captured and eventually exploited, not re-generated later or left unrecognized. PLM systems also help identify and communicate sources of knowledge.(Swink, 2006)

2.2   Technology Adaptation Theories

DeSantics and Poole (1994) proposed Adaptive Structuration Theory (AST) which is a framework with the focal point on the impact made to use behavior from technology, work practice, organization environment, and group structuration appropriation. If it matches with the technology design idea, it then creates higher decision making performance.

Leonard-Barton (1988) proposed that new technology adoption is a kind of innovation and such adoption is a mutual adjustment between the technology and the adopting company. Thus, the misalignment between the technology and the adopting company commonly happens during the adaptation process. The alignment can be observed and analyzed through the dimensions of technical, deliver system, and performance criteria. The center of Leonard-Barton's (1988) idea is the large and small adaptive cycles.

From the behavior theory perspective, Tyre and Orlikowski (1994) characterize adaptation as a highly discontinuous process rather than gradual and continuous. In the adaptation process, discrepancy events occur continuously, especially in the early stage. Discrepant events represent subsequent changes often occurred in an episodic manner and trigger adaptations in terms of use, leading to new new discoveries on the part of users. Later, the degree of routinization becomes high and the modification of process and technology become restricted, discrepant events are reduced. Finally, a balance is reached between discrepant events and new ideas.

Majchrzaket al. (2000) further combined the theories of Leonard - Barton(1988), Tyre and Orlikowski (1994), and Desantics and Poole (1994) that are correspondingly named as mutual adaptation model of Technology and organization, discrepancy events of technology adaptation. and Adaptive Structuration Theory.

Majchrzaket et al. (2000) believes that the models differ in the continuity presumed to occur in the adaptation process. Leonard-Barton's model proposes that adaptations occur continuously in response to misalignments, gradually leading to a successful alignment. In contrast, Tyre and Orlikowski characterize adaptation as a highly discontinuous process, where discontinuities occur during brief windows of opportunity which open the constraint set. This difference may reflect different conditions in the field rather than invariant theoretical conclusions.

2.3   As-Is and To-Be Analyses in BPR

During the IT implementation process, business process reengineering (BPR) is frequently taken by organizations. As-Is and To-Be analyses are performance during BPR. The main objective of As-Is analysis is to identify disconnects that prevents the process from achieving desired results. Having identified the potential improvements to the existing processes, the development of the To-Be models is done using the benchmarking technique, which is the comparing of both the performance of the organization's processes and the way those processes are conducted with those relevant peer organizations to obtain ideas for improvement. After setting the To-Be model, discrepant events occur frequently in the organization which means a turning away from the planned scenario as set in the To-Be model. When this happens, how can we make adjustment in order to achieve the organizational objectives? The technology adaptation literature has different views to deal with this problem. One group believes that the organization will make adjustment automatically (Leonard-Barton, 1988), and some believe a solution plan will be emerged through the interaction between user and the technology (Majchrzak et al. 2000). Yet, a clear adjustment process is missing in both explanations.

3   Research Framework

Successful product development involves relatively autonomous problem solving by cross- functional teams with high communication and the organization of work according to the demands of the development task. This perspective also highlights the role of project leaders and senior management in giving problem solving a discipline-a product vision. There is an emphasis on both project and senior management, on the one hand, to provide a vision to the development efforts and yet, on the other hand, to provide autonomy to the team. Thus, we may portrays product development as a balancing act between product vision developed at the executive level and problem solving found at the project level. The balance actions must be achieved through the use behavior of PLM.

Since the package software implementation has strong impact on organizations and the adjustment process is critical for system implementation, we proposed a goal-driven technology adaptation model (Figure 1) to understand the adjustment behavior, solution steps, and methods in the adaptation between organization environment and group structure. The Adaptive Structuration Theory (AST) is also embedded in this model so as to provide the basis to help businesses implement PLM.

This research adopts the PLM solution from PTC, which is called Product Development System (PDS) and it can be used to manage the interdependences of all kinds of information generated from product development process. Participants in the process can understand opinions from others easily and how ideas make impact to the product..

PDS include three parts including Pro/Engineer Wildfire, Windchill PDMLink, and Windchill ProjectLink. Pro/Engineer Wildfire can be used by engineers to examine product functions and identify product usage model in the early stage. CAE application lets engineers realize the design performance in the early stage of product design in order to enhance product quality, reduce costs, and time. With the aid of network technologies, organizations can use Windchill PDMLink to share accurate product information sets and latest information, to manage product design changes automatically, to reduce error cost, and to help organizational information elicitation and product collaboration. Windchill ProjectLink provides the functions for team members to control project document in the real-time basis, to increase team integration, to control progress, and to assist organizations in integrating product information generated in interorganizational activities.

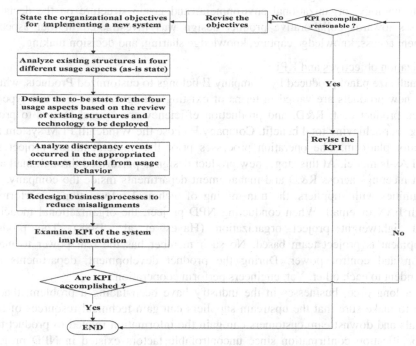

Fig. 1. Goal-driven technology adaptation model

4   Case Study of PLM Implementation

4.1   Case Description

This research adopts Company E as the studying case, which produces small and medium size LCD with a turnover of 3.8 billion Taiwanese dollars and 110% annual growth rate. The revenue model of the company includes 33% on cellphone related products, 14% on consumer electronics, 40% on industrial equipment, and 13% on medical applications. Company E adopts OEM business model, and is one of the best performed LCD makers in Taiwan. The barrier for entering the LCD industry is very high. In addition, the customers are various kinds in the industry. The products are highly customized. Thus, Company E has to communicate with customers and suppliers through the collaborative design platform in the early stage of product development in order to provide more customer support. In terms of OEM business model, product development time and cost are critical to profit making. Thus, Company E utilized PLM to control NPD project time schedule and build a platform for project technical and knowledge sharing in order to satisfy customer needs, achieve operational objectives, and establish technical support.

4.2   Case Analysis

4.2.1   As-Is and To-Be analysis

This study analyzed PLM implementation in two phases: process analysis ("as-is model") and process design ("to-be model"). The impact of contextual factors such as PLM technology, organizational environment, and group structure on the decision processes towards collaborative product design was investigated in four aspects: document access, knowledge capture, knowledge sharing, and decision making.

Organization objectives and KPI

The small size panel produced by Company E belongs to customized Products, which means new products are varied in terms of existing configurations. For this type of product, product cost, R&D, and production efficiency are critical factors to profit earning. For achieving total benefit, Company E chose the Windchill PLM system for the major platform. The operation processes prior the implementation project are termed As-Is model. At this stage, new product design is performed through mail and project meetings across R&D and management departments inside the company. To communities with suppliers, the transmitting of technical information is performed through FAX or email. When conducting NPD project, the organizational model is termed lightweight project organization (Hayes et al., 1988). The product development is project team based. No such member has the real power to make decision and control power. During the product development, departments are independent to each other. Yet, engineers perform cooperation.

For a long time, businesses in the industry have been facing a problem, that is unable to make sure that the upstream suppliers can gain technical resources of raw materials and downstream customers can gain the information related to product test and specification confirmation since uncontrolable factors existed in NPD project. Under a strong competition pressure, Company E implemented PLM platform to improve NPD design environment and integrate resources of up and down streams in order to achieve knowledge sharing and deliver products earlier. After BPR process and the implementation of PLM, it is called To-Be model. The following narrative organizes the model by the DeSanctis and Poole (1994) classification of organizational environment, group structure, and technology. Space considerations permit us only to highlight the most critical details for understanding these structures. Table 1 shows appropriated Structures from As-is model and To-be model.

4.2.2   Discrepant event Analysis and Business Process Redesign

Company E uses PLM system to be the basis of the best operation process. Through BPR, the To-Be model of operations is developed. Yet, when implementing the model, the discrepant events occur among organization, group, and technology. This means that a gap exists between the planned and actual actions of the To-Be model. The causes for creating the gap include the prevention of providing much technical information from the project members, being uncomfortable with the interface, or disintegration of system resources. Thus, the To-Be model is not performed as expected.

Table 1. Appropriated Structures from As-is model and To-be model

Stage	Structure	As-is model	To-be model
Document access	Organization Environment	During the project process, engineers call for meeting every week, and record the meeting minutes and hand them over to the CEO in paper version or through E-mail after meeting. The CEO can understand project content and progress through these materials.	By using the project integration meeting, the project manager makes detailed record of meeting content and project condition. During the project audit process, the CEO controls the project condition at any time through the notification functionality of the system.
	Group Structure	The project engineer discusses project content and progress by using paper document in regular meeting weekly. When the stage task is complete, a full copy of document is passed to the engineer in the next stage. This process repeats until the project is complete.	The project engineer uses Pro/E system for design visualization interface. The project manager uses Pro/E system to collaborate with the project engineer for design progress discussion, modification, and document revision.
	Technology	The design activities are performed through the use of AutoCAD software. The audit activity is performed by using paper document and email.	Once the project starts, the project manager and project engineer use Pro/E system to discuss the project blueprint and manage the versions of products.
Knowledge capture	Organization Environment	The project engineer collects product design drawing, cost analysis, and test results on returned products to RD manager and CEO in paper reports. The CEO gives comments directly on the reports.	In the project close meeting, the project manager uses the online function of Pro/E system to link technical reports, and explain product technical specifications and cost structure.
	Group Structure	The project engineers discuss the technical problems of products in the face-to-face basis. The departments coordinate and review in meeting. When considering technologies integration, it is recorded after the meeting and waited for RD's review and further instruction.	The project engineer uses Product View function in PLM to discuss the document of collaborative design with customers and suppliers and to discuss product design specifications during the meeting.

Table 1. (*continued*)

Knowledge sharing	Technology	Product design drawing and project document are sent out in hard copy or file.	Product design drawing and project document are stored in project database. The project engineers use Web interface to retrieve the document.
	Organization Environment	The engineer tracks the CEO's key requirements in the meeting, and provides project progress to the CEO by using weekly and monthly reports.	The project manager develops specialized control process to manage project progress, and the CEO can use subscription functions of the system to understand project progress.
	Group Structure	The engineers exchange technical document in paper version or file copies in the work place, and review project problems and the technique improvement cooperatively.	The project engineer provides search function to others to use the database on the preset roles and authorities, and also to share latest version information about products.
Decision making	Technology	The project document is transmitted and shared via paper document or file copies.	The system provides search and backup functions to the database based on the authority of roles.
	Organization Environment	The engineer receives evaluation opinions from the CEO, and the CEO makes important decisions in the project close meeting.	The project manager calls for online meeting by using Product View. The engineer explains project progress and cost framework through the use of Pro/E, and the CEO can understand the project status and develops product development and marketing strategies.
	Group Structure	The engineer has to prepare and organize project design document before the project meeting, and make notes on meeting decisions.	The engineers discuss and modify design drawing over the online functions of Pro/E. During the project meeting, the internal opinions are quickly integrated, the time for documentation is shortened, and the information in document is efficiently actualized in the product development process.
	Technology	The Microsoft Office is used during the project process to collect technical information and such information is shared through files or paper document.	The online function of Product View is used to solve the communication problem, and help make decisions and records.

Table 2. The adaptation of discrepant events

Stage	Structure	Discrepant events	Causes for process redesign	Emergent structure
Document access	Organization Environment	Because the need to solve problem immediately, the RD manager asks the project engineer to check project document. Yet, the project engineer may avoid some problems which are difficult to solve. The CEO cannot understand the project accurately.	Project members tend to avoid receiving responsibility. The integrity of project document is lacking, and it has reduced the professionality of the document.	The CEO makes annual assessment policy for checking the integrity of document and developing annual assessment report for the organization.
	Group Structure	The project engineer does not update the version after the project meeting. This has caused confusion on document's current version and wrong record in the system.	Employees believe their development and capabilities are important. They care less about project tasks and document, but more on personal skills for their promotion and competition.	The project manager emphasizes on the integrity of the document used in every stage. The manager collects all performance records of project engineers, and then develops annual KPI for engineers.
	Technology	The project engineer treats the problem solving process as personal asset. The complete technical document is not uploaded to the platform. Project document include only pictures and diagrams.	PLM system integrates the document in every stage. Yet, the records in the system are sometimes fragmented. This has caused low integrity on the document.	The audit process on document takes the format of ISO standard. The project manager audits the content of document and ensures the integrity of knowledge.
Knowledge capture	Organization Environment	The project manager lacks willingness of being good in management. The management of technical document is weak and the CEO cannot understand the current situation and full benefits of the project.	The project engineer has to deal with ISO format, and the document is recorded under a simplified structure.	The CEO considers the integrity of recording project solutions as an important annual assessment item. The CEO uses the document to provide feedback to the project managers.
	Group Structure	The RD manager communicates with the project manager through face-to-face or phone. The project	Project members communicate face-to-face in order to exchange ideas. The technical document is	The project manager provides bonus based on the KPI of the project engineer and the

Table 2. (*continued*)

Decision				
	Knowledge sharing	manager communicates with the project engineer according to opinion category and project designs. The improvement ideas are not recorded in the database.	the record in the document that belongs to personal asset.	contribution on creating intellectual property.
	Technology	The file size of design drawing is large and hard to be shared and searched over the network.	The project engineer downloads project document from the database and this has reduced the quality of network of the organization.	The system provides a security mechanism for controlling document safety and managing knowledge database. It offers online reading function only. Once the document is away from the network, it cannot be read.
	Organization Environment	Since many online projects are discussed online, the design time becomes longer. The CEO cannot give complete auditing comments by reading fragmented data.	The content of project document is not reached organizational level of knowledge standard. Knowledge cannot be maintained and shared. The CEO cannot understand the full picture of the project by simply assessing the project document.	The CEO needs to see the quality of project document, and the document needs to show assessment results of the project team.
	Group Structure	The project engineer does not maintain the version of document appropriately and this causes a confusion of document version.	The version of document of project evaluation process is controlled by the project engineer, and it causes an inconsistency on document versions in the database.	Once the change is made on document version, the change date and reason have to be recorded. The project manager finds the version of document based on the change date.
	Technology	The project engineer misuses the authority to copy project document and to provide it to external persons. This causes the problem of disclosure information.	The engineer copies project information into his or her own computer through the misuse of the authority.	The knowledge database has the encryption system (DRM) to protect information. Once the information is outside the organization, it cannot be read.
Decision	Organization	The CEO focuses on organizational	Regarding the problems in the	The CEO provides quick

Table 2. (*continued*)

making			
Environment	benefits and does not emphasize on the relationship between products. The product related analysis is conducted when the sample is set in product line, and this causes the delay of produce delivery.	project, the project manager reports to the CEO and RD manager face-to-face or through meetings. Professional document is described in outline model. The content recorded in the professional document does not contain much knowledge value.	suggestion to the closing stage of the project via Product View, and discusses and solves problems with customers and suppliers in order to establish strong supply chain relationship.
Group Structure	During the improvement online meeting, the major members in the project are not involved on time. This causes that the conclusion in the meeting is not implemented well.	Project members commonly communicate with suppliers and customers face-to-face or through the telephone. The engineer depends on his or her own habits to complete project document, and the quality of the document is not stable.	The project manager reviews the technical problems with customers and suppliers in online meeting, and makes final decision to ensure the outcome of product development.
Technology	During the project auditing process, the RD manger discuss with the project manager orally in most of time. Product View is used to check format only. The solution is not store in the document database.	For dealing with system auditing function, the engineer avoids the automatic judgment function on document of the system. The project produces useless document.	The system automatically checks the document format, keeping invalid document from storing in the system database.

We will discuss the gap existed in the To-Be model based on three aspects: organization, group, and technology, and discuss possible strategies and ways to minimize with the gap. After the rectification from BPR, the emerged structure comes out, which can be used to improve the incompatibility existed between the To-Be model and systems. The emergent structure will be the best model for using PLM in organizations. Table 2 presents the adaptation of discrepant events and the emergent structures in the case

4.2.3   Emergent Collaborative Processes in Product Development

Table 3 shows a summary of BPR actions resulting in emergent structures for dealing with discrepant events. After the system implementation, the gap emerges which is caused by the interaction of usage behavior including top-down and bottom-up. These two are relevant to organizational performance.

Table 3. Summary of BPR actions resulting in emergent structures

Usage Behavior	Gap adjustment
Document access	The project manager summarizes the important items given by engineers and makes a report. The CEO sets annual evaluation standards for the project based on the integrity and contribution of the report. The organization establishes system control documentation for PLM, which is conformed to ISO standards. The system includes a function to check the document in the project.
Knowledge capture	During the annual evaluation of engineers' project contribution, awards are given based on the number of patterns received. The system provides security mechanism to protect document and ensure the databased cannot be accessed outside the business network.
Knowledge sharing	The project manager audits the dates to manage the version of document, and the system uses encoding system (DRM) to protect project information.
Decision making	Product View function is used to solve the technical problem occurred between suppliers, buyers, and the organization. The online auditing function is able to record the process of making solutions and manage the document versions, and protect the organization's intellectual property.

As shown in Table 4, the case company uses the online function of PLM system to perform collaborative meeting with customers and suppliers over the Internet. The system provides pictures, diagrams, and simulation function to help locate problems and make solution efficiently.

There are also interactions between performance metrics in supporting product development decisions (Krishnan and Ulrich, 2004). Table 6 presents performance impacts results from implementing BPR actions as summarized in Table 4.

Table 4. Project outcomes in terms of organizational KPI

Item		KPI	
Six potential interaction		Pre-existing	Technology spirit
Development Time	Development Cost	103 Days	80 Days
		Shorten 23 days (eliminate 22.3%)	
Development Cost	Product Cost	Manage benefit : 98.7 million/ year	
Development Cost	Product Performance	Product benefit : 122 million/ year	
Product Performance	Product Cost	Process benefit : 29.8 million/ year	

(Unit : NT$)

5 Conclusion

Our research results as depicted in Figure 2 indicate that there are three contextual factors (technology, group structure, and organizational environment) affecting the usage of PLM. Through interactions between multilevel users, modification of system features and process flows leads to the achievement of organizational objectives.

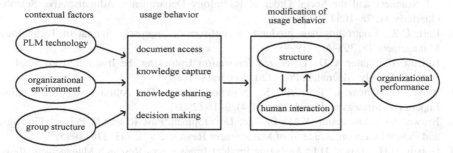

Fig. 2. The impact of contextual factors on PLM system usage

Prior research has suffered from versatile and complicated organizational flows in product development decisions. Only multilevel studies can resolve conflicting results produced in different levels because only they examine the linkages between levels, such as discovering how individual contributions generate and sustain communities (Burton-Jones and Gallivan, 2007). Scholars have long proposed that CEO's product vision such as ranking project priorities influences product development performance (Brown and Eisenhardt, 1995). Yet, how beliefs of top management affect individual team members to achieve organizational goals remains to be addressed. The multilevel research concept and supporting IT infrastructure to conduct multilevel PLM system usage research have only recently matured to allow this in-depth case study.

There are different functional perspectives of product development: concept development, supply chain design, product design, product strategy and planning, (Krishnan and Ulrich, 2001). The focus of this study is essentially a combination of product strategy and PLM system implementation, investigating how strategic goals can realize through system usage behavior. As the use of PLM has come across almost every aspect of product development, great research opportunities exist in applying this system usage concept to other areas of product development. For example, a firm may have intensive product development collaboration with its supply chain partners like customers or suppliers. How would the usage of PLM system affect the firm in managing relationships with its partners? Addressing this issue will require theoretical support from supply chain management, IS implementation, and system usage.

References

1. Aaby, N.E., Discenza, R.: Strategic marketing and new product development. Marketing Intelligence & Planning 13(9), 30–35 (1995)
2. Alavi, M., Carlson, P.: A Review of MIS Research and Disciplinary Development. Journal of Management Information Systems 8(4), 45–62 (1992)
3. Allport, C.D., Kerler III, W.A.: A Research Note Regarding the Development of the Consensus on Appropriation Scale. Information Systems Research 14(4), 356–359 (2003)
4. Barley, S.R.: Technology as an Occasion for Structuring: Evidence from Observation of CT Scanners and the Social Order of Radiology Departments. Administrative Science Quarterly 31, 78–108 (1986)
5. Bart, C.K.: Controlling new products: a contingency approach. Journal of Technology Management 18, 395–413 (1999)
6. Bidault, F., Butler, C.D.: Leveraged Innovation Unlocking the Innovation Potential of Strategic Supply. Macmillan Press Ltd., London (1998)
7. Blake, D., Cucuzza, T., Rishi, S.: Now or Never: The Automotive Collaboration Imperative. Strategy and Leadership 31(4), 9–16 (2003)
8. Brown, S.L., Eisenhardt, K.M.: Product Development:Past Research, Present Findings, and Future Direction. Academy of Management Review 20(2), 343–378 (1995)
9. Gobeli, D.H., Brown, D.J.: Analyzing Product Innovations. Research Management 30(4), 25–31 (1987)
10. Jacobs, R., Bendoly, E.: Enterprise resource planning: Developments and directions for operations management research. European Journal of Operational Research 146(2), 233–240 (2003)
11. Ulrich, K.T., Eppinger, S.D.: Product Design and Development. McGraw-Hill (2004)
12. Kosaka, T., Fitzgerald, G.: Research in Progress: Structurational analysis of comparative study of EIS between the United Kingdom and Japan. In: ICIS Proceedings, Atlanta, GA, pp. 496–497 (1997)
13. Krishnan, V., Ulrich, K.T.: Product Development Decisions: A Review of the Literature. Management Science 47(1), 1–21 (2001)
14. Kuczmarski, T.D.: Success Isn't Always Its Own Reward – Big Bucks Help. Marketing News 22(24), 10 (1988)
15. Leonard-Barton, D.: Implementation as Mutual Adaptation of Technology and Organization. Research Policy (17), 251–267 (1988)
16. Liberatore, M.J., Stylianou, A.C.: Toward a Framework for Developing Knowledge-based Decision Support Systems for Customer Satisfaction Assessment, an Application in New Product Development. Expert Systems with Applications 8(1), 213–228 (1995)

17. Majchrzak, A., Rice, R.E., Malhotra, A., King, N.: Technology Adaptation: The Case of a Computer-Supported Inter-Organizational Virtual Team. MIS Quarterly 24(4), 569–600 (2000)
18. Markus, M.L., Robey, D.: Information Technology and Organizational Change: Causal Structure in Theory and Research. Management Science 34(5), 583–598 (1988)
19. Tyre, M.J., Orlikowski, W.J.: Windows of Opportunity: Temporal Patterns of Technological Adaptation in Organizations. Organization Science 5(1), 98–118 (1994)
20. McGrath, M.E., Romeri, M.E.: From experience the R&D effectiveness index: A metric for product development performance. Journal of Product Innovation Management 11(3), 213–220 (1994)
21. Grieves, M.: Product Lifecycle Management: driving the next generation of lean thinking. McGraw-Hill (2006)
22. Swink, M.: Building Collaborative Innovation Capability. Research Technology Management 49(2), 37–47 (2006)
23. Orlikowski, W.J.: The Duality of Technology: Rethinking the Concept of Technology in Organizations. Organization Science 3(3), 398–427 (1992)
24. Orlikowski, W.J., Robey, D.: Information Technology and the Structuring of Organizations. Information Systems Research (2), 143–168 (1991)
25. Orlikowski, W.J., Iacono, C.S.: Research commentary: Desperately seeking the "IT" in IT research - A call to theorizing the IT artifact. Information Systems Research 12(2), 121–134 (2001)
26. Sands, S., Warwick, L.M.: Successful Business Innovation: A Survey of Current Professional Views. Californian Management Review 20(2), 5–16 (1977)
27. Serrano, V., Fischer, T.: Collaborative innovation in ubiquitous systems. Journal of Intelligent Manufacturing 18(5), 599–615 (2007)

Flipped Learning: Integrating Community Language Learning with Facebook via Computer and Mobile Technologies to Enhance Learner Language Performances in Taiwan

Paoling Liao

Foreign Language Education Centre, National Kaohsiung Marine University,
No.142, Haizhuan Rd., Nanzi Dist., Kaohsiung City 81157, Taiwan
{Paoling Liao,efl301}@webmail.nkmu.edu.tw

Abstract. This study investigated how a Community Language Learning (CLL) approach, when utilised via new technology with the social network Facebook, can be most effective in a flipped EFL classroom. Curran [2] claims his CLL approach can reduce learner anxiety and insecurity. One feature of CLL is that trusting relationships are established not only between students and teachers but also among students themselves, and a learning community based on trusting relationships is claimed as the key element for the successful learning of foreign languages. Meanwhile, computer and mobile technologies have grown increasingly influential in many areas, including education. Flipped learning, a learning model first proposed by Sams and Bergmann [11], is becoming widespread globally, and mobile technology makes flipped classrooms both feasible and moveable. Liao and Fu's study [14] showed that task rehearsal in computer-mediated communication does enhance learners' performances. Thus, this study focused on learner anxiety and language performance issues.

Keywords: Anxiety, Community Language Learning, EFL, Facebook, Flipped Learning, Mobile Technology, Task Rehearsal.

1 Introduction

As a result of continuing advances in information and technologies, the globalisation of business has become a trend around the world. Accordingly, a common language for business activities has become more important than ever, to the point of being required in many cases. In this context, English has becomes a clear favourite as a second/foreign language for non-native speakers. However based on the author's personal experiences and previous research [1], it is clear that in spite of how much time some students have spent learning English, most of them remain very uncomfortable using English in the classroom, with spoken and written English being particularly difficult for most. In short, many students lack motivation and show high anxiety when they are actually using English as a foreign language (EFL).

L.S.-L. Wang et al. (Eds.): MISNC 2014, CCIS 473, pp. 92–101, 2014.
© Springer-Verlag Berlin Heidelberg 2014

Curran [2] has claimed that his Community Language Learning (CLL) approach can reduce the learner's anxiety and increase his/her feelings of security. Curran believes that learning not only involves intellectual interactions but also relationships between teachers and students, and among students themselves. Learning, according to Curran, is actually made possible by the quality and structure of these personal relationships. Samimy [3] is one of the followers of the Counselling Learning approach in teaching a second language, which is itself derived from Curran's counselling learning theory. Samimy [3] explains the approach as follows:

> The Counselling Learning Approach: One of the most basic ingredients for using CL was to bring personal qualities to the teacher-learner relationships and to learning activities. In order to help establish a personal atmosphere in a foreign language classroom, Curran provides a guideline by introducing six interrelated elements in learning. The acronym SARD symbolizes these six elements: Security, Attention-Aggression, Retention-Reflection, and Discrimination. (p. 171)

Along these lines, Zou [4], consistent with Samimy [3], emphasises the importance of language as well as the need for a supportive relationship and meaningfulness in second language learning. As Kristjansson [5] explains, "In his (Zou's) view, the prerequisite for effective teaching and meaningful learning is the establishment of trusting relationships. This is the basis for successful interaction in another language and for empowerment." In her own STAR ESL programme, Kristjansson [5] focuses on the links between second language learning, human relationships and community.

From the above descriptions, it can be seen that Curran [2], Samimy [3], Zou [4] and Kristjansson [5] all focus on trusting relationships in the learning community. Such trusting relationships, moreover, are established not only between students and teachers but also among students themselves. The quality of the relationships between teachers and learners facilitates and enhances learning and is, therefore, prior to and more important than any particular methodology. In this view, a learning community based on trusting relationships is claimed to be the key element for successful learning of a foreign or second language.

Moreover, the American psychiatrist M. Scott Peck [6], in his book *The Different Drum*, describes community as a place where no one is attempting to heal or convert you, but where you are best placed to heal and convert yourself. This is possible, he argues, because people are called to wholeness, to be the best that they can be. However, healing and converting need to happen in a community, for we are also called to recognise our limitations and cannot be whole without each other. This book implies at least two states, namely, joy and sorrow. The joy comes from the realisation that, yes, community is possible, while the sorrow comes from the acceptance of the fact that we must "die" to achieve community, where "die", in this context, means the necessary act of emptying ourselves of our prejudices, our need to control, our need to convert, our theology, etc. According to Peck [6], community is emphasised as a true, global meaning in a technological society more than a false society that people usually regard as a true one. With love and tolerance, we can start to transform false society into a true community.

As Curran [7] states, "The community means a common group commitment would imply both the self-involved and self-committed dynamics. The community implies genuine communication. It would be an open trustworthiness, which is essential to one's freedom to communicate his whole self in a group." In this way, Curran's community is associated with the true meaning of Peck's community. One common feature of both of their visions of community is a genuine and trusting society.

In contrast with such a true community, the traditional classroom seems to create more competition than trust among learners. Traditional approaches may thus result in learner anxiety and become an obstacle to learning. Curran's CLL approach, in contrast, may bring a promising solution for the English learning problems of Taiwanese college students.

In addition, technological advances all around the world, and especially in developed countries, have achieved substantial influences on many areas, including education. Computers, smart phones and social networks in Taiwan are quite popular. Almost every student has a computer and a mobile phone in his or her university residence or at home, and they are very interested in and adept at using computers, mobile devices and social networks such as Facebook, Twitter, and others. A few years ago, I created the website Pony's EFL Café [8] and related Facebook groups [9] for my students to enjoy and flip their English learning without the imposition of time limits or constant teacher control. Many of the students showed great interest in English learning via computer and mobile technology. In particular, some of the students who had previously received low marks in English showed more interest than they had before. These changes among my students were very similar to findings observed by other researchers. For example, as Warschauer and Healey [10] point out:

> Some studies show that students tend to like using computers, even when they may not make much progress (Stenson et al., 1992) and when they may feel that computers do not necessarily improve their language learning (Schcolnik et al., 1995/96).   (p.62)

Moreover, a new teaching approach proposed by Sams and Bergmann [11] termed the "flipped classroom" has seen emerging interest in recent years. A flipped classroom helps instructors shift direct instruction onto a more learner-centred approach. In this approach, learners watch video lectures at home which are made by the teacher beforehand rather than coming to the classroom to watch the instructor's lectures. A flipped classroom allows learners to do homework in class (see Appendix 1) and allows instructors to use their in-class time most effectively with learners. Learners can propose questions for discussion during class time individually or in groups, and instructors can utilise their face-to-face time to help learners solve their problems based on their different proficiency levels. By implementing the flipped classroom approach, instructors can focus on all the learners in a class in spite of their different learning curves. Furthermore, social networks supported by new technologies are now making flipped classrooms more widespread than ever. For example, smartphones and wearable devices such as Google glasses make flipped classrooms portable, and diffusive. Students can watch lecture videos whenever it is comfortable and convenient to do so. For example, they may watch videos when they are waiting for a bus or standing in a queue. In other words, learners are not limited to watching videos at home; instead, they can do so at anytime, anywhere they would like to. In one recent research study on new technology, Liao [12] proposes a three-stage game-based innovative mobile-mediated communication task (see Appendix 2) to help learners to enhance their interest in English as a foreign language (EFL) learning and their language performances. Mobile learning (ML) is becoming a new paradigm of education.

Gallagher [13] has conducted an evaluation of CLL compared to a traditional approach. The results of that study showed that students in a CLL group have more positive attitude changes than students in a traditional group. The computer-assisted language learning (CALL), ML and CLL approaches have all been proven to be capable of increasing a learner's positive attitude toward learning. Therefore, three approaches have the same features and links among them.

Warschauer and Healey's study [10], together with my own classroom observations, has encouraged me to use computer and mobile technologies as an aid to the CLL approach in the EFL classroom. Liao and Fu's study [14] has shown that task rehearsal in computer-mediated communication does enhance the performance of learners with regard to syntactic variety and lexical complexity in terms of Skenhan's [15] complexity, accuracy and fluency (CAF) model. This study integrates Facebook groups (see Appendix 3) as social networks into a trusting community (CLL theory) via computer and mobile technologies. The research that I propose to undertake will investigate both the theory and practice of CLL, CALL and ML from the past decades up to the present. It will also focus on how to reduce the student's anxiety. The issues of anxiety in CLL, CALL and ML will be discussed and argued respectively. The research will be conducted in a university in Taiwan. There will be one control group and one experimental group. They will be taught, respectively, in one traditional classroom not using the CLL, CALL and ML approaches, and in another in which the CLL, CALL and ML approaches are applied. The methodologies I adopt will be both quantitative and qualitative, such as questionnaires, interviews, classroom observations, etc.

Finally, from the results of this research, I should be able to prove whether or not CLL with CALL and ML can reduce learner anxiety and increase communicative competence. Cziko and Park's [16] study reports that "there have been several projects that have used the Internet to link second language learners with native speakers, but for the most part these efforts have been limited to text communication." Little research has been done in the field on using synchronous, computer-mediated audio communication (SCMAC), which provides access to interact with native speakers of the target language. My research will thus seek to more fully explore

SCMAC's major impact on second/foreign language learning and teaching. Those findings will be a new contribution to the field of second/foreign language learning via Internet audio communication.

2   Rationale of This Study

The rationale of this study is to determine the effectiveness of Community Language Learning (CLL), Computer-Assisted Language Learning (CALL), Mobile Learning (ML) and Flipped Learning (FL).

3   Main Objectives of Research

This research aims to investigate and examine how a Counselling Learning in a Second Language approach (CLL) together with Computer-Assisted Language Learning (CALL) and Mobile Learning (ML) can be used most effectively in the EFL classroom. To that end, the research seeks to answer the following questions:

1. Would the adoption of aspects of CLL, CALL and ML be compatible with established university curricula and could they bring significant changes in students' affective variables such as anxiety, as well as in their communicative and linguistic competence?
2. What further research is needed concerning the subtle interplay between the personal and pedagogic factors that determine successful acquisition?
3. In Curran's research, the CLL approach was applied to learners whose mother tongues are very close to the target languages. What will happen, in contrast, if the students must learn different grammatical systems or different kinds of phonemic features with which they are not familiar? For example, how affective will the approach be for Chinese students learning English as a foreign language in Taiwan?

4   Main Research Procedures/ Methodology

To achieve the objectives stated above, the following approach is envisaged: The target group for the research will be 48 students with Basic English proficiency according to their scores on the Test of English for International Communication (TOEIC) for global test takers administered by ETS Taiwan. Their TOEIC scores, in other words, will range between 350 and 400. All the participants will be divided into two groups, one control group and one experimental group. The control group will be taught using a traditional approach while the experimental group will be taught using an approach combining CLL with CALL and ML. The measure of anxiety developed by Macintyre and Gardner [17] will be administered using three scales at the input, processing and output stages. The results will then be analysed and interpreted.

Moreover, whether the use of CLL with CALL and ML in a foreign language setting can improve learner communicative competence or not will be investigated.

The setting for the data collection will principally be my present workplace, National Kaohsiung Marine University, Taiwan. The data will be collected via the following methods: 1) anecdotal records; 2) classroom diagrams and maps; 3) discussion; 4) archival data: documents and student work; 5) feedback cards; 6) making and transcribing audio recordings; 7) making and transcribing video recordings; 8) observation/field notes; 9) teacher and student journals; 10) lesson plans and teaching logs; 11) sociograms; and 12) face-to-face interviews. These methods will be carefully designed in order to gain meaningful results.

Beatty [18] has suggested that traditionally, much research on CALL has focused on whether or not students learn better with a computer. Research questions now include *how* computers and mobile devices should be used and *for what purpose*. He also proposes eight methodological approaches for research into the CALL approach. They are the literature review, a pilot study, corpus linguistics, error analysis, the experiment, a case study, the survey and the ethnographic approach. Those research approaches are worthy of further studies, and some of them may fit into my research project.

A more comprehensive, necessary/advised literature review will also be undertaken, from the general to the specific. Having critically reviewed this literature with respect to my proposal, further in-depth desk-research is also to be expected.

After thorough desk-research, the objectives of this research will first be revisited to ensure that it is properly targeted. Hypotheses will then be established in relation to this desk-research, and the hypotheses will be tested against the results of the above data collection.

5   Analysis, Interpretation and Results

Both quantitative and qualitative analyses are envisioned to investigate this research proposal. I intend to name any key ideas/concepts/group categories that arise and discover significant relationships, patterns and themes. How to decrease learners' anxiety levels in English classrooms and enhance their language performances is the main concern of this study. The results will be interpreted and presented along with recommendations for future research. Since this study is still in development, it is worth mentioning that the draft results showed that the experimental group taught with an approach which integrated CLL with CALL and ML did exhibit enhanced language performances (according to Covington's [19] propositional idea density (P-density) measure) in comparison to the control group for which a traditional approach was used.

In sum, this study implies a feasible integration of an old theory (CLL) with new technology (CALL/ML) for future language classrooms. The recommendation for further research studies is that the application of newly developed wearable devices such as Google glasses and smart watches to educational domains will be vital for researchers, teachers and learners.

References

1. Liao, P., Fu, K.: Theory-Based Evaluation of an Innovative CMC Task. In: The International Symposium on Education and Psychology, pp. 109–124. Knowledge Association of Taiwan, Taipei (2013)
2. Curran, C.A.: Counselling-Learning in Second Languages. Apple River Press, Apple River (1976)
3. Samimy, K.K.: A Comparative Study of Teaching Japanese in the Audio-Lingual Method and Counselling- Learning Approach. Modern Language Journal 73(2), 169–177 (1989)
4. Zou, Y.: Rethinking Empowerment: The Acquisition of Cultural, Linguistic, and Academic Knowledge. TESOL Journal 7(4), 4–9 (1998)
5. Kristjánsson, C.: Whole-Person Perspectives on Learning in Community: Meaning and Relationships in Teaching English as a Second Language (Unpublished doctoral dissertation). University of British Columbia, Vancouver, Canada (2003)
6. Peck, M.S.: The Different Drum: Community-Making and Peace. Arrow Books, London (1990)
7. Curran, C.A.: Counselling Learning a Whole-Person Model for Education. Apple River Press, Apple River (1972)
8. Pony's EFL Café, http://goo.gl/dsBJ/
9. Facebook Groups, https://www.facebook.com/groups/485856841542177/
10. Warschauer, M., Healey, D.: Computer and Language Learning: an Overview. Language Teaching 31(2), 57–71 (1998)
11. Sams, A., Bergmann, J.: Flip Your Students' Learning. Educational Leadership 7(6), 16–20 (2013)
12. Liao, P.: Beyond "Spiderman", Teacher's Design Thinking: Theory-Based Evaluation of an Innovative Mobile Mediated Communication Task. In: The 7th Conference on College English, Trends and Issues in Teaching English for Academic Purposes, p. 34. National Chengchi University, Taipei (2013)
13. Gallagher, R.M.: An Evaluation of a Counselling-Community Learning Approach to Foreign Language Teaching or Counselling-Learning Theory Applied to Foreign Language Learning. Final report. National Inst. of Education, Washington, DC (1973)
14. Liao, P., Fu, K.: Effects of Task Repetition on L2 Oral (in Written Form) Production in Computer-Mediated Communication. The International Journal of Humanities and Arts Computing 8, 221–236 (2014)
15. Skehan, P.: Modelling Second Language Performance: Integrating Complexity, Accuracy, Fluency, and Lexis. Applied Linguistics 30(4), 510–532 (2009)
16. Cziko, G.A., Park, S.: Internet Audio Communication for Second Language Learning: a Comparative Review of Six Programs. Language Learning &Technology 7(1), 15–27 (2003)
17. MacIntyre, P.D., Gardner, R.C.: The Subtle Effects of Language Anxiety on Cognitive Processing in the Second Language. Language Learning 44(2), 283–305 (1994)
18. Beatty, K.: Teaching and Researching Computer-Assisted Language Learning. Pearson Education, London (2003)
19. Covington, M.A.: Idea Density — A Potentially Informative Characteristic of Retrieved Document. In: Southeastcon IEEE, pp. 201–203. IEEE Press, Atlanta (2009)

Appendix 1: This Is an English Short Film Assignment Completed by Students in and outside of Class Time. It Is Available at http://www.youtube.com/watch?v=nq1PEBx-vMM&feature=youtu.be

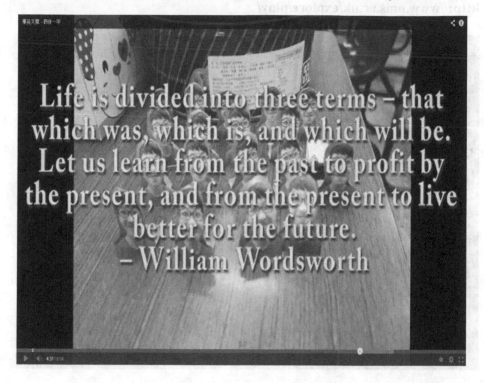

Appendix 2: This Is a Games and Adventures Website Which Can Be Played on Computers and Smartphones. Copyright Owned by Museum of Scotland. It Is Available at http://www.nms.ac.uk/explore/play/

Appendix 3: A Facebook Group Created by the Author of This Study. It Is Available at https://www.facebook.com/groups/485856841542177/

Elucidating the Continual Use of Mobile Social Networking Web Sites: Case Study of Facebook Mobile

Chuan-Chun Wu and Ching-Kuo Pu

Department of Infomation Management, I-Shou University
miswucc@isu.edu.tw, ckboo@kcg.gov.tw

Abstract. By selecting mobile social networking sites as the research focus, this study examines perceived values that affect the behavioral intention of users to use mobile social networking sites continuously. Exactly how external variables are considered as internal/psychological values is also addressed to provide further insight into the interactions between external variables and the internally perceived values as well as the extent to which such interactions impact behavioral intention to continuously use. Results of this study significantly contribute to the efforts of social network developers and researchers to more thoroughly understand mobile social network users.

Keywords: Mobile Social Networking Sites; Facebook.

1   Introduction

The fact that the computer (rather than a human) was TIME Magazine 1993 Person of the Year strongly suggests that the Internet will be the most influential entity of the 21st century, owing to its significant impact on human life. The Internet has dramatically transformed daily life and social network models - whether reserving a dinner table, purchasing tickets or participating in political events, in which social networks gather social support in opposition to governments [1].

A virtual community comprises individuals communicating mainly via the Internet. Such group members tend to be familiar with each other, as well as share information and maintain their relationships by participating continuously in a virtual community [2-3].

Social network sites (SNS) flourish with the pervasiveness of online usage. A wide variety of social networks are available, including Friendster, Meetspot, Sconex, Myspace, Foursquare, Twitter and Facebook, with the latter one especially popular [4]. Facebook reached one billion users on October 2012, or 1/7th of the global population [5]. According to Facebook's IPO released at the end of 2012, approximately 4.25 billion of Facebook's 8.45 billion users have logged onto the site via their mobile devices or the mobile version [6], reflecting the popularity of social networks via mobile devices.

L.S.-L. Wang et al. (Eds.): MISNC 2014, CCIS 473, pp. 102–116, 2014.
© Springer-Verlag Berlin Heidelberg 2014

While the relation between virtual communities and consumers has received considerable attention, most of those studies examine either factors supporting continuous participation of virtual community members [7-9] or the extent to which community member interaction impacts knowledge sharing [10-12]. However, mobile networking communities have seldom been studied. While focusing on the mobile version of Facebook (i.e. the most popular mobile network site), this study examines how key perceived values influence the behavioral intention of users to continuously use mobile devices for social networks.

This study also describes how external variables (e.g., self-efficacy, website quality, user involvement, and social influence) affect perceived usefulness (PU) and perceived ease of use (PEOU). Exactly how these psychological factors affect user attitudes and behavior is also explained. Moreover, whether the behavioral intention of users to continuously use mobile devices for social networks can affect the stickiness and word-of-mouth nature of Facebook Mobile is also addressed.

2   Related Concepts and Hypotheses

2.1   Technology Acceptance Model (TAM)

The technology acceptance model (TAM) is characterized by the PU and PEOU concepts. While PU focuses on individual perceptions toward using a specific information technology for increasing work efficacy, PEOU concerns itself with individual perceptions toward the ease of using a specific information technology for increasing work efficacy, PEOU, concerns itself with individual perceptions toward the ease of using a specific information technology. Despite the extensive use of TAM in studying various information technologies, subsequent empirical studies have demonstrated that in addition to PEOU and PU, external factors also significantly impact the user intention of consumers. Therefore, many later studies have included other dimensions or external variables to more thoroughly investigate the behavioral intention of users [13-16]. This study examines factors affecting the behavior and intentions of mobile network users by using TAM2 and four external variables (i.e. system quality, self-efficacy, social influence and user involvement).

2.1.1   System Quality

The information system quality of a website can affect the perceived usefulness (PU) and perceived ease of use (PEOU) of website users. In particular, reliable system quality or information quality can generate improved satisfaction and level of use [17-18]. This study investigates whether Facebook Mobile information system quality significantly affects users' PEOU and PU. We thus hypothesize the following:

H1: Website quality of Facebook Mobile can positively and significantly affect PEOU.

H2: Website quality of Facebook Mobile can positively and significantly affect PU.

2.1.2   Self-efficacy

Self-efficacy refers to an individual's capabilities to perform a specific task, allowing one to evaluate an individual's level of confidence by assessing an individual's self-efficacy [19]. According to previous empirical studies, self-efficacy can positively affect PU and PEOU [20-21].

This study also examines how users' self-efficacy affect their PEOU and PU. We thus hypothesize the following:

H3: Self-efficacy of Facebook Mobile users can positively and significantly affect PEOU.

H4: Self-efficacy of Facebook Mobile users can positively and significantly affect PU.

2.1.3   Social Influence

The social life of humans begins at birth. Social influence refers to the pressure from other individuals or groups in shaping an individual's behavior or thinking.

Many empirical studies have demonstrated that social influence significantly affects user behavior [22-24]. While assuming that peer influence among youth significantly affects consumer participation in mobile social networking sites, this study examines whether peer and social influence can significantly affect users' PEOU and PU by including peer factor. We thus hypothesize the following:

H5: Peer and social influence of Facebook Mobile users can positively and significantly affect PEOU.

H6: Peer and social influence of Facebook Mobile users can positively and significantly affect PU.

2.1.4   User Involvement

An individual with a great ego is less likely to accept constructive criticism. Moreover, individuals with a great ego not only approve only of opinions similar to theirs, but also accentuate those opinions that are similar to theirs [25].

Previous studies have demonstrated that involvement levels can significantly and positively affect personal purchase behavior, involvement intention, and service satisfaction [26-29]. Therefore, for information system use, user involvement can affect the positive attitudes of users toward information system use, while the attitude can subsequently affect users' reuse intention [30]. Cumulatively, according to our results, the level of user involvement can significantly affect their PEOU and PU. We thus hypothesize the following:

H7: User involvement of Facebook Mobile users can positively and significantly affect PEOU.

H8: User involvement of Facebook Mobile users can positively and significantly affect PU.

Additionally, PEOU and PU mentioned in TAM2 impact continuous use intention. We thus hypothesize the following:

H9: PEOU can positively and significantly affect PU.

H10: PEOU can positively and significantly affect continuous use behavior.

H11: PU can positively and significantly affect continuous use behavior.

2.2   Stickiness

A successful virtual website depends on the ability of the user interface to attract users in order to browse the website and, more importantly, encourage them to return to the website in the future. To assess the performance of a website, stickiness is a good indicator. While referring to a user's loyalty to a specific website, stickiness depends on whether a user revisits a website and remains there for a relatively long time [31].

Walczuch et al. proposed three dimensions of the contents of stickiness, content depth, content breadth and update frequency [32]. A website is considered to have stickiness when a user frequently visits a website or spends a considerable amount of time on a website.

As is assumed here, continuous use intention of Facebook Mobile can positively and significantly affect stickiness of a website. We thus hypothesize the following:

H12: Behavioral intention to continuously use a website can positively and significantly affect the stickiness of Facebook users.

2.3   Word of Mouth

Word-of-mouth is an informal communication behavior between message senders and receivers; the messages are normally concerned with a specific product or service [33]. Consumers will trust this information source since it does not have a commercial intention [34]. Cumulatively, word-of-mouth has multiple functions, including offering information and influencing the ideas of other individuals [35]. Positive messages can also improve the rating from individuals and, therefore, positive messages are more easily to be accepted by others [36].

The emergence of various network platforms has ensured that word-of-mouth is no longer limited to individuals within a social circle but expanded to an entire virtual community [37].

Based on the above literature review, this study also examines whether the behavioral intention of Facebook users to continuously use this social network can positively and significantly affect the word-of-mouth of Facebook. We thus hypothesize the following:

H13: Behavioral intention to continuously use this social network can positively and significantly affect the word-of-mouth of Facebook among its users.

3   Methods

TAM2 is used as the theoretical framework of this study, in which four external factors (i.e. website system quality, self-efficacy, peers and social environment, and user involvement) are assumed to affect PEOU and PU of TAM. Consequently, based on PEOU, PU, external factors, and behavioral intention, the research dimension of perceived values is formed to examine how perceived values influence behavioral intention to continuously use Facebook.

Assume that website stickiness plays a major role in determining the success of a website. While word-of-mouth can significantly alter consumer behavior, this study incorporates stickiness and word-of-mouth to examine how they impact the behavioral intention users to continuously use Facebook Mobile. Figure 1 illustrates the research model.

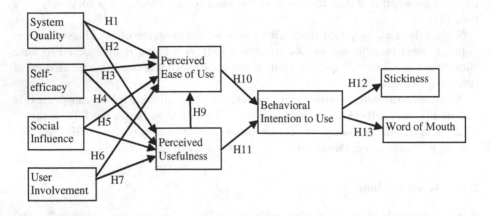

Fig. 1. Research Model

3.1   Measures

This study designed the questionnaire using the five-point Likert scale, in which the scores are 1 (highly disagree), 2 (disagree), 3 (neutral), 4 (agree), and 5 (highly agree). The formal questionnaire consisted of two sections. Section 1, the main part of the questionnaire, contains the nine research dimensions (i.e. system quality, self-efficacy, peers and social influence, user involvement, PEOU, PU, continuous use intention, stickiness, and word-of-mouth). The second part of the questionnaire collected personal information regarding the respondents for analyzing their characteristics.

3.2   Sample

3.2.1   Descriptive Statistics on the Sample

The validity and consistency of the questionnaire were ensured using a questionnaire survey method. A pre-test was performed by distributing the questionnaires among individuals who have used mobile devices for accessing the Internet. Reliability test results suggested that only two question items, system quality (4) and peer and social influence (7), had a Cronbach's α lower than 0.40. All other items had a Cronbach's α higher than 0.700, implying that the research variables had an excellent internal consistency.

Reliability of each item was tested by factor loading, and the value should be higher than 0.5. If the value is lower than 0.5, the question item should be eliminated [38]. Besides the two above-mentioned question items, all other question items had a factor loading value higher than 0.5 (p<0.05), implying an excellent convergent validity [39].

Descriptive statistical analysis was performed based on collected personal information of the study respondents. The study sample variables consisted of gender, age, education level, and duration for using mobile Internet service. The sample distribution was examined by preparing a frequency table, based on the acquired data; the frequency was also presented in percentage form. Next, the level of agreement was rated, i.e. ranging from highly disagree (1) to highly agree (5), by using the five-point Likert scale. The mean and SD of each variable were calculated, which represent the intention of respondents to continuously use the mobile community. The mean denotes the level in which the respondents agree with the question item, and a higher mean refers to a situation in which the respondents highly agree with the question item. SD denotes whether respondents' viewpoints on a question item were consistent with each other. A smaller SD implies highly consistent viewpoints.

3.2.2   Gender Analysis
The gender of website visitors was analyzed using the independent t-test. Those results revealed a lack of significance of gender on PU (t=-1.22, p>.05), PEOU (t=-1.22, p>.05), website system quality (t = -1.46, p>.05), peers and social influence (t=1.95, p>.05), self-efficacy (t=-1.87, p>.05), user involvement (t=-1.54, p>.05), behavioral intention to continuous use (t=-1.12, p>.05), word-of-mouth (t=-1.69, p>.05) or stickiness (t=-1.45, p>.05). Restated, gender cannot significantly affect PU, PEOU, website system quality, peers and social influence, self-efficacy, user involvement, behavioral intention to continuous use, word-of-mouth or stickiness.

3.2.3   Age Analysis
The age of Facebook Mobile users was analyzed by one-way ANOVA. Those results suggested that age was a significant factor for Facebook Mobile users' PU (F=139.92*, *p<.05), PEOU (F=116.36*, *p<.05), website system quality (F=5.87*, *p<.05), peers and social influence (F=78.17*, *p<.05), self-efficacy (F=74.64*, *p<.05), user involvement (F=89.07*, *p<.05), users' behavioral intention to continuous use (F=84.25*, *p<.05), word-of-mouth (F=87.00*, *p<.05), and stickiness (F=79.23*, *p<.05). This finding indicates that Facebook Mobile users' PU, PEOU, website system quality, peers and social influence, self-efficacy, user involvement, users' behavioral intention to continuous use, word-of-mouth and stickiness differed significantly from each other, depending on their age. Additionally, Scheffe's test was performed for post-hoc comparison, indicating that those Facebook Mobile visitors aged 15-17 displayed a significantly higher PU than those aged 18-25, 26-35, 36-49, or above 50 (including 50). Similarly, those Facebook Mobile visitors aged 18-25 or 26-35 also showed a significantly higher PU than those aged 36-49 or above 50 (including 50). Finally, Facebook Mobile visitors aged 36-49 showed a significantly higher PU than those aged 50 or older.

3.2.4   Education Level Analysis
The effect of education level of website visitors was analyzed using one-way ANOVA. According to those results, educational level significantly affected Facebook Mobile users' PU(F=57.83*, *p<.05), PEOU (F=59.84*, *p<.05), website system quality (F=5.73*, *p<.05), peers and social influence (F=48.71*, *p<.05), self-efficacy (F=50.29*, *p<.05), user involvement (F=55.14*, *p<.05), behavioral intention to continuous use (F=60.42*, *p<.05), word-of-mouth (F=51.79*, *p<.05), and stickiness (F=59.92*, *p<.05). This finding suggests that educational level can significantly affect Facebook Mobile users' PU, PEOU, website system quality, peers and social influence, self-efficacy, user involvement, behavioral intention to continuous use, word-of-mouth, and stickiness. Additionally, the data was further analyzed using Scheffe's test for post-hoc comparison. Analysis results indicated that Facebook Mobile users with general/vocational high school education rated PEOU and PU higher than those with education of junior high school or under, college/university, or post-graduate school or above did. Additionally, Facebook Mobile users with college/university or post-graduate school or above education rated PEOU and PU higher than those with a junior high school or lower education did.

3.3   Analysis Criteria

3.3.1   Reliability Analysis
Reliability test evaluates the trustworthiness or stability of the scores of an inventory. This test represents the level of consistency of answers from the same respondent of the same inventory, but given at different time intervals. A higher inventory reliability implies an improved stability. In this study, internal consistency was determined using the most commonly used reliability coefficient Cronbach's α. According to Cronbach [40], an inventory with a reliability coefficient higher than 0.7 has an excellent reliability

3.3.2   Variance Analysis
Independent sample one-way ANOVA can determine the significance of an average of more than three groups of population. This study assessed whether age, education, and length of mobile Internet usage can significantly affect use intention.

3.3.3   Structural Equation Modeling
While consisting of measuring modeling and structural modeling, structural equation modeling (SEM) explores the correlation between latent variables and observed variables as well as the correlation among latent variables. SEM includes path analysis and confirmatory factor analysis tools. This study also analyzed the correlation among the research dimensions using SEM in order to determine the direction and level of influences of these dimensions.

4 Result

4.1 Reliability

For constructing reliability and average variance extracted, Fornell and Larcker [41] suggested that a construct reliability higher than 0.600 implies a better construct reliability of the observed variables for the latent variables. For an average variance extract to exceed 0.50, the measurement error of the observed variables for the latent variables should be smaller than 50%, implying that the observed variables have a high accuracy in determining the latent variables.

In this study, the values of construct reliability of PU, PEOU, user involvement, peers and social influence, behavioral intention to continuous use, self-efficacy, website system quality, stickiness, and word-of mouth ranged from 0.73 to 0.891 (all higher than 0.600). Therefore, the study has an excellent internal consistency. Moreover, the average variance extracted of each latent variable was between 0.605 and 0.748 (all higher than 0.500), indicating that the measurement errors of observed variables for latent variables were less than 50%. Restated, the latent variables of this study have an excellent reliability and convergent validity.

Table 1. Analysis of Construct Reliability and Average Variance Extracted (AVE)

Variables	Construct Reliability	AVE
PU	0.891	0.672
PEOU	0.857	0.667
System Quality	0.859	0.605
Social Influence	0.730	0.575
Self-efficacy	0.898	0.746
User Involvement	0.794	0.659
Continued Use Intention	0.794	0.659
Word-of-Mouth	0.855	0.748
Stickiness	0.834	0.626

1. Component reliability = (Sum of standardized factor loading)2/[(Sum of standardized factor loading)2 + Sum of Errors
2. Average variance extracted $=\Sigma$(Multiple correlation)2/Number of factors

4.2 Validity

Wixom and Watson [38] indicated that convergent validity uses confirmatory factor analysis for modularization to estimate the correlation between question items and research variables. The factor loading of recommended convergent validity should exceed 0.50 and statistically significant.

The factor loadings of latent variables of PU, PEOU, user involvement, peers and social influence, behavioral intention to continuous use, self-efficacy, website system quality, stickiness, and word-of-mouth ranged from 0.607 to 0.883 (higher than 0.50, which is statistically significant). This finding suggests that the study's inventories on PU, PEOU, user involvement, peers and social influence, behavioral intention to continuous use, self-efficacy, website system quality, stickiness, and word-of-mouth have an excellent convergent validity.

Table 2. Convergent Validity Analysis

Variables	Measuring Items	Factor Loading	Squared Multiple Correlations	t-Value
PU	PU6	.832	.692	—
	PU5	.817	.667	37.500*
	PU 3	.786	.618	35.289*
	PU 1	.694	.481	29.474*
PEOU	PE6	.757	.574	—
	PE4	.761	.579	29.771*
	PE2	.819	.671	32.187*
System Quality	SQ6	.667	.445	—
	SQ4	.607	.369	19.284*
	SQ3	.729	.531	21.159*
	SQ2	.715	.511	20.871*
Social Influence	SI 5	.731	.534	—
	SI 2	.775	.601	29.907*
Self-efficacy	SE5	.827	.683	—
	SE 4	.843	.711	39.458*
	SE 2	.883	.779	42.681*
User Involvement	UI 4	.762	.581	—
	UI 1	.785	.616	31.240*
Continued Use Intention	CI 2	.879	.772	—
	CI 1	.789	.623	38.856*
Word-of-Mouth	WM 7	.737	.543	—
	WM 3	.810	.657	30.885*
	WM 2	.822	.676	31.260*
Stickiness	ST 4	.895	.801	—
	ST2	.795	.632	38.960*

4.3   Structural Modeling Analysis

4.3.1   Goodness of Fit of the Model

Structural modeling analysis consists of analyzing both the goodness-of-fit of the model and the explanatory power of the overall research model. Based on the results of previous studies[42-44], this study selected seven indicators for the goodness-of-test of the overall model: the ratio between χ^2 and degree of freedom (χ^2 /df), adjusted goodness of fit index (AGFI), normed fit index (NFI), non-normal fit index (NNFI), comparative fit index (CFI), relative fit index (RFI) and root mean square error of approximation (RMSEA).

Analysis results suggested that in addition to having a $\chi2/d.f.$ higher than the recommended value, all other indices satisfied the values recommended by majority studies. Although $\chi2/d.f.$ of the study was 6.012 (>3), Hair et al. suggested that for a sample size greater than 200, $\chi2$ and a situation in which the degree of freedom may become over-sensitive, the ratio may be higher than 3 and other goodness-of-fit indices should be used [45]. Owing to the 576 samples in this study, the research model may still be acceptable, i.e. the study model and observed data had a reasonable goodness-of-fit.

Table 3. Goodness-of-Fit test of the Model

Goodness-of-Fit Indicator	Recommended Value	Test Value	Test Result
X2/df	≦ 3.00	6.012	Nearly Qualified
AGFI	≧ 0.80	.900	Qualified
NFI	≧ 0.90	.947	Qualified
NNFI	≧ 0.90	.944	Qualified
CFI	≧ 0.90	.955	Qualified
RFI	≧ 0.90	.933	Qualified
RMSEA	≦ 0.08	.058	Qualified

4.3.2   Pathway Analysis Results

Based on structural modeling, this study analyzed PU, PEOU, user involvement, peers and social influence, behavioral intention to continuous use, self-efficacy, website system quality, stickiness, and word-of-mouth. Those results suggested that website system quality significantly affected PEOU (β=.09*, *p<.05). Restated, the website system quality of Facebook Mobile can significantly and positively affect PEOU. The effect of self-efficacy on PU (β=.11*, *p<.05) was also statistically significant, indicating that the self-efficacy of Facebook Mobile can positively and significantly affect PU. Namely, self-efficacy of Facebook Mobile can positively and significantly affect PU. The effect of peers and social influence significantly impacted PEOU (β=.46*, *p<.05). Restated, peers and social environmental of Facebook Mobile can significantly and positively affect PEOU. The effect of user involvement was also significant on PEOU (β=.74*, *p<.05). Therefore, user involvement of Facebook Mobile can significantly and positively affect PEOU. Moreover, PEOU significantly affected PU (β=.92*, *p<.05). Therefore, PEOU of Facebook Mobile can significantly and positively affect PU. Both PU (β=.30*, *p<.05) and PEOU (β=.74*, *p<.05) significantly and positively affected behavioral intention to continuous use. Finally, behavioral intention to continuous use can positively and significantly affect stickiness (β=.92*, *p<.05) and word-of-mouth (β=.92*, *p<.05).

Table 4. Structural Modeling Pathway Analysis

Pathways	Pathway Value	t-Value	Test Result
Website system quality —> PU	.01	2.562	Accepted
Website system quality —> PEOU	.09	2.894*	Accepted
Self-efficacy —> PU	.11	3.501*	Accepted
Self-efficacy —> PEOU	.23	5.870*	Accepted
Peers and social influence —> PU	.05	2.718	Accepted
Peers and social influence —> PEOU	.46	11.238*	Accepted
User involvement —> PU	.03	2.488	Accepted
User involvement —> PEOU	.74	16.490*	Accepted
PEOU —> PU	.92	17.397*	Accepted
PEOU —> Behavioral intention to continuous use	.30	8.535*	Accepted
PEOU —>Behavioral intention to continuous use	.74	15.827*	Accepted
Behavioral intention to continuous use —> Stickiness	.92	32.455*	Accepted
Behavioral intention to continuous use —> Word-of-mouth	.92	33.979*	Accepted

5 Conclusion

The study examined the correlation between PU, PEOU, user involvement, peers and social influence, continuous use behavior intention, self-efficacy, website system quality, stickiness, and word-of-mouth of Facebook Mobile. The research model and questionnaire were examined using descriptive statistics, independent t-test, one-way ANOVA, and structural equation modeling.

The study sample displayed a medium or higher level of approval of PU, PEOU, user involvement, peers and social influence, behavioral intention to continuous use, self-efficacy, stickiness and word-of-mouth of Facebook Mobile. In terms of website system quality, the study sample displayed a medium level of approval.

Additionally, this study demonstrated that gender does not significantly affect the approval of various Facebook Mobile variables. In terms of age, analysis results indicated that a younger population would more highly approve of these variables than an older one would. Individuals between 15 and 17 years old in this study displayed the highest approval. As for education, individuals with a higher educational level in this study approved more of the study variables than those with a lower one. Individuals with general/vocational high school, college/university, or post-graduate school degrees showed the highest approval rating. As for cell phone usage experience, respondents with a longer cell phone use experience more highly approved of the study variables than those with a shorter experience. The population with one to two years of cell phone usage experience gave the highest approval rating.

Moreover, SEM analysis in this study indicated that 1) PEOU was positively and significantly related to the website system quality, self-efficacy, peers and social influence and user involvement of Facebook Mobile; 2) PU was positively and significantly related to self-efficacy; 3) PU was positively and significantly related to PEOU of Facebook Mobile; 4) PU and PEOU of Facebook Mobile were positively and significantly related to behavioral intention to continuous use, and 5) behavioral intention to continuous use of Facebook Mobile was positively and significantly related to stickiness and word-of-mouth.

Moreover, SEM analysis in this study indicated the following: 1) PEOU was positively and significantly related to the Web site system quality, self-efficacy, peer and social influence, and user involvement in Facebook Mobile; 2) PU was positively and significantly related to self-efficacy; 3) PU was positively and significantly related to the PEOU of Facebook Mobile; 4) the PU and PEOU of Facebook Mobile were positively and significantly related to behavioral intention to use Facebook Mobile continually; and 5) the behavioral intention to use of Facebook Mobile continually was positively and significantly related to stickiness and word of mouth.

6   Limitations and Implications

6.1   Limitations

Despite its contributions, this study has certain limitations. The true population of cell phone users changes constantly, and the population is huge. Consequently, verifying the representativeness of the sample must rely on those with a more acceptable means. Owing to the impossibility of obtaining a list of the entire population of cell phone users in Taiwan, this study adopted a randomized sampling approach for the questionnaire survey. We recommend that future researcher develop improved sampling methods for the mobile business sector.

Additionally, this study performed cross-sectional analysis to explain the interactions among dimensions of this research model. Owing to possible oversimplification of the research process, we recommend that future research include theoretical, longitudinal evidence to ensure a more comprehensive explanation.

Furthermore, many studies have indicated that involvement level can affect consumers' acceptance of technological products. Therefore, consumers' technological product involvement level (high vs. low) can affect their willingness for using technological products. We recommend that subsequent studies discuss whether consumers' product involvement level (high vs. low) can affect modeling pathway association.

6.2   Implications

This study has several managerial implications for researchers and practitioners. Managers should strive to maintain the high approval ratings from younger or more educated consumers and from consumers with more experience. For other consumer segments, researchers should identify the causes for their low approval, especially those older consumers since they have higher spending power and development potential. Therefore, researchers should more closely examine how to enhance older consumers' approval of mobile network sites.

Additionally, this study demonstrated that only website system quality was negligibly associated with PU. This finding suggests the importance of having easy operation for Facebook Mobile users. Restated, mobile social networking users do not value complex functions. Given rapid advances in information technology, easy operation should be of priority concern.

7   Discussion

Although this study did not discuss the mediating effect among variables, many studies have demonstrated that mediators play a critical role in this mediating effect. Therefore, some mediators should be considered when examining the correlations among the variables.

Additionally, owing to the lower overall score of website system quality in this study, we recommend further elucidating website system quality in future research. We further recommend adopting IPA to further analyze components warranting further improvement.

References

1. Ben, W.: Push-button-autocracy in Tunisia: Analysing the role of Internet infrastructure, institutions and international markets in creating a Tunisian censorship regime. Telecommunications Policy 36(6), 484–492 (2012)
2. Rheingold, H.: The Virtual Community: Homesteading on the Electronic Frontier. Addison-Wesley Publishing Co., MA (1993)
3. Hagel, J., Armstrong, A.G.: Net gain: Expanding markets through virtual communities. Harvard Business School Press, Boston (1997)
4. Anderson, B., Fagan, P., Woodnutt, T., Chamorro-Premuzic, T.: Facebook psychology: Popular questions answered by research. Psychology of Popular Media Culture 1(1), 23–37 (2012)
5. Checkfacebook.com (2012), http://www.socialbakers.com/blog/897-socialbakers-congratulates-facebook-on-1-billion-active-users
6. DIGITIMES (2012), http://www.digitimes.com.tw/tw/dt/n/
7. Hsiao, C.C., Chiou, J.S.: The effect of social capital on community loyalty in a virtual community: Test of a tripartite-process model. Decision Support Systems 54(1), 750–757 (2012)
8. Tsai, H.T., Pai, P.: Explaining members' proactive participation in virtual communities. International Journal of Human-Computer Studies 71(4), 475–491 (2013)
9. Zhao, K., Stylianou, A.C., Zheng, Y.: Predicting users' continuance intention in virtual communities: The dual intention-formation processes. Decision Support Systems 55(4), 903–910 (2013)
10. Lea, B.R., Yu, W.B., Maguluru, N., Nichols, M.: Enhancing business networks using social network based virtual communities. Industrial Management & Data 106(1), 121–138 (2006)

11. Cheolho, Y., Erik, R.: Knowledge-sharing in virtual communities: familiarity, anonymity and self-determination theory. Behaviour & Information Technology 31(11), 1133–1143 (2012)
12. Chen, Y.J., Chen, Y.M.: Knowledge evolution course discovery in a professional virtual community. Knowledge-Based Systems 33, 1–28 (2013)
13. Mun, Y.Y., Hwang, Y.J.: Predicting the use of web-based information systems: self-efficacy, enjoyment, learning goal orientation, and the technology acceptance model. International Journal of Human-Computer Studies 59(4), 431–449 (2003)
14. Daniel, J.M., Diane, H.: Adding contextual specificity to the technology acceptance model. Computers in Human Behavior 22(3), 427–447 (2006)
15. Teo, T., Su Luan, W., Sing, C.C.: A cross-cultural examination of the intention to use technology between Singaporean and Malaysian pre-service teachers: an application of the Technology Acceptance Model (TAM). Educational Technology & Society 11(4), 265–280 (2008)
16. Lee, D.Y., Lehto, M.R.: User acceptance of YouTube for procedural learning: An extension of the Te chnology Acceptance Model. Computers & Education 61, 193–208 (2013)
17. Liu, C., Arnett, K.P.: Exploring the factors associated with web site success in the context of electronic commerce. Information and Management 38(1), 23–34 (2000)
18. DeLone, W.H., McLean, E.R.: The DeLone and McLean Model of Information System Success: A Ten-Year Update. Journal of Management Information Systems 19(4), 9–30 (2003)
19. Bandura, A.: Social Foundations of Thought and Action. Prentice-Hall, Englewood Cliffs (1986)
20. Petrus, G., Nelson, O.N.: Borneo Online Banking: Evaluating Customer Perceptions and Behavioural Intention. Management Research News 29(1/2), 6–15 (2006)
21. Ong, C.S., Lai, J.Y.: Gender Differences in Perceptions and relationships Among Dominants of E-learning Acceptance. Computers in Human Behavior 22(5), 816–829 (2006)
22. Lu, J., Liu, C., Yu, C.S., Yao, J.: Acceptance of wireless Internet via mobile technology in China. Journal of International Technology and Information Management 14(2), 117–130 (2005)
23. Wu, Y.L., Tao, Y.H., Yang, P.C.: Using UTAUT to explore the behavior of 3G mobile communication users. In: Proceeding(s) of the IEEE International Conference on Industrial Engineering and Engineering Management, pp. 199–203 (2007)
24. Hsu, C.L., Lin, J.C.C.: Acceptance of blog usage: The roles of technology acceptance, social influence and knowledge sharing motivation. Information & Management 45(1), 65–74 (2008)
25. Hanna, N., Wozniak, R.: Consumer Behavior. Prentice-Hall, Inc., New Jersey (2001)
26. Sanchez-Franco, M.J.: The Moderating Effects of Involvement on the Relationships Between Satisfaction, Trust and Commitment in e-Banking. Journal of Interactive Marketing 23(3), 247–258 (2009)
27. Gutierrez, S.S.M., Izquierdo, C.C., Cabezudo, R.S.J.: Product and channel-related risk and involvement in online contexts. Electronic Commerce Research and Applications 9(3), 263–273 (2010)
28. Chang, H.H., Chuang, S.S.: Social capital and individual motivations on knowledge sharing: Participant involvement as a moderator. Information & Management 48(1), 9–18 (2011)

29. Park, N., Oh, H.S., Kang, N.: Factors influencing intention to upload content on Wikipedia in South Korea: The effects of social norms and individual differences. Computers in Human Behavior 28(3), 898–905 (2012)
30. Swanson, E.B.: Management-information-systems: appreciation and involvement. Management Science 21(2), 178–188 (1974)
31. Holland, J., Baker, S.M.: Customer Participation in Creating Site Brand Loyalty. Journal of Interactive Marketing 15(4), 34–45 (2001)
32. Walczuch, R., Verkuijlen, M., Geus, B., Ronnen, U.: Stickiness of Commercial Virtual Communities, MERIT-Infonomics Research Memorandum series (2001)
33. Harrison-Walker, L.J.: The Measurement of Word-of-Mouth Communication and an Investigation of Service Quality and Customer Commitment as Potential Antecedents. Journal of Service Research 4(1), 60–75 (2001)
34. Richins, M.L.: Negative word-of-mouth by dissatisfied consumers: a pilot study. Journal of Marketing 47(1), 68–78 (1983)
35. Richins, M.L., Root-Shaffer, T.: The Role of Involvement and Opinion Leadership in Consumer Word-of-Mouth: An Implicit Model Made Explicit. Advances in Consumer Research 15(1), 32–36 (1988)
36. Eagly, A.H., Chaiken, S.: The psychology of attitudes. Harcourt Brace Jovanovich, Fort Worth (1993)
37. Dellarocas, C.: Strategic Manipulation of Internet Opinion Forums: Implications for Consumers and Firms. Management Science 52(10), 1577–1593 (2003)
38. Wixom, B.H., Watson, H.J.: An empirical investigation of the factors affecting data warehousing success. MIS Quarterly 25(1), 17–41 (2001)
39. Bock, G.W., Zmud, R.W., Kim, Y.G., Lee, J.N.: Behavioral Intention Formation in Knowledge Sharing: Examining the Roles of Extrinsic Motivators, Social-Psychological Forces, and Organizational Climate. MIS Quarterly 29(1), 87–111 (2005)
40. Cronbach, L.J.: Coefficient Alpha and the internal structure of tests. Psychometrika 16(3), 297–334 (1951)
41. Fornell, C., Larcker, D.F.: Evaluating structural equation models with unobservable and measurement error: A comment. Journal of Marketing Research 18(3), 39–50 (1981)
42. Joreskog, K.G., Sorbom, D.: LISREL 8: User's reference guide. Scientific Software International, Chicago (1996)
43. Bentler, P.M.: Comparative Fit Indexes in Structural Models. Psychological Bulletin 107(2), 238–246 (1990)
44. Bentler, P.M.: On the Fit of Models to Covariance and Methodology to the Bulletin. Psychological Bulletin 112(3), 400–404 (1992)
45. Hair, J.F., Black, W.C., Babin, B.J., Anderson, R.E., Tatham, R.L.: Multivariate data analysis, 6th edn. Prentice Hall, Upper Saddle River (2005)

Do Social Network Services Successfully Support Knowledge Transfer in Organizations?

Jong-Chang Ahn[1] and Soon-Ki Jeong[2,*]

[1] Department of Information Systems
[2] Department of Electronics Computer Engineering, Hanyang University,
Industrial Technology Center 208, 222 Wangsimni-ro, Seongdong-gu, Seoul 133-791, Korea
{ajchang,skjeong}@hanyang.ac.kr

Abstract. Social Network Service (SNS) has affected large sections of society. After Web 2.0 appeared, SNS caused a revolution in information flow with explosive diffusion. Many organizations have invested millions in their current Knowledge Management Systems (KMS). However, some limitations to KMS usage exist. Retrieval of optimal information and knowledge from repository systems can be difficult. This leads to avoidance of knowledge sharing and hinders knowledge transfer. This article demonstrates how horizontal communication structure and information diffusion created by SNS can affect the knowledge transfer mechanism. Our case study targets a representative IT-service enterprise in Korea. Our analysis reveals that SNS could provide a complementary technology for knowledge transfer activation because it affects organizational structure and cultural flexibility. However, SNS cannot substitute for KMS. This article provides a theoretical and experiential foundation for future research on the relationship between SNS and knowledge management.

Keywords: Social network service (SNS), Knowledge management (KM), Knowledge management system (KMS), Knowledge transfer.

1 Introduction

Social Network Services (SNS) have accelerated changes in the knowledge information society. Web 2.0 represented "participation and openness." It contributed to information technology and knowledge development (e.g., Wikipedia, Blogs, RSS, SNS) because of user participation [1]. Social Network Sites are virtual communities. Users create individual public profiles, interact with real-life friends, and meet others who share interests [2]. A social network consists of a set of actors and a set of ties that connect relationships between these actors [3]. Use of collaborative technologies such as blogs and SNS can lead to the development of instant online communities where people can communicate rapidly and conveniently. SNS is a flexible and convenient platform for individuals to form and maintain online friendships [4].

---

* Corresponding author.

L.S.-L. Wang et al. (Eds.): MISNC 2014, CCIS 473, pp. 117–133, 2014.

Social interactions that occur in SNS because of these social relationships may increase knowledge sharing intentions.

Firms have invested in knowledge management systems (KMS) to solve problems and create value by knowledge reuse. KM's early stages were characterized by the development of KMS that stored pieces of knowledge uploaded by employees [5]. However, accumulation of knowledge assets can lead to increases in management costs and decline in the quality of available knowledge.

Firms may intentionally attempt to activate knowledge reuse in order to increase productivity by internal KMS usage. This decision may be related to practical uses of SNS outside firms. Organizations realize that SNS can be used to establish horizontal relationships between members through a centralized organization management. Therefore, organizations believe that social interactions that occur in SNS may promote knowledge sharing and collaboration by horizontal communication [6]. A horizontal organization is well suited to knowledge sharing because sub-units are decentralized to facilitate collaboration that may activate knowledge flux [7].

Generally, scholars believe that SNS usage fostered the development of trust-based social interactions. These social interactions exert positive effects on information usage [2, 8]. Jeong et al. [9] empirically that horizontal organizational structures positively affect knowledge sharing and knowledge transfer activation. If we suppose that SNS meaningfully affects the horizontal relationship formation, we might conclude that SNS bears a close relationship to KMS.

Benbya and Van Alstyne asserted that the knowledge trade market could develop possibilities for knowledge usage in organizations [10]. We hope to discover how SNS affects the knowledge trade market and the knowledge culture of the knowledge transfer mechanism. We also hope to discover whether SNS can extend its effects to an analysis of the knowledge transfer mechanism. However, limited studies have examined the impact of SNS on KM and KMS. In this study, we analyze how SNS interacts with knowledge trade and knowledge culture.

We analyzed the knowledge trade market and knowledge sharing culture framework based on social network theory. We performed an analysis of a case study. Our results revealed that SNS may serve as a complementary factor in the activation of KMS usage because the knowledge trade market is in its early stages. An optimized strategy is needed to reveal ways to integrate SNS characteristics into each organizational structure to strengthen the link between SNS and KMS. A knowledge transfer framework mediated by SNS might provide a theoretical and practical basis for the application of KM with SNS.

In the next section, we provide a literature review on KMS and SNS. We then provide an overview of our research methodology and the case study what is the effects of SNS to knowledge transfer framework. Finally, we provide an outline and discussion for readers and suggestions for future research.

2   Related Literature

2.1   Knowledge Transfer and SNS (Is SNS an Accelerator for Knowledge Transfer?)

According to Nonaka [11], knowledge creation converts tacit knowledge to explicit knowledge based on the spiral SECI model. Tacit knowledge is acquired by interactions between individuals that are shared experiences. Social interaction between individuals then provides an ontological dimension to the expansion of knowledge.

Knowledge transfer is a process by which tacit knowledge is transformed into explicit knowledge by an organic combination of internal knowledge and socialization. Knowledge transfer could be created by two dimensions. Sharing and internalization of tacit knowledge is mapped to knowledge sharing by an internal mechanism. The externalization of explicit knowledge is essentially a knowledge trade, and the market in which it occurs can thus be known as a knowledge trade market. Knowledge trade represents how knowledge can emerge as an externalization of tacit knowledge. Figure 1 (modified from Nonaka's knowledge spiral model) demonstrates that knowledge transfer can be affected by two different viewpoints of knowledge trade and knowledge culture. We will discuss how SNS relates to knowledge trade and knowledge culture in the next section.

* Source: [11] Nonaka I., Lewin A.Y., the figure was modified.

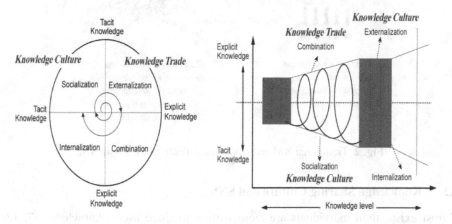

Fig. 1. Two Dimensions of Knowledge Transfer: Internal Dimension (Knowledge Culture) and External Dimension (Knowledge Trade Market)

A social network is a network connected by relationships between individuals, groups, organizations, or entire societies. Social networks are determined by these interactions. Boyd [12] defined SNS as web-based services that allow individuals to:

1.   construct public or semi-public profiles within a bounded system
2.   articulate lists of other users with whom they share connections
3.   view and traverse lists of connections and connections made by others within the system

Social relationships within SNS are connected by profiles. The visibility of a profile varies by site and user discretion. Users in SNSs are prompted to identify other users who they have relationships with [12]. In SNS, individuals search for others with whom they can form and maintain close relationships. Social relationships within SNS are connected by profiles. Some individuals serve as social members in SNS. This could explain why some individuals attempt to gain "recognition" within social relationships.

Problem solving and knowledge sharing depends on the knowledge possessed by professionals in KMS. The knowledge transfer model with SNS depends on the rapid exchange of collective intelligence. This SNS application allowed the SECI knowledge spiral model devised by Nonaka [11] to grow quickly. SNS's two-directional communication structure caused by the diffusion of information across the network may explain why SNS is so effective. In KMS, knowledge is shared by 1:N. However, in SNS, knowledge may be spread on the structure of an N: N personal network. Figure 2 shows comparisons of traditional KM and KM from the SNS perspective.

* Source: [10] Benbya BH, Van Alstyne M, The figure was modified.

Fig. 2. Traditional KM and Knowledge Trade Market with SNS

2.2   Knowledge Sharing Culture and SNS

If trust exists, then individuals are more willing to share useful knowledge [7, 13]. Trust is an essential attribute in organizational culture. We can assume that trust exists between members in organizations that possess cultures that rely on good knowledge transfer [14]. SNS is a technology that might strengthen trust-based social relationships. Social ties between members that use SNS may affect knowledge transfer intentions. This differs from the idea that the SNS usage may directly affect knowledge transfer intention because trust is essential for SNS implementation into organizational culture. Trust accumulation leads to the solidification of relationships between members. Enriched relationships activate horizontal communication, and this helps flatten the organizational structure. Knowledge culture develops best within a balanced horizontal organization structure that includes liberal communication [15].

KM scholars stated that rewards (economic and social) can positively affect knowledge sharing [16, 17, 18]. A literature review on rewards revealed that social rewards are more important than economic rewards [19]. If this is so, what is the relationship between SNS application and possible rewards for knowledge transfer? A balance between social and economic rewards is important for successful knowledge transfer. Users share their knowledge via SNS because they hope to improve their reputations and gain recognition in social networks where they maintain relationships. Individuals may share knowledge in social networks because human beings are social animals who enjoy recognition. Individuals' desires to gain positive identities in online society exert positive effects on knowledge transfer.

A knowledge sharing culture is an important condition that can lead to knowledge transfer. However, it can be the most difficult part of KM because KM is a complex socio-technical system that includes various forms of knowledge generation, storage, representation, and sharing [20]. Information systems support data interaction, processes to support daily operations, and problem solving. These are critical success factors for organizational culture and knowledge sharing [14]. The term "socio-technical" emphasizes the interrelatedness of social and technological function [21]. SNS is composed of socially N: N networks where users connect autonomically. N: N networks in SNS are diffused by autonomous knowledge sharing culture. Organizational knowledge is created by socio-technology within KMS. Knowledge is socially constructed and shaped by the interplay between technology and organizational factors. Pan [21] believed the creation of knowledge in KMS is controlled by information flow. He also believed that KM is divided into three parts. The socio-technical perspective of KM consists of infrastructure, info-structure, and info-culture. Infrastructure enables physical/communicational contact between network members. Info-structure consists of formal and informal rules that govern exchanges between actors. Info-culture for knowledge sharing is created by social relations [21]. From a social-networking point of view, KMS' socio-technical perspective can be represented as follows:

- infrastructure: KMS
- info-structure: SNS
- info-culture: knowledge sharing culture

Figure 3 shows where SNS is positioned in KMS. A rapid formation of social relationships accelerates knowledge flow in KMS.

A good understanding of SNS's characteristics requires integration of SNS into organizational culture. Although SNS can help knowledge sharing, the purpose of knowledge sharing must be clear. Marouf [22] believed that knowledge sharing intention relates more to business than to social interactions. Thus, SNS application to knowledge transfer must be managed differently for business and personal purposes.

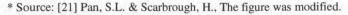

* Source: [21] Pan, S.L. & Scarbrough, H., The figure was modified.

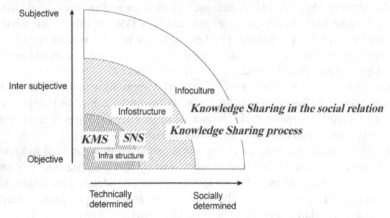

Fig. 3. Positions of SNS and KMS: a Socio-technical perspective on KM

2.3   Knowledge Market and SNS

Based on social exchange theory, people make decisions and act based on individual satisfaction levels within relationships. The knowledge market contains the precepts of social exchange theory. Exchange theory depicts individuals as rational profit seekers who must choose between alternative actions to obtain the greatest value at the lowest possible costs [23]. Emerson [24] stated that the market could be perceived in three ways:

1.   the market is a location where people can assemble to engage in transactions;
2.   the market is a location where people can find specific goods or services;
3.   the market process.

The knowledge market is a virtual location where intangible knowledge goods are traded. In knowledge market systems, members act as buyers (seekers) and/or sellers (possessors). Knowledge transactions result from knowledge trade in the knowledge market. Knowledge trade contributes to equilibrium by its effects on the balance between economic rewards and social rewards. Collaboration in the market creates a complicated social system [25]. A complex social system develops from collaboration within the organization.

To graft a suitable SNS onto the internal knowledge-trade market in organizations, strategies must be devised that recognize how this type of SNS differs from general SNS. Marouf [22] proposed that the strength of fact connection differs between business and social relationships. He stated that people congregate in communities for purposes other than knowledge sharing. Thus, internal users may use SNS to find business knowledge rather than general knowledge. We must consider the optimist process when we consider the adoption of SNS for KMS. It can be difficult to connect knowledge producers and knowledge consumers. A two-sided network might connect different markets because a mutual benefit is offered [10]. Platforms supported by two-sided networks must apply a strategy to provide mutual benefits to each other's customers. The expansion of network externalities may lead to the improvement of valuable services. The process of "subsidization" is used to allow maximization of network externalities. "Deciding which market to subsidize depends on the relative network externality benefits [26]." A practical application of KMS with SNS and two-sided networks may lead to the addition of value and quantitative and qualitative expansion of the knowledge shared.

Interactions that occur through SNS include: (a) consolidation of relationships; (b) intention to share knowledge; (c) knowledge inflow from outside; and (d) intention to access current knowledge. Knowledge consumers' desires adrenalize knowledge-production by knowledge producers. Figure 4 shows a conceptualization of the network externalities that may result from KMS with SNS.

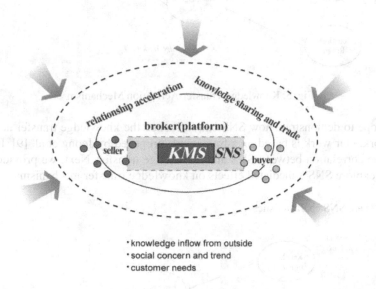

* knowledge inflow from outside
* social concern and trend
* customer needs

Fig. 4. Network Externalities that may result from KMS and SNS

3 Methodology

Jeong et al. [9] proved empirically that knowledge transfer could be activated by knowledge sharing and knowledge trade in the knowledge market. Knowledge organization and knowledge strategy are positive factors that affect knowledge sharing. KMS and knowledge reward are positive factors that affect the knowledge trade market. We show the proven platform for knowledge transfer activation in Figure 5 below.

* Source: [9] Jeong, S.K., Ahn J.C., Rhee, B.H.

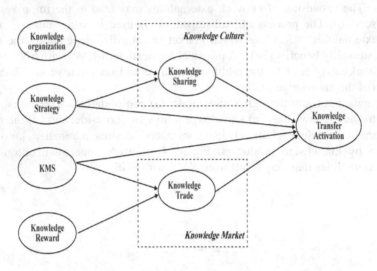

Fig. 5. Knowledge Transfer Activation Mechanism

We hope to demonstrate how SNS can influence the knowledge transfer activation framework. Our work is based on a framework proposed by Jeong et al. [9]. Figure 6 shows the correlation between SNS and knowledge transfer. Next, we provide a case study to explore SNS's mediated effects on knowledge transfer mechanisms.

Note: *s are SNS characteristics

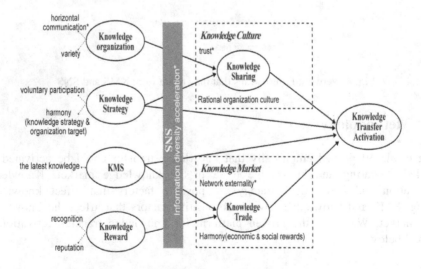

Fig. 6. Knowledge Transfer Activation Mechanism with SNS

4 Case Study

Here, we discuss factors of knowledge transfer related to SNS. We demonstrated how each factor could be influenced by mediated SNS. Table 1 shows the probable results along with the relationship between factors based on analysis and literature review of SNS.

Table 1. Knowledge Transfer Activation Factors with SNS

Main Construct Factors	Construct Factors	Relationships for Knowledge Transfer	With SNS
Knowledge Sharing Culture	Knowledge Organization	Although knowledge is transferred by vertical hierarchy in traditional KM, an evolution to knowledge organization leads to horizontal organization.	SNS may activate horizontal communication.
	Knowledge Strategy	Knowledge strategy and planning harmonized with management targets can exert positive effects on knowledge sharing culture.	For an institutional SNS to succeed for firms, it must be differently optimized with commercial SNS.
Knowledge Trade Market	KMS	Knowledge is transferred via 1:N network operations in traditional KM. Knowledge trade leads to peer-to-peer knowledge exchanges in knowledge trade market.	The efficiency of knowledge transfer may be developed by KMS supplementation by the diffusion of information flux through SNS.
	Knowledge Reward	The balance between economic rewards and social rewards is important for knowledge transfer.	People share knowledge in social networks not to get economic rewards but to gain recognition and better reputations.

The case study method can be used to discover the "how" and "what" that operate in a research topic. Although p level < 0.05 (below 95% confidence level) is significant in statistical analysis, a case study must designate a different verification standard [27]. For case study verification, we planned our research based on the standards proposed by Yin [27]. Table 2 shows case study methodology standards.

Table 2. A Methodology for the Verification of a Case Study

Verification	Case study strategy	Methodology for this case study
Construct Validity	Utilization of various sources. Creation of chain of evidence. Gain the services of an individual to examine a draft report.	Utilization of various sources: Firm's accounting report, articles, and direct interviews.
Internal Validity	Utilization of pattern matching. Explanation of a causal relation. Settlement of competition hypothesis. Utilization of logical model.	Utilization of the Figure 6 framework that was proposed to explain SNS effectiveness for knowledge transfer.
External Validity	Utilization of theory in single case study. Utilization of replication-study logic.	This research is a single study based on theoretical frame of reference that analyzed prior research on knowledge transfer and SNS.
Reliability	Utilization of case study protocol. Development of database for case study.	

* Source: [27] Yin, R., The characteristics were summarized.

A social network is a set of people (or organizations or other social entities) connected by a set of social relationships, such as friendship, co-working or information exchange [28]." Large connections of SNS form huge networks on the Internet. Recently, organizations that incorporated SNS into internal organization structures have grown because they anticipated the activation of horizontal communication by members and real-time knowledge sharing via SNS.

We chose to study Organization A, a firm that successfully operates KM in Korea. We performed both qualitative and empirical analyses to discover whether SNS influences knowledge transfer. Organization A is a good representative of KM users because it provides IT and consulting services. KMS is widely used by many IT services and technology consulting organizations. Organization A is a traditional IT service and technology consulting enterprise. Professional IT enterprises' characteristics help them integrate recent IT trends into KM. Organization A was founded in 1987. As of 2011, it had $ 2.7 billion in revenue and 10,000 employees. It invested millions of dollars in KMS even prior to the 2000s when KMS was widely introduced in Korea. SNS was introduced in 2009 and is now in active use. Members of Organization A simultaneously use self-developed and commercial SNS such as Facebook and Twitter.

The knowledge-transfer framework has organically conjoined many factors: knowledge organization, knowledge strategy, KMS, knowledge rewards, knowledge trade market, and knowledge sharing culture. We analyzed how SNS influences each factor within the knowledge-transfer framework. We based our questions on literature reviews and professional reports. We show our questionnaire about the influences between SNS and knowledge-transfer in an Appendix. We conducted interviews to analyze how SNS might influence knowledge transfer. In August 2012, we interviewed five persons during primary and secondary interviews. These individuals hold sales and technical positions located near the headquarters of Organization A and its customer site. Each interviewee possessed a specific view of SNS depending on his/her position. We interviewed a manager and senior manager who made significant contributions to the organization. They have used KMS for many years (five years ~ twelve years). The interviews lasted about 1 ½ hour. Interviewees that held technical positions expressed strong interests in knowledge of the latest IT trends. Interviewees that held sales positions expressed strong interests in the creation of internal human-resource networks based on SNS.

Knowledge organizations that allow smooth distribution of knowledge are more likely to be horizontal rather than vertical organizations. Communication flows smoothly in horizontal organizational structures. Interviewees recognized that their organization was horizontal and described the organization as "informal". Real time feedback is provided based on knowledge inquiry within the internal KMS to help find solutions to problem. SNS usage within the organization aids organizational flexibility. SNS users share their concerns and hobbies. The CEO actively participates in SNS activity. This provides momentum for the development of closeness among employees. The organization's SNS usage reveals that the organizational structure could easily convert from a structure based on formality and authority to a horizontal, more flexible structure where employees feel a sense of closeness.

A knowledge strategy that allows smooth knowledge flow must develop from an organization's core values. Organization A's employees recognized that their organization used certain strategies. Organization A presents the core value of organization "DNA 3.0" to its members as its overriding organizational concern. It emphasizes employee education, communication, and member cohesion. Employees compose yearly projects based on themes provided at the beginning of each year. They suggest ideas continuously in regular offline and online meetings. Excellent outcomes are relayed to the top-level of the management hierarchy. These outcomes are converted into money-making ventures. Groups ("sCoPs") are created for temporary projects. sCoPs consist of general learning teams, learning teams across teams, and mentoring teams. The company pays for expenses incurred during learning activities. Thus, we can see that knowledge strategies must be practiced as management activities. They help employees believe that the organization's culture belongs to them and that their duty is to maintain it. In reality, interviewees possessed good understanding of the core value: Organization A believes that organizational sustainability results from endless learning.

KMS allows collection, storage, and transfer that can change tacit knowledge to explicit knowledge [29]. Various developments in IT technologies have raised questions about the limitations of KMS. Organization A adopted many IT technologies into their KM. All employees maintain blogs and all systems (SNS, KMS, and groupware) are logically and technologically connected. Employees directly access to all systems via groupware. Their ability to connect with mobile technology anytime and anywhere has greatly improved accessibility. Individuals can find desired knowledge if upper-bracket knowledge in KMS is ranked in a repository system. Thus, information stored in SNS can flow in real-time. This will lead to faster transfer of current knowledge. Good utilization of SNS's advantages can activate the transfer of knowledge in KMS.

Scholars have emphasized social, rather than economic rewards, for knowledge transfer [19, 22]. In KMS' early stages, organizations that employed strategies that solely emphasized economic rewards found them helpful for the quantitative expansion of knowledge. However, knowledge quality did not expand. Experienced members who might have shared knowledge imbued with core values did not do so. Thus, members had difficulty finding current knowledge. The expansion of barely-adequate knowledge resulted in a decline in KMS usage. To remedy the disadvantages of current KMS, organizations might consider a balance between social and economic rewards [30]. "Recognition" and "reputation" represent the social rewards of knowledge transfer [10, 31, 32].

Trust formation in knowledge cultures that share knowledge autonomically is vitally important. Opportunities to transfer knowledge in the organizational atmosphere will not occur without trust. Trust must be accumulated in informal meetings. Informal meetings offer spaces where people with similar concerns can form knowledge and human networks. SNS facilitates communication-based trust between members. Members within Organization A tend to place unconditional trust in internal SNS. Therefore, members believe that other members would not upload knowledge that might negatively affect their recognition within the organization. Appropriate SNS utilization supports culture formation that can activate knowledge transfer by active participation in communication.

The knowledge trade market remains in the early stage of institution. We would not describe Organization A as a situation of the knowledge trade market because the knowledge trade market in Korea is apparently in the early stage of institution. (In January 2012, we conducted with 223 respondents who work for Korean representative knowledge organizations. Respondents stated that the price of the knowledge trade market was a little flexible by 11.7% and virtual currency was in effect by 27.4%). The knowledge trade might be autonomously controlled by economic behaviors that exchange knowledge about price between producers and consumers.

5   Conclusion

The intention to participate in SNS positively affects the activation of horizontal communication in KMS and in knowledge transfer. Relationship consolidation in SNS influences informal horizontal communication. SNS does not directly influence knowledge transfer. Certain factors such as knowledge organization, knowledge strategy, KMS, knowledge rewards, knowledge culture, and knowledge market affect knowledge transfer activation.

The effects of SNS can be described as follows: First, a knowledge organization can be directly affected by knowledge transfer activation via SNS. For knowledge transfer activation to occur in a knowledge organization, a horizontal organization is more desirable than a vertical organization. SNS's flexible communication architecture serves as a conditioner for communication activation. A collaborative atmosphere and communication activation occurs between members who maintain relationships in SNS. Second, organizations attempt to keep their knowledge strategies in step with KM architecture. The bases of knowledge strategies must be prepared and translated into action along with management architecture so they can serve as connection factors that influence patterns of behavior such as organizational vision. Interviewees expressed sympathy with their organization's strategy that emphasizes knowledge sharing. Third, SNS is not a substitute for KMS. However, SNS could provide alternatives for knowledge transfer activation. SNS could provide neural networks that might effectively support knowledge transfer within organizations. The results of our case study demonstrated that SNS was not directly connected to KMS. Fourth, people desire social rather than economic rewards for knowledge sharing. These include reputation and recognition within organizations. Indirect social rewards such as the expansion of human networks within organizations can be achieved by SNS use. Fifth, trust formation is important for the creation of a knowledge culture that will transfer knowledge. The typical characteristic of SNS-based relationships leads to the creation of trust-based human networks. This characteristic exerts a fundamentally positive impact on knowledge culture creation. Sixth, organization A was in the early stage of institution of the knowledge trade market. The factors of knowledge transfer mentioned previously must be connected organically.

Further research is needed on knowledge transfer. We must analyze how mediated SNS can influence the knowledge trade market. Further empirical and quantitative research on the knowledge transfer framework with SNS is needed. Studies should be conducted on how the sociality of SNS related to economic feasibility of the knowledge trade market can be employed to synthetically establish the knowledge transfer mechanism. Therefore, further studies (based on the extent of the progress of the knowledge trade market) should be conducted to concretely investigate the correlation between SNS and the knowledge trade market.

References

1. Tredinnick, L.: Web 2.0 and Business: A pointer to the intranets of the future? Business Information Review 23(4), 228–234 (2006)
2. Kuss, D.J., Griffiths, M.D.: Online social networking and addiction: a review of the psychological literature. International Journal of Environmental Research and Public Health 8(9), 3528–3552 (2011)
3. Brass, D.J., Butterfield, K.D., Skaggs, B.C.: Relationships and Unethical Behavior: a Social Network Perspective. The Academy of Management Review 23(1), 14–31 (1998)
4. Fu, F., Liu, L., Wang, L.: Empirical analysis of online social networks in the age of Web 2.0. Statistical Mechanics and its Applications 387(2-3), 675–684 (2008)

5. Sasson, J.R., Douglas, I.: A conceptual integration of performance analysis, knowledge management, and technology: from concept to prototype. Journal of Knowledge Management 10(6), 81–99 (2006)
6. Forkosh-baruch, A., Hershkovitz, A.: A case study of Israeli higher-education institutes sharing scholarly information with the community via social networks. The Internet and Higher Education 15(1), 58–68 (2012)
7. Tsai, W.: Social structure of "coopetition" within a multiunit organization: Coordination, competition, and intraorganizational knowledge sharing. Organization Science 13(2), 179–190 (2002)
8. College, B., Hill, C., Borgatti, S., Cross, R.: A Relational View of Information Seeking and Learning in Social Networks. Management Science 49(4), 432–445 (2003)
9. Jeong, S.K., Ahn, J.C., Rhee, B.H.: Knowledge Transfer Activation Analysis: Knowledge Trade Perspective. Journal of Computer Information Systems 53(2), 47–55 (2013)
10. Benbya, B.H., Van Alstyne, M.: How to Find Answers within Your Company. MIT Sloan Management Review 52(2), 65–75 (2011)
11. Nonaka, I., Lewin, A.Y.: Dynamic Theory Knowledge of Organizational Creation. Organization Science 5(1), 14–37 (1994)
12. Boyd, D.M., Ellison, N.B.: Roles of Extrinsic Motivators, Social Network Sites: Definition, History, Psychological Forces and Scholarship. Journal of Computer-Mediated Communication 13(1), 210–230 (2007)
13. Chow, W.S., Chan, L.S.: Social network, social trust and shared goals in organizational knowledge sharing. Information & Management 45(7), 458–465 (2008)
14. Al-Alawi, A.I., Al-Marzooqi, N.Y., Mohammed, Y.F.: Organizational culture and knowledge sharing: critical success factors. Journal of Knowledge Management 11(2), 22–42 (2007)
15. Goh, S.C.: Managing effective knowledge transfer: an integrative framework and some practice implications. Journal of Knowledge Management 6(1), 23–23 (2002)
16. Milne, P.: Motivation, incentives and organizational culture. Journal of Knowledge Management 11(6), 28–38 (2007)
17. Vorakulpipat, C., Rezgui, Y.: Value creation: the future of knowledge management. The Knowledge Engineering Review 23(03), 283–294 (2008)
18. Xing, W.: Knowledge capitalism put into practice as an operational mechanism. Journal of Knowledge Management 10(1), 119–130 (2006)
19. Bock, G.W., Zmud, R.W., Kim, Y.G., Lee, J.N.: Behavioral Intention Formation in Knowledge Sharing: Examining the Roles of Extrinsic Motivators, Social-Psychological Forces and Organizational Climate. Management Information Systems 29(1), 87–111 (2005)
20. Ardichvili, A., Maurer, M., Li, W., Wentling, T., Stuedemann, R.: Cultural influences on knowledge sharing through online communities of practice. Journal of Knowledge Management 10(1), 94–107 (2006)
21. Pan, S.S., Scarbrough, H.: Knowledge Management in practice: An exploratory case study. Technology Analysis & Strategic Management 11(3), 359–374 (1999)
22. Marouf, L.N.: Social networks and knowledge sharing in organizations: a case study. Journal of Knowledge Management 11(6), 110–125 (2007)
23. Eschenfelder, K., Heckman, R., Sawyer, S.: The Distribution of Computing: The Knowledge Markets of Distributed Technical Support Specialists. Information Technology & People 11(2), 84–103 (1998)
24. Emerson, R.M.: Social Exchange Theroy. Annual Review of Sociology (2), 335–362 (1976)
25. Deng, P.S., Tsacle, E.G.: A market-based computational approach to collaborative organizational learning. Journal of the Operational Research Society 54(9), 924–935 (2003)

26. Parker, G.G., Van Alstyne, M.: Two-sided Network Effects: A Theory of Information Product Design. Management Science 51(10), 1494–1504 (2005)
27. Yin, R.: Case Study Research: design and methods, Thousands Oak, California (2002)
28. Garton, L., Haythornthwaite, C., Wellman, B.: Studying online social networks. Journal of Computer-Mediated Communication 3(1) (1997),
   http://onlinelibrary.wiley.com/doi/10.1111/j.1083-6101.1997.
   tb00062.x/full?sms_ss=facebook&at_xt=4da62ce480237b90,0
29. Alavi, M., Leidner, D.: Review:Knowledge Management and Knowledge Management Systems: Conceptual Foundations and Research Issues. Management Information Systems 25(1), 107–136 (2001)
30. Bénabou, R., Tirole, J.: Incentives and prosocial behavior. National Bureau of Economic Research Cambridge, USA (2005)
31. Brachos, D., Kostopoulos, K., Soderquist, K.E., Prastacos, G.: Knowledge effectiveness, social context and innovation. Journal of Knowledge Management 11(5), 31–44 (2007)
32. Wasko, M.M., Faraj, S.: Why Should I Share? Examining Social Capital and Knowledge Contribution in electronic networks of practice. MIS Quarterly 29(1), 35–57 (2005)

Appendix

I. Related KM

	Question
	What is your role in the department?
	What is your position in the company? (manager/employee)
Knowledge Management Status	How many years your company has operated KMS? - Is knowledge sharing activity actively operated? - Is there any knowledge trade selling and buying with virtual point in your company?
	Who is the manager that controls your company's knowledge contents? (a knowledge dedicated team or per headquarters)
	Have your company utilized SNS in your company? - How many years have been SNS operated in your company?
	How could SNS influence on knowledge sharing and trade?
Knowledge Organization	What type of your company's organization structure do you think? (vertical, horizontal, with authority and with collaborative) - So, do you think that the utilization of SNS is related to horizontal communication? - How SNS does influence on organization structure?
	What effects the utilization of SNS influence on communication between superiors or inferiors?
	Do you have any previous experience of feeling by authorities like superior's one-directional order?

	There are so many types of employees in organization. Is SNS effective for merging employees' diversified opinions? So, why do you think like this?
Knowledge Strategy	Do you have the importance of knowledge included in your organization vision like creative person?
	Do you think that your knowledge strategy is well related to company's vision?
	What types of knowledge strategy for activating the utilization of KMS of your company do you have? - For example, cultivating human resources based learning - Change management system like a just estimation and rewards - Connectivity between CoP practices and management activity - Connectivity between knowledge sharing and process
	How could your company's knowledge strategy influence on employee's knowledge activity?
	Is there any commercial SNS prohibited to access?
KMS	What type of system related KMS is?
	What type of IT technology for activating the utilization of KMS does your company operate?
	Is connected your company's SNS to KMS physically and logically?
Knowledge Rewards	What type of rewards for knowledge sharing does your company operate?
	What type of rewards is effective for knowledge sharing? For example, economic rewards or social rewards like recognition and reputation.
	Is your company's knowledge rewards properly given to your contribution for knowledge sharing?
	What type of additional rewards do you have by SNS?
Knowledge Culture	Do you think what factors contribute for activating knowledge transfer in your organization?
	Do you participate in informal community in your organization?
	What is your role of informal community in your company? (knowledge transformer, passive visitor, community leader etc.)
	Have you experienced the motivation to participate in informal community or to intend transfer knowledge by the utilization of SNS?
	What type of knowledge is shared in your company? (project experiences, technology trend, personal technology, market trend, external reports, etc.)
	Do you think that trust is the important factor for knowledge transferring?
	Do your employees in your organization share autonomously knowledge?

	Do you depend on people who have relations on SNS?
Knowledge Trade	Is there any activity of knowledge exchange with virtual currency (mileage, currency)?
	How points are given? (static or flexible, for example, 10 points per one uploading event or comment, flexible points by count of 'like')
	How does your company set the price for shared knowledge? (estimation of knowledge dedicated team, good feeling of users, unconditional allocation for uploading knowledge)
	How does your company manage price liquidity? (a self-regulating system in knowledge market, central control)
	Do you think that feedbacks of customers have a positive impact on improving knowledge?
	Does your company have what route customer needs inflow? (marketing, FGI(focus group interview), and SNS)

II. Organization Changes with the Utilization of SNS

We want to know how SNS utilization influences on knowledge sharing and trade in your organization. Could you explain there are what impacts on the utilization of knowledge by SNS?

Relationships between SNS and KM	What purpose do you use SNS in company? (keeping closeness, obtaining information, etc.)
	What types of SNS are used in your company? (Facebook, Twitter, self-developed SNS) - Do you think current using SNS is proper to your company's KM? If not, is there any supplementation?
	How many persons do you related in SNS?
	How fast does information flow by using SNS as compared the past?
	Do you think that the utilization of SNS is effective to transfer from outside trend and knowledge?

Evolution of Social Networks
and Depression for Adolescence

Hsieh-Hua Yang[1], Chyi-In Wu[2], Yi-Horng Lai[1], and Shu-Chen Kuo[3]

[1] Department of Health Care Administration, Oriental Institute of Technology
58, Sec 2, Sihchuan Rd., Pan-Chiao Dist., New Taipei City 22061, Taiwan
{FL008,FL006}@mail.oit.edu.tw
[2] Institute of Sociology, Academia Sinica, Nankang, Taipei 11529, Taiwan
ssslciw@gate.sinica.edu.tw
[3] Center for General Education, National Defense Medical Center, Taipei 114, Taiwan
kuochen1327@gmail.com

Abstract. The goal of the present study is to analyze the evolution of adolescent friendship network and depression. A network survey was carried out in classrooms of high schools. The participants are 93 boys and 82 girls. Sociometric data were collected by having each student nominate up to 16 intimate classmates. Mandarin Chinese version of the center for epidemiological studies-depression scale (MC-CES-D) was adopted as the measurement of depression. Panel data was collected across 3 semesters from Sep. 2008 to Jan. 2010. The program SIENA was applied to estimate the models for the evolution of social networks and depression. The result showed that gender had effect in the beginning, and depression had effect during the 2nd semester. The results and implication are discussed.

Keywords: Social networks, depression, adolescence, evolution.

1   Introduction

Depressive disorders with a steadily increased lifetime risk and earlier onset in successive birth cohorts have been documented in several epidemiologic studies [1-3]. Lifetime prevalence of depressive disorders increases with age. The prevalence of adolescents' major depressive disorder in previous studies ranged from 0.4% to 8.3% [4]. In Taiwan, epidemiological studies have found that 0.5-4.4% of adolescents aged 13-15 had major depressive disorder [5,6]. In another study, the rate of adolescents who had significant depression was 12.3% [7].

Social networks, close relationships and perceived social support have all been linked to the healthy psychological development of adolescents [8]. Prior studies of adolescent depression and friendships can be divided into two categories: (1) studies that examine the importance of a student's social integration with their peers, and (2) studies that examine whether a student's friends' characteristics are related to his/her risk of depression. The first group of studies finds that students who are more socially integrated in school are less likely to be depressive [9]. Students who have few friends

L.S.-L. Wang et al. (Eds.): MISNC 2014, CCIS 473, pp. 134–144, 2014.

[10], are less popular, and report low peer support and changes in peer acceptance [11-17] are more likely to be depressive.

A longitudinal network behavior models indicated that gender similarity and perceived popularity are influential in the formation of social ties, and the associations between distressed mental health and students' network have been tested [18]. Social network analysis is uniquely suited for measuring and understanding the behavior of peers because it provides a formal means for "mapping" friendships and measuring properties of those friendships [19]. In a majority of applications of social network analysis, there is interdependence between network structure and the individual characteristics of the actors. McPherson, Smith-Lovin and Cook [20] proposed a homophily principle to explain the network autocorrelation, which stands for the argument that it is easier or more rewarding for an actor to interact with a similar other than with a dissimilar other. An alternative explanation of the same phenomenon is the assimilation principle according to which network actors adapt their own individual characteristics to match those of their social neighborhood [21,22].

The maintenance of friendship needs more focused similarities. A friendship is more likely to come into existence if each individual perceives the other as attractive, responsive, and in particular, similar in a variety of ways [23,24]. Friendship network formation is an evolution process. During the evolution process, strangers are converted to acquaintances and acquaintances to friends. In the initial stage, a friendship between two people can emerge only when they meet in the same place; in the later stage, the individual should be of the same taste. van Duijn et al. [25] developed a theory to explain how changes in the network structure. The theory posits that proximity plays a dominant role in the formation of networks, visible similarity plays a role in the beginning of the friendship development process, and invisible similarity will have effect at later stages of the friendship development. The theory tried to differentiate the main effects during this evolution process, but unfortunately, no significant effects of invisible similarity are found in his study. In van Duijn's study, the invisible similarity variables are activities and interests important to students in general. We argue that the invisible similarity should be something invisible, such as depressive symptoms. In this research, the depression similarity is the focus to be tested during friendship network evolution.

2   Method

2.1   Participants

Studying the friendship network evolution, it must be assumed that the group is closed, the boundary of the friendship network is clear. The data was collected from high school in Taiwan. The participants were 175 high school students of 93 boys and 82 girls.

2.2   Measures

Panel data was collected 10 waves during 3 semesters from Sep. 2008 to Jan. 2010. Sociometric data were collected by having each student nominate up to 16 intimate classmates. Depression was measured in the beginning of each semester, namely at wave 1, 5, and 8.

Mandarin Chinese version of the center for epidemiological studies-depression scale (MC-CES-D) was adopted as the measurement of depression. The 20-item MC-CES-D is a self administered four-point evaluation scale assessing frequency of depressive symptoms in the preceding week, with scores ranging from 0 (none or very few) to 3 (always) [26,27]. Higher CES-D scores indicated more severe depression. A person with total score less than 16 is not depressed. The psychometric of the MC-CES-D for assessing depressive symptoms among non-referred adolescents in Taiwan has been examined in the previous study [5]. And the previous study using the MC-CES-D in a two-phase survey for depressive disorders among non-referred adolescents in Taiwan found that the adolescents with total MC-CES-D scores >28 were more likely to have major depressive disorder with or without functional impairment. In the present study we defined those adolescents whose MC-CES-D score was higher than 28 as having significant depression. The Cronbach alpha for the MC-CES-D in the present study is 0.89.

2.3   Data Analysis

The program SIENA (Simulation Investigation for Empirical Network Analysis) was applied to estimate the models for the evolution of social networks according to the dynamic actor-oriented model of Snijders [28-30]. For the estimation, sociometric data was transformed into adjacency matrices. Boy was coded as 1 and girl as 2. Depression was classified as not depressed (<16), subclinically depressed (16-28), and clinically depressed (>28) by the sum score of 20-item MC-CES-D measures.

In the study, the evolution model components include density, reciprocity, transitivity, gender similarity, and similarities of depression. Density effect is defined as the number of outgoing ties. Reciprocity effect is defined the number of reciprocated ties. A transitive triplet for actor i is an ordered pairs of actors (j, h) for which $i \rightarrow j \rightarrow h$ and also $i \rightarrow h$. Depression- alter represents the effect on the actor's popularity to other actors, a positive parameter implies the tendency that the in-degree of actors with higher values of depression increase more rapidly. Depression-ego represents the effect on the actor's, a positive parameter implies the tendency that actors with higher values of depression increase their out-degrees more rapidly. A positive parameter of depression-similarity implies that actors prefer ties to others with similar depression.

The program can estimate all possible effects simultaneously, and is suitable to test our hypotheses. The SIENA program is included in the StOCNET system, and can be downloaded from http://stat.gamma.rug.n1/stocnet/ [31].

3   Result

The analyses were run on the 175 students who were present at all ten measurement points. At 3 observation points, the majority of students were not depressed. The rates were 48.6%, 48.0%, and 46.3% respectively at wave 1, 5, and 8. Depressive students at wave 1 were related to that at wave 5 ($\chi^2 = 67.79, p < .001$). During wave 1 and wave 5, 61.7% of students remained in the same depressive status at both measurements, whereas among 18.3% the depression increased, and among 20% depression decreased. Depressive students at wave 5 were related to that at wave 8 ($\chi^2 = 55.09, p < .001$). During wave 5 and 8, 62.8% of them remained in the same depressive status, whereas among 20.6% the depression increased and among 16.6% depression decreased.

The average degree at beginning is 7.489, and then decreased to 4.966 at wave 8. The change counts are indicated in Table 2. The total number of changes between consecutive observation moments is 740 in the first period, and 520 to 627 in all further periods.

Table 1. Density and degree

Wave	1	2	3	4	5	6	7	8	9	10
Density	0.043	0.036	0.039	0.032	0.033	0.031	0.033	0.028	0.029	0.031
Average degree	7.489	6.328	6.747	5.626	5.845	5.379	5.718	4.966	5.034	5.345

Table 2. Change frequencies for the periods of wave 1 to wave 10

Wave	w1-2	w2-3	w3-4	w4-5	w5-6	w6-7	w7-8	w8-9	w9-10
0 to 0	28878	29017	29075	29146	29184	29183	29207	29281	29287
0 to 1	269	332	201	325	249	331	248	305	287
1 to 0	471	259	396	287	330	272	379	293	233
1 to 1	832	842	778	692	687	664	616	571	643

Out-degree, in-degree, and density of boys and girls are shown as Fig. 1 to 2. Out-degree and in-degree of boys and girls have decreasing tendency during 10 waves. Boys' out-degree and in-degree are higher than girls' in the beginning. At wave 2, both out-degree and in-degree decreased rapidly, and were lower than girls'. After wave 8, both out-degree and in-degree were increased until the last observation, and closing to girls'. Density kept rather steady, as Fig. 3. Boys' density is lower than girls' during 10 waves.

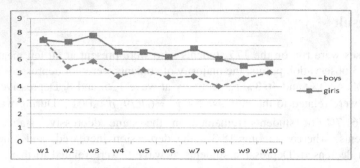

Fig. 1. Out-degree of boys and girls during 10 waves

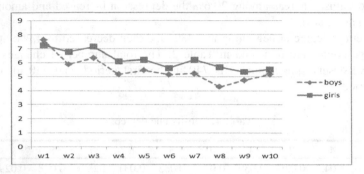

Fig. 2. In-degree of boys and girls during 10 waves

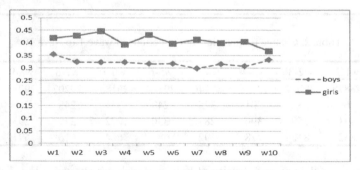

Fig. 3. Density of boys and girls during 10 waves

For network evolution, the rate function describes the average number of changes in network ties between measurement points. The highest rate parameter is in the beginning, and then decreases. The rate is higher between 2 semesters than during the same semesters. The endogenous network effects are all statistically significant. The density, reciprocity and transitive triplets have effect during all the stages. The reciprocity effect indicates a preference for reciprocating relationships. The transitivity effect indicates a preference for being friends with friends' friends.

The effect of gender similarity was significant during all waves. The positive parameter implies that actors prefer ties to others with same gender. The effect of gender ego was significant during w1-2, w5-6, and w6-7. It implies the tendency that girls increase their out-degrees more rapidly than boys. The effect of gender alter was significant during w2-3, w3-4, w5-6, w6-7, w7-8, w8-9. The negative parameter implies the tendency that the in-degree of boys increases more rapidly than girls.

Table 3. Estimates of the evolution model (wave1 to wave4)

Variable	w1-2	w2-3	w3-4
	Estimate(se)	Estimate(se)	Estimate(se)
Rate parameter	6.965(0.268)	5.409(0.235)	5.541(0.243)
Network effects			
Density	-2.035(0.059)*	-1.910(0.064)*	-2.218(0.071)*
Reciprocity	0.944(0.085)*	1.225(0.094)*	1.348(0.098)*
Transitive triplets	0.104(0.008)*	0.128(0.010)*	0.108(0.011)*
Gender alter	-0.042(0.080)	-0.284(0.088)*	-0.254(0.092)*
Gender ego	0.466(0.081)*	0.137(0.085)	0.181(0.101)
Gender similarity	0.424(0.085)*	0.395(0.093)*	0.443(0.099)*
Depression alter	-0.052(0.051)	-0.011(0.056)	-0.086(0.057)
Depression ego	-0.063(0.055)	-0.0671(0.061)	0.011(0.061)
Depression similarity	-0.060(0.114)	0.035(0.123)	-0.251(0.131)

* $p<.05$

Table 4. Estimates of the evolution model (wave4 to wave7)

variable	w4-5	w5-6	w6-7
	Estimate(se)	Estimate(se)	Estimate(se)
Rate parameter	5.889(0.260)	5.665(0.265)	6.213(0.278)
Network effects			
Density	-2.075(0.064)*	-2.149(0.069)*	-1.905(0.059)*
Reciprocity	1.076(0.099)*	1.412(0.102)*	1.177(0.100)*
Transitive triplets	0.175(0.012)*	0.127(0.012)*	0.134(0.012)*
Gender alter	-0.170(0.094)	-0.411(0.099)*	-0.197(0.087)*
Gender ego	-0.017(0.094)	0.299(0.101)*	0.246(0.089)*
Gender similarity	0.615(0.095)*	0.410(0.096)*	0.522(0.086)*
Depression alter	0.170(0.057)*	-0.055(0.058)	-0.086(0.056)
Depression ego	0.209(0.060)*	-0.226(0.070)*	-0.024(0.058)
Depression similarity	0.301(0.126)*	0.034(0.133)	0.119(0.118)

* $p<.05$

For the first semester, the non-significance of the depression alter, depression ego, and depression similarity effects shows that there is no evidence for depression related differences in the tendency to have friends in the beginning of friendship formation. For the second semester, the effect of depression similarity was significant only during w4-5. The positive parameter during w4-5 implies that actors prefer ties to others with similar values of depression. The effect of depression ego was significant during w5-6, and w7-8. The positive parameter during w7-8 implies the tendency that actors with higher value of depression increase their out-degrees more rapidly.

Table 5. Estimates of the evolution model (wave7 to wave10)

variable	w7-8	w8-9	w9-10
	Estimate(se)	Estimate(se)	Estimate(se)
Rate parameter	6.508(0.306)	6.518(0.340)	4.976(0.263)
Network effects			
Density	-2.215(0.063)*	-1.996(0.052)*	-1.884(0.063)*
Reciprocity	1.536(0.100)*	1.358(0.094)*	1.136(0.108)*
Transitive triplets	0.146(0.011)*	0.140(0.011)*	0.134(0.013)*
Gender alter	-0.198(0.094)*	-0.233(0.084)*	-0.105(0.089)
Gender ego	0.055(0.097)	-0.112(0.091)	0.025(0.094)
Gender similarity	0.270(0.085)*	0.458(0.086)*	0.612(0.090)*
Depression alter	0.020(0.054)	0.051(0.052)	-0.046(0.059)
Depression ego	0.163(0.059)*	0.067(0.055)	0.022(0.062)
Depression similarity	0.073(0.118)	0.070(0.110)	0.218(0.124)

$*\ p<.05$

4   Discussion and Conclusion

Longitudinal network data can yield important insights into social processes. Methodologically, the assessment of friendship evolution needs to be based on longitudinal designs that include multiple measurement waves. Poulin and Chan [32] indicated that such detailed analysis of stability will allow a better understanding of the dynamic processes by which friendships change over time and affect children's and adolescents' psychosocial development. Ten waves' observations may offer rich materials for understanding adolescents' development.

4.1   Friendship Network Evolution

The descending individual degree leads to the decreasing network density during time 1 to time 10. And there is no gender difference for indegree and outdegree. But, studies have found more social participation on the part of girls. Girls are found to make or receive more friendship choices [33,34]. Since there are more boys than girls in this class and gender plays a key factor of nomination, the result is reasonable.

The rate parameter is up and down, then decreasing at the last observation. It represents the friendship network is dynamic but not steady. The network effects, including density, transitive triplets, and reciprocity are important during all stages. The results are the same as the results of van Duijn et al. [25].

4.2   Gender is the Visible Demographic Characteristics and Has Effect on Friendship Formation from the Beginning

Gender is a powerful organizer of peer relationships throughout development. Heterosex friendships represent unique opportunities both for healthy development and adverse development [35]. Other-sex friendships may aid youth in learning how to interact productively with individuals with interests, experiences, and backgrounds that differ from their own. However, adolescents report feeling closer to their same-sex friends than to their other-sex friends [36,37]. And friends who provide psychological closeness are beneficial for adolescents' socio-emotional development [38,39]. Thus, same-sex friendships are an important resource for the development of psychological health during adolescence.

In our findings, positive parameter of gender ego and negative parameter of gender alter indicate that girls are more active than boys to make friends. The effect is significant almost during all the stages. It is suggested that girls have better interpersonal skills than boys. Girls language is relatively more likely to be collaborative and affiliative [40,41], while boys' language is relatively more assertive, controlling, and competitive and incorporates more demands [42]. Girls are more likely than boys to spend their time in small intimate groups, usually made up of two or three people [43-45]. Most of all, gender is the visible demographic characteristics and has effect on friendship formation in the beginning and during all the stages.

4.3   Depression Is Invisible and Has Effect in Later Stage

Friendship network is initiated by chance; in the early stages the proximity and visible similarity variables determined dyadic relations. In later stages, dyadic relations will be strengthened by invisible similarities. Our findings indicated the invisible role of depressive symptom and its effect in later stage of friendship network evolution.

The basic idea of the model is that the actors in the network may evaluate the network structure and try to obtain a "pleasant" configuration of relations, a configuration that increases their social well-being. The actors will evaluate the network structure in order to obtain a rewarding pattern of relationships. The evaluation takes place on the basis of visible or invisible factors. The visible factor can be seen in the beginning and has effect on making friends. The invisible factors will become visible after a period of interaction. In the findings, depressive symptoms had effect at the beginning of the 2nd semester. The positive parameter of depression ego changed from positive to negative. Coyne (1976) found that the normal female students who had spoken to depressed patients were significantly more depressed, anxious, hostile, and rejecting. The decreased tendency of out-degree of higher depressive students may be due to rejection or hostility. It needs further exploration.

4.4   Conclusion

The purpose of this article is testing the similarity effect of depressive symptoms on friendship network evolution. Similarity breeds connection. By the time adolescent enter high school, they have learned that gender is a notable personal characteristic. The effect of gender similarity is significant in the beginning and for all the three semesters. However, the similarity effect of the depressive symptoms is significant during the second semester.

Compared with the studies of van Duijn et al. [25], the network dynamic characteristics are the same, but not invisible similarity. Our finding revealed the depression similarity may exert effect in later stage. The results are entirely consistent with the theory proposed by van Duijn.

Acknowledgments. The work is supported by the National Science Council (NSC101-2410-H-161-002).

References

1. Wickramaratne, P.J., Weissman, M.M., Leaf, P.J., Holford, T.R.: Age, Period and Cohort Effects on the Risk of Major Depression: Results from Five United States Communities. J. Clin. Epidemiol. 42, 333–343 (1989)
2. Lavori, P.W., Klerman, G.L., Keller, M.B., Reich, T., Rice, J., Endicott, J.: Age-period-cohort Analysis of Secular Trends in Onset of Major Depression: Findings in Siblings of Patients with Major Affective Disorder. J. Psychiatr. Res. 21, 23–35 (1987)
3. Keyes, K.M., Nicholson, R., Kinley, J., Raposo, S., Stein, M.B., Goldner, E.M., Sareen, J.: Age, period, and cohort effects in psychological distress in the United States and Canada (March 31, 2014), doi:10.1093/aje/kwu029
4. Birmaher, B., Ryan, N.D., Williamson, D.E., et al.: Childhood and Adolescent Depression: A Review of the Past 10 Years. Part I. J. Am. Acad. Child Adolesc. Psychiatry 35, 1427–1439 (1996)
5. Yang, H.J., Soong, W.T., Kuo, P.H., Chang, H.L., Chen, W.J.: Using the CES-D in a Two-phase Survey for Depressive disorders among Nonreferred Adolescents in Taipei: A Stratum-specific Likelihood Ratio Analysis. J. Affect. Disord. 82, 419–430 (2004)
6. Gau, S.S., Chong, M.Y., Chen, T.H., Cheng, A.T.: A 3-year Panel Study of Mental Disorders among Adolescents in Taiwan. Am. J. Psychiatry 162, 1344–1350 (2005)
7. Lin, H.C., Tang, T.C., Yen, J.Y., Ko, C.H., Huang, C.F., Liu, S.C., Yen, C.F.: Depression and Its Association with Self-esteem, Family, Peer and School Factors in a Population of 9586 Adolescents in Southern Taiwan. Psychiat. Clin. Neuros. 62, 412–420 (2008)
8. Kawachi, I., Berkman, L.: Social Ties and Mental Health. J. Urban Health 78, 458–467 (2001)
9. Millings, A., Buck, R., Montgomery, A., Spears, M., Stallard, P.: School connectedness, peer attachment, and self-esteem as predictors of adolescent depression. J. Adolescence 35, 1061–1067 (2012)
10. Sund, A.M., Larsson, B., Wichstrom, L.: Psychosocial correlates of depressive symptoms among 12-14-year-old Norwegian adolescents. Journal of Child Psychology and Psychiatry 44, 588–597 (2003)
11. Birmaher, B., Bridge, J.A., Williamson, D.E., Brent, D.A., Dahl, R.E., et al.: Psychosocial functioning in youths at high risk to develop major depressive disorder. J. Am. Acad. Child Adolesc. Psychiatry. 43, 839–846 (2004)
12. Lewinsohn, P.M., Gotlib, I.H., Seeley, J.R.: Depression-related psychosocial variables: Are they specific to depression in adolescents? J. Abnorm. Psychol. 106, 365–375 (1997)

13. Olsson, I.G., Nordstrom, M.-L., Arinell, H., von Knorring, A.-L.: Adolescent depression and stressful life events. A case-control study within diagnostic subgroups. Nord. J. Psychiat. 53, 339–346 (1999)
14. Rose, A.J., Rudolph, K.D.: A review of gender differences in peer relationship processes: potential trade-offs for the emotional and behavioral development of girls and boys. Psychol. Bull. 132, 98–131 (2006)
15. Caldwell, M.S., Rudolph, K.D., Troop-Gordon, W., Kim, D.Y.: Reciprocal influences among relational self-views, social disengagement, and peer stress during early adolescence. Child Dev. 75, 1140–1154 (2004)
16. Levendosky, A.A., Okun, A., Parker, J.G.: Depression and maltreatment as predictors of social competence and social problem solving skills in school-age children. Child Abuse Neglect 19, 1183–1195 (1995)
17. Vernberg, E.M.: Psychological adjustment and experiences with peers during early adolescence: reciprocal, incidental, or unidirectional relationships? J. Abnorm. Child Psych. 18, 187–198 (1990)
18. Pachucki, M.C., Ozer, E.J., Barrat, A., Cattuto, C.: Mental Health and Social Networks in Early Adolescence: A Dynamic Study of Objectively-measured Social Interaction Behaviors. Soc. Sci. Med. xxx, 1–11 (2014)
19. Ennett, S.T., Bauman, K.E.: Adolescent social networks: school, demographic, and longitudinal considerations. J. Adolescent Res. 11(2), 194–215 (1996)
20. McPherson, J.M., Smith-Lovin, L., Cook, J.M.: Birds of a feather: homophily in social networks. Annu. Rev. Sociol. 27, 415–444 (2001)
21. Friedkin, N.: Norm formation in social influence networks. Soc. Networks 23, 167–189 (2001)
22. Oetting, E.R., Donnermeyer, J.F.: Primary socialization theory: the etiology of drug use and deviance. Subst. Use. Misuse 33, 995–1026 (1998)
23. Cramer, D.: Close Relationships: The Study of Love and Friendship. Arnold, London (1998)
24. Duck, S.W.: Friends for Life: The Psychology of Personal Relationships. Harvester, New York (1991)
25. van Duijn, M.A.J., Zeggelink, E.P.H., Huisman, M., Stokman, F.N., Wasseur, F.W.: Evolution of sociology freshmen into a friendship network. J. Math. Sociol. 27, 153–191 (2003)
26. Chien, C.P., Cheng, T.A.: Depression in Taiwan: Epidemiological Survey Utilizing MC-CES-D. Psychiatria et Neurologia Japonica 87, 335–338 (1985)
27. Radolff, L.S.: The CES-D Scale: A Self-report Depression Scale for Research in the General Population. Appl. Psychol. Meas. 1, 385–401 (1977)
28. Snijders, T.A.B.: Stochastic Actor-oriented Models for Network Change. J. Math. Sociol. 21, 149–172 (1996)
29. Snijders, T.A.B.: The Statistical Evaluation of Social Network Dynamics. In: Sobel, M.E., Becker, M.P. (eds.) Sociological Methodology. Basil Blackwell, Boston (2001)
30. Snijders, T.A.B., Steglich, C., Schweinberger, M.: Modeling the Coevolution of Networks and Behavior. In: van Montfort, K., Oud, J., Satorra, A. (eds.) Longitudinal Models in the Behavioral and Related Sciences. Lawrence Erlbaum Associates, New Jersey (2007)
31. Snijders, T.A.B., Steglich, C., Schweinberger, M., Huisman, M.: Manual for SIENA version 3, University of Groningen: ICS. University of Oxford, Department of Statistics, Oxford (2007), http://stat.gamma.rug.nl/stocnet
32. Poulin, F., Chan, A.: Friendship stability and change in childhood and adolescence. Dev. Psychol. 30, 257–272 (2010)

33. Urberg, K.A., Degirmencioglu, S.M., Tolson, J.M., Holliday-Scher, K.: The structure of adolescent peer networks. Dev. Psychol. 31(4), 540–547 (1995)
34. Berndt, T.J., Hoyle, S.G.: Stability and change in childhood and adolescent friendships. Dev. Psychol. 21, 1007–1015 (1985)
35. Sippola, L.K.: Getting to know the "other": the charracteristics and developmental significance of other-sex relationships in adolescence. J. Youth. Ado. 28(4), 407–418 (1999)
36. Bukowski, W., Sippola, L., Hoza, B.: Same and other: Interdependency between participation in same- and other-sex friendships. J. Youth Adolesc. 28, 439–459 (1999)
37. Lundy, B., Field, T., McBride, C., Field, T., Largie, S.: Same-sex and opposite-sex best friend interactions among high school juniors and seniors. Adolescence 33, 279–289 (1998)
38. Hartup, W.: The company they keep: Friendships and their developmental significance. Child Dev. 67, 1–13 (1996)
39. Hartup, W., Stevens, N.: Friendships and adaptation in the life course. Psychological Bulletin 121, 355–370 (1997)
40. Leaper, C.: Influence and involvement in children's discourse: Age, gender, and partner effects. Child Dev. 62, 797–811 (1991)
41. Strough, J., Berg, C.A.: Goals as a mediator of gender differences in high-affiliation dyadic conversations. Dev. Psychol. 36, 117–125 (2000)
42. Leaper, C., Smith, T.: A meta-analytic review of gender variations in children's language use: Talkativeness, affiliative speech, and assertive speech. Dev. Psychol. 40, 993–1027 (2004)
43. Fabes, R., Martin, C., Hanish, L., Anders, M., Madden-Derdich, D.: Early school competence: The roles of sex-segregated play and effortful control. Developmental Psychology 39, 848–858 (2003)
44. Maccoby, E.E.: Gender and group process: A developmental perspective. Current Directions in Psychological Science 11, 54–58 (2002)
45. Moller, L., Hymel, S., Rubin, K.: Sex typing in play and popularity in middle childhood. Sex Roles 26, 331–353 (1992)
46. Coyne, J.C.: Depression and the response of others. Journal of Abnormal Psychology 85, 186–193 (1976)

Classification of Terrorist Networks
and Their Key Players

Fatih Ozgul

Turkish National Police Academy, Faculty of Security Sciences, Golbasi, Ankara, Turkey
fatih.ozgul@istanbul.com

Abstract. Due to the interest by public audience and academic research, there has been a great interest in Terrorist Networks by the academicians, analysts and criminologists. Either to learn how to disrupt or to prevent their activities, structure of these networks are investigated. The final conclusion about their structure and topology came to the fact that they do not resemble each other, but there are categories of them. In this paper, we categorized these networks into six because of their ideologies and common practices. Topologies of these six categories are observed and importance of key players (leaders, financiers, propaganda units and armed units) are compared based on these categories.

Keywords: terrorist networks, network topology, community structure, hierarchy, key players, degree, closeness, betweenness.

1   Introduction

After the September 11th, there has been a great interest in Terrorist Networks by the academicians, analysts and criminologists in the field. Sageman [8] as the pioneer of the research worked intensively on 'homegrown' terrorist and developed "bunch of guys" theory which refers to multi cell-based small self-organized terrorist networks where these terrorists have little or no contact before with known terrorist groups. Sageman [9] says that they are usually anonymous groups of disaffected amateurs who self-radicalize, then independently plan and execute their own operations with no outside guidance. As a result, they are very difficult to detect, infiltrate and prevent. But these characteristics are more resemble to Al-Qaida type networks where later we would call them Shura based networks. But how about the other ideologies and terrorist groups such as extreme-left, ethnic and anarchist groups? Can we characterize other terrorist groups concerning their cultural and practices within topologies? Can we also predict important key players within those terrorist networks in order to prevent or disrupt them?

2   Terrorist Networks

Terrorist networks are criminal networks. But the main differences between terrorist networks against other criminal networks are; firstly they have certain

L.S.-L. Wang et al. (Eds.): MISNC 2014, CCIS 473, pp. 145–157, 2014.

ideologies where they build aims around it. For spreading ideology, they have propaganda units, books, web sites, radio and TV channels. Secondly they have armed units to reach their aims by using violence. Thirdly they have financiers in order to supply resources for terrorist activities. Fourthly they take great care about their secrecy during meetings and interactions in order to hide themselves from surveillance of the police and other legal institutions. They sometimes operate globally all over the world in small cells by using means of communications such as mobile phones and internet. Fifth, some of the terrorist networks are protected and financed by governmental or non-governmental organizations (NGOs) so they use the capabilities of governmental resources for their political aims.

Rapoport [12] describes terrorism historically in four waves: In the 1880s, an initial "Anarchist Wave" appeared which continued for some 40 years. Its successor, the "Anti-Colonial Wave" began in the 1920s, and by the 1960s it disappeared. The late 1960s witnessed the birth of the "New Left Wave," which continued till the 90s leaving a few groups still active. The fourth or "Religious Wave" began in 1979, and, if it follows the pattern of its predecessors, it still has twenty to twenty-five years to run.

In many countries criminal codes terrorism is defined as a criminal activity. Although there is no standard definition of terrorism in the world, but in general there are three requirements of a terrorist group. In order to define a terrorist network, they need an ideology, a hierarchy, and using violence and terror for reaching their political aims. Contrary to organized crime networks, the primary motivation of terrorism is not economic profits but the destabilization of political, constitutional, economic or social structures. They also ensure their continuity of their life-cycle by secrecy even if they had to minimize or disrupt themselves after police operations or inactivation by their own choice.

Despite the ideological and purpose differences between organized crime and terrorist groups, they share the same loosely connected and fluid ad hoc organizational principles. Hence, social ties and connections are still to a large extent crucial determinant for the performance, sustainability and success of both criminal and terrorist organizations [13]. Most of the international terrorist networks get support from legal companies, institutions, governments, NGOs, and from the public for their finance, recruitment and materials. Sometimes terrorist networks have their own companies, businesses, charities, clubs, premises on behalf of other people and social institutions. In some cases supporters of terrorist networks only seek to harm their common enemies. They simply need seek adventure based on an ideology, using each other's skills and capabilities for harming their common enemies. They see terrorism as a family business. For instance Sageman [8, 9] defines home-grown terrorists as citizens of the western countries, they were born in western countries, educated there and they are non-immigrant. Sageman points them as more dangerous than others who are born in their originated country, but migrated to western countries after they are grown-up, and they conceal their ideology while living in. On the other hand, 'home-grown' terrorist networks [8, 9] are steered by political, philosophical, ideological, racial, ethnic or religious motivations all of which can be called as ideology. Systematically accumulating knowledge about the structural knowledge of criminal activity gives us an understanding of their functioning and ways to counteract and disrupt terrorist networks. Most members of those terrorist networks are not aware of other cells which are geographically very close to them. They share the same ideology, meet each other on internet, but they only know each other with nick-names. That is why they are more difficult to follow, more complex, and unpredictable.

3   Counter-Terrorism Priorities

Memon and Larsen [10] refers that counter-terrorism often focus on characteristics of the network structure in order to gain insight into the following questions when analyzing terrorist networks; Which terrorist is highly/less connected? Which terrorists are connected to highly connected terrorists? Who is depending on whom? On whom do many terrorists depend? What is the efficiency of the network? Who is the most important person in the network? What are the various roles of terrorist in the network? Which terrorists are key players? How can police use (often incomplete and faulty) network data to disrupt and destabilize terrorist networks? Knowledge of these structural characteristics helps in revealing vulnerabilities of terrorist networks and may have important implications for investigations.

Counter-terrorism methods may even include the use of assassination to disrupt terrorist networks. But in many cases taking out the leadership may not cause disruption of terrorist network. In fact, leadership removal may make the network denser to future analysis given the emergence of new leadership that may not be known. Carley [11] proposed a well-constructed set of criteria for requirement of taking out the key player or disrupting terrorist network as: Ineffective terrorist organization means either the rate of information flow through the network has been reduced to zero, or the network, as a decision making body, cannot reach a consensus or the ability of the network to accomplish tasks is impaired.

4   Analysis of Terrorist Networks

There are three aspects that are important when analyzing terrorist networks [5]; the first aspect is structural analysis which analyzes the characteristics of the network structure and topology. What kind of terrorist network is it? What is their ideology? Are they structured hierarchically or more informal? Structural analysis of criminal networks gives a general about the type of network. The second aspect [4] is positional analysis of network members. Such as the position of a particular member occupies in the network structure, or answering the question of who is the key player in the network. Are there any members who are information rich about all the activities? Are there any brokerage roles that can interact between subgroups or cells? The third aspect [4] is relationship analysis between the network and the people. Relationship analysis of criminal networks gives us the ideas for who is a-friend- of- who in the network or getting recent information about the acquaintances, friends, and participants of criminals. For instance, Krebs [7] analyzed all members in notorious 9/11 attackers network members in great detail and found interesting relationships.

5   Topology of Terrorist Networks

Since there are advantages and disadvantages of specific network topologies, they prefer according to their needs [2,3,6]. They mostly shape themselves in order to maximize their gains but also hide their activities from the enemy [1]. Another important reason for topology changes are the need of trust between network members and cliques. They may seek for immunity and protection from breaches and may not trust others, so they shape the network structure to this end. For instance, Wasserman [6] focused on star shaped networks and triads where star shaped networks are simply concentrate on distribution of power on a single member whereas triads divide power equally to three members in a network. One-man leadership prefer star shaped networks where all the information flows into the leadership and authority is in his hand.

Historical practices of ideologies can also be reflected as inheritance of particular network topology choices. Based on the Rapoport's [12] definition of four waves of terrorism, anarchist wave of terrorism mostly preferred lonely wolf or muti cell-based topologies. Anti-colonial wave of terrorism preferred ethnic groups which were shaped as corporate based or brokerage based terrorist networks. New-left wave of terrorism preferred politburo based terrorist network topologies. Religious wave preferred to shura based or multi cell based terrorist network topologies. For instance, most of the Marxist-Leninist networks historically choose politburo topology. Based on this approach network topologies can quickly be identified based on their ideologies and historical practices. As a result we classified terrorist networks into six topologies; corporate based terrorist networks, central committee (politburo) based terrorist networks, shura based terrorist networks, multi-cell based terrorist networks, and brokerage based terrorist networks, lonely-wolf based terrorist networks.

- Terrorist groups acting with racist / ethnic / minority communities mostly choose corporate based network topology.

- Terrorist groups with an extreme left Marxist ideology choose politburo based network topology.

- Terrorist groups with abused Islamic approach or with middle-east origin choose Shura based network topology.

- International terrorist networks choose multi-cell based network topology no matter what ideology they have.

- Secret organizations or group of individuals who seek to work with terrorist or organized crime networks choose brokerage based network topology.

- Individual activists who decide to use violence as means of environmentalism, animal rights, or in some cases extreme racist/religionists choose lonely wolf based network topology for their terrorist activities.

5.1   Corporate Based Terrorist Networks

Corporate based terrorist networks are evolved in time so that they include armed-subgroup, propaganda-subgroup, finance-subgroup, youth& women-subgroup, and leaders of the network acts like board of directors. These type of networks have a long history and tradition and they converted themselves from a couple of members into multi-functioning enterprise. Every terrorist network wants to reach a political aim but most of the corporate based networks end up either as organized crime network or a political party with a concrete supporter network. The need for sustaining logistics& weapon need of armed subgroup pushes the network to develop their connections in illegal fertile money making activities such as drug dealing and arms smuggling. As similar, the need for increasing the number of members pushes the network to recruit newcomers thereby running propaganda machinery to convince them about the arguments of terrorist ideology. In some cases, they create their television and satellite networks as well as newspapers, magazines, and web sites. To accommodate their supporters and members they start running businesses, camps, and clubs and guest houses. Some members of the network is either killed, prosecuted or put into prison. That is why these members and their families need financial and emotional support from their supporters. So these terrorist organizations create and run charities and hire lawyers to support their members and their families.

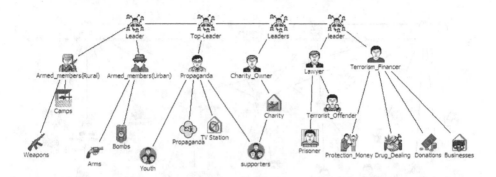

Fig. 1. Conceptual Representation of a Corporate based Terrorist Network

If corporate based terrorist networks cannot produce expected political gains, chains of command might convert the capabilities of the terrorist network into legitimate money making business and armed – members into organized crime networks. If corporate based terrorist networks produce some of expected political gains, then some part of the terrorist network create their own political party to achieve more political gains within target countries' political system. The bigger the network gets the more they tent to corrupt and the more they execute their own members by judging them as corrupted traitors. As similar, a clique in the network can declare war against another clique for various reasons and they start to destroy themselves.

Corporate based terrorist networks don't direly need money or weapon but they need prestige and new recruits to better themselves. In order to disrupt corporate based terrorist networks it is better to publish their corruption and show the leaders of network with their ideological pitfalls. Another way of disrupting corporate based terrorist network is waiting for passing away of first generation leaders and ideologists. Following generation of network members will rule the network out of ideological and moral values which would result criticisms within terrorist networks.

Most of the corporate based terrorist networks aim to own piece of land and declare independent state where they assume the terrorist network to be a government. They also learn politics where the network evolves over time. In some cases, they get support from political groups and government.

5.2   Politburo (Central Committee) Based Terrorist Networks

Many extreme left terrorist groups shape themselves in politburo based networks, where they collected all the authority into the hands of central committee. Each member of the central committee governs one function in particular but all members of the central committee decides all strategy of the terrorist network. When compared to corporate based networks, central committee works less specialized and they are closer to front end members. Most of the members are armed units and they operate within link to all central committee members. Armed units are much more cell based and they have no idea about each other's existence. One of the committee member acts as periodical leader of the organization and called as secretary general. This structure was firstly successfully created in Russian Communist Revelation and therefore adopted by many extreme left-wing Marxist/Leninist terrorist networks.

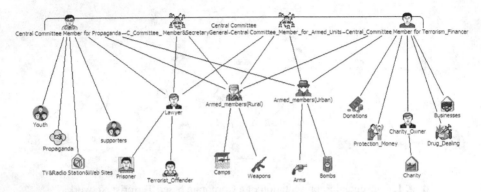

Fig. 2. Conceptual Representation of a Politburo based Terrorist Network

The advantages of politburo structure are secrecy and power are in the hands of central committee so the other members always see each central committee member as equally important and respect all of them. Propaganda and financing is also the business of central committee so they have to decide financial issues and propaganda as well. In case of disclosing a secret within the lower level of cells and sympathizers there is less risk of disruption.

In case of an operation made into an armed unit or other cells, it is less likely to disclose all names of the central committee because less number of members only know more than two central committee members. There are also disadvantages of politburo based terrorist networks. One disadvantage is due to the fact they decide most of the decisions together slowly and there is a lot of discussion goes on between committee members in every detailed issue, they tend to divide and act in fraction of the organization in time because of their different approaches. The second dilemma of politburo based is in case of operation against most of the central committee members together the organization can totally be disrupted. Recovery after such operations can be impossible because other members cannot act as central committee easily simply because they have less capacity to manage and never performed as rulers in the past.

5.3   Shura Based Terrorist Networks

Another topology is Shura based terrorist network topology. *Shura* is an Arabic word meaning is an Arabic word for "consultation". It is encouraged for Muslims to decide their affairs in consultation with those who will be affected by that decision. Shura is a praiseworthy activity, and often used in organizing the affairs of a group. Terrorist networks who are of Asian or middle-eastern origin use this structure. In Shura based networks each member of the Shura governs all functions and they also know every member within the terrorist network. Shura is somewhat similar to Central Committee apart from Shura members are also spiritual leaders of whom manage his own cell. Members of the cell are different, some members are military apparatus, some of them working on ideology and propaganda and some members manage the finance of the network.

Shura members have deep knowledge about the ideology and accepted as authority by the network. Each Shura member manages all resources in micro level and direct members and sympathizers with his own initiative. Since each Shura member is also an expert in ideology, they create their own learning circle in the process of teaching the ideology. They recruit some of the members as armed unit, as finance unit, and propaganda/ideology unit. The advantage of Shura based topology is each member has the potential to govern the whole network. Since they are both spiritual and practical leaders, they deeply affect their members and they always have organizational and spiritual links in closer contact. One member of Shura is accepted as the leader of the network who is also accepted more spiritually superior than other Shura members. The leader must have the virtues and spiritual life otherwise there should be conflict about who will be the leader. Since every Shura member has his own ideological/spiritual circle, they are open to the public and they don't much about secrecy. Therefore, one disadvantage of Shura based network is its vulnerability for intrusion. Dishonest or pragmatist members can easily disclose information in such cases. Shura based networks might be targeted for prosecution but they can easily recover afterwards. Even after each Shura member is prosecuted followers or members who are showing spiritual leadership and virtues can easily be promoted as Shura members.

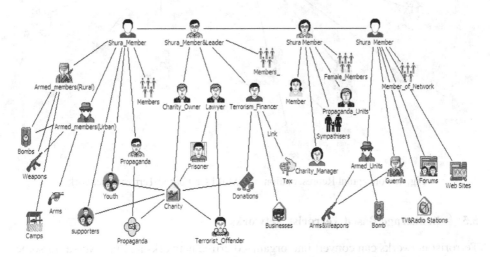

Fig. 3. Conceptual Representation of a Shura based Terrorist Network

5.4   Multi Cell (Bunch of Guys) Based Terrorist Networks

Multicell-based terrorist networks are much similar to star-shaped networks. As Sageman [8, 9] refers them as 'bunch of guys' where each of the members known in a long time within a small world such as peers around a school, kids in the same neighborhood, small number of terrorists met in a training camp. In the center of a cell, there is a strong leader who manages all the business and around him there are other network members. At the beginning, most of the other terrorist networks start their life-cycle as multi cell based networks. In the course of time they convert themselves other topologies. But some of the networks deliberately continue to be multi cell based because of concerns about power and secrecy. International terrorist organizations who operate globally choose to be acting as multi cell based networks. The advantage of multi cell based network is they are always handing power in terms of money, armed-members, and logistics so it is easy to govern the whole network with a high secrecy level. The disadvantage of multi cell based terrorist networks are in case of operation or prosecution, it is impossible to continue the life-cycle of network without strong leadership.

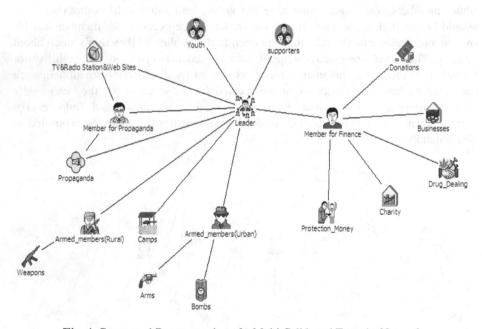

Fig. 4. Conceptual Representation of a Multi Cell based Terrorist Network

5.5   Brokerage Based Terrorist Networks

Terrorist networks can convert into organized crime networks and vice versa. In some cases, legitimate networks such as societies, unions, communities need terrorism and violence to reach their aims. This aim might be destabilizing a country, putting economic in crisis, creating a sense of hatred to a particular community, forcing governments to change their policies. Sometimes legitimate and illegitimate terrorist networks works hand in hand against the common enemy. In such occasions, organizing power who acts as brokerage and invite legitimate networks to work with illegitimate terrorist networks must ensure that this cooperation is legal and fully trusted. On the other side of the network, terrorist networks are also guaranteed that they cooperate with friends who are not enemy according to terrorist networks' ideology. Brokerage involvement is so important that only the organizing power must see the whole picture of business. Secrecy and trust is dealt with brokerage members and other members or institutions never see the other side of the business. Brokerage based networks do exist in many countries all over the world but it is too difficult to disclose all cells and members because they are never aware of themselves as acting cooperation with terrorists. For terrorists also such a relationship is just an order of the leader and other members cannot entirely understand why they act with other legitimate networks. Advantage of brokerage based networks is secrecy so that even the members do not realize that they are a part of terrorist activity. Disadvantage of brokerage based network is the whole network is bounded by one or two brokerage member roles and while they are removed from the system it can take so long to find new brokerage members like older ones who are fully trusted by the leaders of the cells.

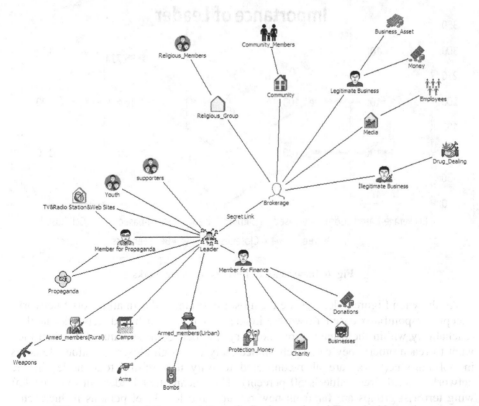

Fig. 5. Conceptual Representation of a Brokerage based Terrorist Network

5.6   Lonely Wolf Based Terrorist Networks

Lonely wolf based terrorist networks are consisting of one member who does everything. In terms of logistics, finance, arms and explosives lonely wolves arrange every piece of detail while they use violence and terror for its ideology. In case of getting help from others, none of the supporter of a lonely wolf can realize that there is an act of terrorism and they might be supporting a person who is a terrorist. Advantage of lonely wolf based network is high secrecy and power in hands of one single member whereas the disadvantage is always the possibility of extinction after the prosecution of lonely wolf terrorist.

6   Importance of Leader, Financier, Armed and Propaganda Units

In order to identify the importance of functions in terrorist networks we used positional analysis of important nodes in the networks. For positional analysis of leader, financier, propaganda and armed units three social network analysis metrics are used as normalized in percentage values: degree, closeness and betweenness.

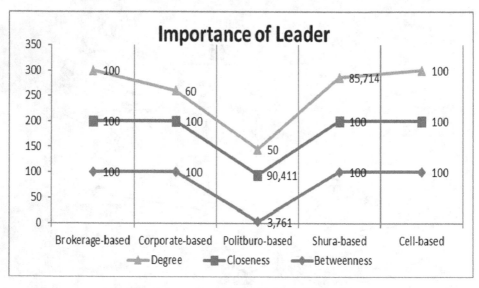

Fig. 6. Importance of Leadership in Networks

As shown in Figure 6, leaders are the most important nodes in all terrorist networks except for politburo based networks. Leaders in politburo based networks neglect reachability within the organization (i.e. very low betweenness value) but when they want to reach others they can reach immediately (i.e. high closeness value). Leaders in politburo networks are also connected to only half of the total nodes in the networks (i.e. degree value is 50 percent). This means that leaders in extreme left wing terrorist groups are far from new recruits and low level persons in hierarchy. Financing terrorism is very important for terrorist networks.

All terrorist groups need finance to carry out their activities. When we look at the importance of financiers of terrorism (Figure 7), financiers are even more important than leaders in politburo based networks. This means that for extreme-left wing terrorist groups, financier of activities is hidden leader. Financiers are also as important as leaders in corporate based networks, brokerage based networks and multi cell-based networks. Corporate based networks are more focused on money matters so it is no surprise to see the person who manages financial matters are of higher importance. Financiers are less important in Shura based networks.

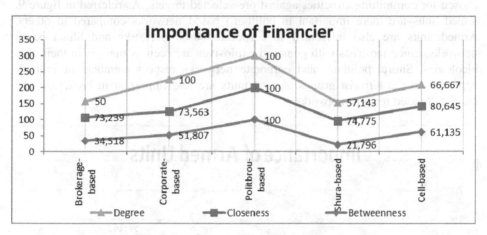

Fig. 7. Importance of Financier in Networks

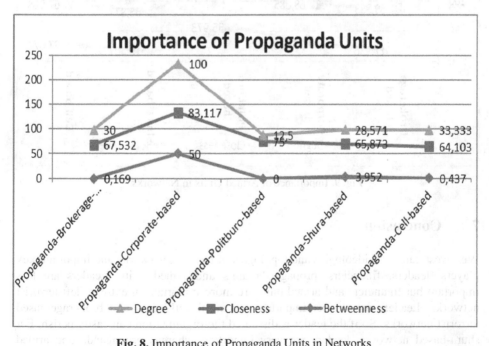

Fig. 8. Importance of Propaganda Units in Networks

Propaganda units can be radio and television companies, site admins, preachers, ideologists, authors, journalists and publisher houses. Propaganda fuels the required ideological approach for the carried out activities. As shown in figure 8, propaganda units in corporate based network are the most influential. Propaganda units in Multi cell based, Shura based and brokerage based networks are connected to one thirds of all members but propaganda units in extreme left-wing politburo based networks are not influential.

Armed units are military apparatus of terrorist networks. They are recruited and chosen for committing atrocities against pre-selected targets. As referred in figure 9, armed units are more important in politburo based networks compared to others. Armed units are also influential in corporate based networks and Shura based networks. Since terrorists with guns and explosives are seen as martyrs in their own ideologies, Shura, politburo and corporate networks respect members of military apparatus in the terrorist groups. Armed units are less important in brokerage and multi cell based terrorist networks.

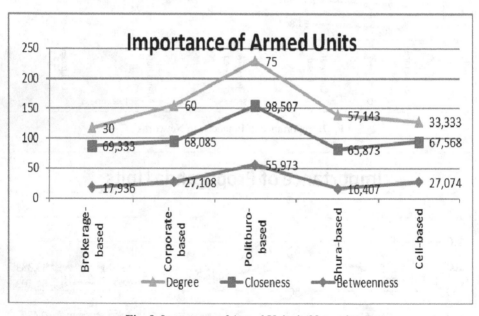

Fig. 9. Importance of Armed Units in Networks

7   Conclusion

We investigated the ideologies and topologies of terrorist networks and important key players: leaders, financiers, propaganda units and armed units. Leaders are less important but financiers and armed units are more important in extreme left terrorist networks. Leaders are utmost important for multi cell-based and brokerage-based terrorist networks. So if the leader is disrupted the organization can easily perish. For shura-based networks, leaders are very important whereas propaganda and armed units are less important. In parallel to common belief, financiers are more important than leaders and propaganda units are more important than armed units.

References

[1] Morselli, C.: Inside Criminal Networks. Springer Science+Business Media LLC, New York (2009)

[2] Scott, J.: Social Network Analysis: A Handbook. Sage Publications (2000)

[3] Finckenauer, J.O.: Mafia and Organized Crime A Beginner's Guide, p. 11. Oneworld Publication, Oxford (2007)

[4] Everett, M.G., Borgatti, S.P.: The Centrality of Groups and Classes. Journal of Mathematical Sociology (1999)

[5] Klerks, P.: The Network Paradigm Applied to Criminal Organisations: Theoretical nitpicking or a relevant doctrine for investigators? Recent developments in the Netherlands. Connections 24(3), 53–65 (2001)

[6] Wasserman, S., Faust, K.: Social Network Analysis Methods and Applications. Cambridge University Press, Cambridge (1994)

[7] Krebs, V.: Mapping networks of terrorist cells. Connections 24(3), 43–52 (2002)

[8] Sageman, M.: Understanding Terror Networks. University of Pennsylvania Press, Philadelphia (2004)

[9] Sageman, M.: Leaderless jihad: Terror networks in the twenty-first century. University of Pennsylvania Press (2011)

[10] Memon, N., Larsen, H.L.: Practical Approaches for Analysis, Visualization and Destabilizing Terrorist Networks. In: Proceedings of the First International Conference on Availability, Reliability and Security, pp. 906–913. IEEE Computer Society (2006)

[11] Carley, K., Dombrowski, M.: Destabilizing Dynamic Covert Networks, 8th International Command and Control Research and Technology, National Defence War College, Washington, DC (2003)

[12] Rapoport, D.C.: The four waves of modern terrorism. In: Attacking Terrorism: Elements of a Grand Strategy, pp. 46–73 (2004)

[13] van der Hulst, R.C.: Introduction to Social Network Analysis as an investigative tool. Trends in Organised Crime (12), 101–121 (2009)

Topic Participation Algorithm for Social Search Engine Based on Facebook Dataset

Hao-Ren Yao and I-Hsien Ting

Department of Information Management
National University of Kaohsiung, Taiwan
ghjk3695@gmail.com, iting@nuk.edu.tw

Abstract. With the rapid growth of users in social networking websites, large amount of data are aggregated. Users are tending to find information through their friends on social network such as Facebook, and this behavior leads to a new search paradigm called social search. However, the traditional search engine like Google cannot handle this kind of search. The data cannot be indexed because of the membership privacy setting and social network relationships. Under this situation, it is harder and harder for users to search information related to their social network.

In this paper, we therefore proposed a system architecture which can deal with this issue and using the data from Facebook as example. An algorithm is also proposed which is the core technique of the system which is called Topic Participation Algorithm (TPA). Furthermore, we will propose a novel implemented social search engine which is developed based on the concept of social network analysis, data mining techniques and searching techniques.

Keywords: Search Engine, Social Search, Social Network Analysis, Data Mining, Web Mining.

1 Introduction

With the rapid growth of the internet and the concept of Web2.0, the web has becoming a popular communication platform. Due to the advances of Web2.0 technology, social network sites like Facebook and Twitter make it possible for users to share their information instantaneously. The interactive nature gives users the chance to communicate and describe their daily life experience in a unique way. Take Facebook as an example, users can publish variety of data on their timeline including posts, photos, videos, and check-ins…, etc. The tagging scheme deeply combines the relationship between object and their friends. The like function can help users to express their attention on certain content such as pages, link sharing, or topics. In this situation, it is easy to figure out the users' preference by their activities on Facebook. [4]

In order to analyze the data on Facebook, social network analysis (SNA) [9] is an important research field due to it focuses on the analysis of social data and social relation. Social network is used to describe the relationship between people and also display what the role of an actor plays in a group. In particularly, the nature of

L.S.-L. Wang et al. (Eds.): MISNC 2014, CCIS 473, pp. 158–170, 2014.
© Springer-Verlag Berlin Heidelberg 2014

network can easily detect a group of people who share similar interests by their interaction such as comments, likes, or tags. [8]

The importance of social search has been considered in Academia and Industrial due to the fast development of social network sites like Facebook and Twitter. Social relationship has been considered as the criteria for ranking the search results by some researchers.

The reason is in past years, traditional search engine rank the search result by keyword similarity or google page rank. These scheme has nothing to do with social relationship since no social interaction has been incorporated. [6][11][5]

In this case, our purpose would like to implement the social ranking approach that proposed by us in this paper called topic participation algorithm by using SNA to analyze the relationship on social network graph built on top of the user query. This can further lead to the discovery of not only explicit but also implicit users who have strong connection to the query. And the magnitude of these connections can be used as a basis of so-called *"Topic Participation Algorithm"*. We then build up a system which implement this approach to evaluate how the social search runs in the real environment and to understand the cost, the requirement, and the presentation.

The rest of the paper is organized as following: In section 2, literatures that related to cloud computing, clustering and social search are reviewed as well as to provide some background of the paper. Then, the system architecture is introduced in section 3 and the explanation of each component in the architecture and the algorithm. In section 4, screenshots of the main functions of the system are presented to show the ability of the social search system. The paper is finally concluded in section 5.

2   Literature Review

2.1   Cloud Computing

Due to the definition from NIST(National Institute of Standards and Technology) of cloud computing, it is a model implementing the ubiquitous, convenient, and on-demand network access to a computing resources which is managed by the service provider. The computing resources typically are a collection of hardware and software which support the implementation of various service model. The infrastructure of the cloud includes broad network access, distributed storage, and a pool of computing server[10].

The service model includes SaaS(Software as a Service), PaaS(Platform as a Service), and IaaS(Infrastructure as a Service). The SaaS is to provide consumer application software usually based on the web by the provider's cloud infrastructure without knowing any detail about the software. The PaaS makes consumer possible to deploy their own application on the cloud infrastructure by the tools, or programming language provided by the service provider without knowing any detail about the platform. Finally, the IaaS gives consumer the usage on the cloud infrastructure including network, computing servers, or storage without managing or controlling the underlying infrastructure on their own[1].

Since the IaaS service model let consumer run their application without the need of server pools, lots of big data analysis task are migrated to cloud especially the development of Hadoop. Hadoop is an open-source software framework for large scale data storage and processing. It includes Hadoop Distributed File System (HDFS) and Map-Reduce computation framework[13]. The HDFS is a distributed file-system that stores data on a cluster of servers. Map-Reduce is a programming model developed by Google software engineer to provide an easy way to process large dataset in parallel and distributed manner.

In our proposed system, Hadoop has been implemented to deal with our large Facebook dataset including data indexing, clustering, and user interaction analysis.

2.2 Cluster Analysis

Cluster analysis or simply called clustering is the process of grouping similar object into a group or cluster. The dissimilar object will lie in different cluster. K-means clustering on the other hand is a type of centroid based clustering algorithm. By the name imply, it partitions a dataset into k clusters. The process is iteratively assign the data object to the closest centroid by user defined distance measure formula. The main drawback of K-means clustering is unpredictable k value[7]. If the dataset is at large scale, the value of k will greatly affects the clustering result. Since our proposed system need to preprocess large amount of data, we develop a clustering method by implementing K-Nearest Neighbors algorithm which is a kind of classification algorithm[3].

2.3 Social Search

What is social search? In traditional searching techniques, keyword similarity is the most used criteria for ranking the results. Some other techniques are applied the ranking method that developed by Google, which is known as PageRank [12] with the concept of keyword similarity and the importance of websites. Therefore, the term social search means the social relationship will be took in to account when ranking the search results[5].

In the past, the researches which add the social relationship are few. Nowadays, there are tags [15] or "like" [14] base on the social search. Most users are connection with others in social networks. When the interactive is frequently, it will become lots of data. Others can get more data though searching [16]. Hence, the relationship of social networks is important for social search [2]. Thus, in the paper, we will focusing on the implementation of social search by using the real world data from Facebook.

3   System Architecture and Algorithm

According to the research background and motivation, we have designed the system architecture to address the related issues. The proposed system architecture is presented in figure 1.

In figure 1, the system can be divided into three parts, the first part is the fetching, indexing, and storage of users' Facebook data. The second part is to conduct the data preprocessing including data clustering and relationship aggregation, and the third part is the social search algorithm that used in our search engine. The process of the three parts will be introduced in detail below.

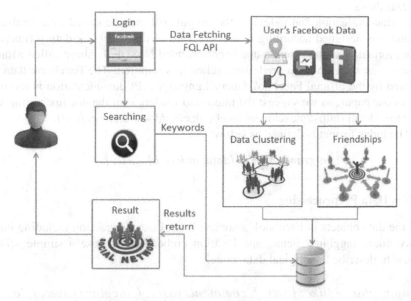

Fig. 1. The system architecture

3.1   Fetching, Indexing, and Storage of Users' Facebook Data

(1) Data Collection

When the user login into our system at first time, we will request the authentication on user's data through Facebook API. Then, the system begins to start the fetching process by using Facebook Query Language API. The data we fetch includes posts, photos, and videos, links which is shared via user, check-ins, pages, and personal interests. All the comments, likes, and tags associate to each kind of the data will be fetched as well.

(2) Indexing

The entire text field in the data we have already collected will be transferred into inverted index which makes query processing more efficient. However, our primary users are located in Taiwan; the text segmentation process is hard since there's no word boundary between Chinese sentences.

Due to the fact that no word boundary exists in Chinese, we use n-gram model to segment the Chinese text. The n-gram model is a contiguous sequence of n characters from a given sentence. All the tokens in our n-gram model are have the same weight. We give an example as below:

Sentence: ABCDEF
1-gram: A B C D E F

2-gram(bigram): AB BC CD DE EF
3-gram(trigram): ABC BCD CDE DEF

(3) Database
After data collection and indexing, the output will then be stored in a database. The database is designed according to the structure of the Facebook data format and the index format. We implement a document-oriented NoSQL database called Mongo DB to store our data since no predefined schema is required. The Facebook data format defined by the official Facebook Query Language API documentation is too many to put in our paper, so we suggest the interested readers visit the documentation website for more detail (https://developers.facebook.com/docs/reference/fql/).

The index format is defined as below:

n-gram token -> [data_object_id...,etc.]

3.2   Data Preprocessing

All the data objects in Facebook associate lots of user interaction including comment reply, likes, tagging scheme, and location embedded. We use a simple schema as below to describe the original data model:

sample_post -> {like: [user,], comment: [user,], tagging: [user,], location: [place,]}

In order to calculate the user interaction, we need to scan through all the data field to check what interaction that user has. Since our search algorithm calculates the user interaction in real time, it is necessary to improve the search speed. We convert the original data model into the following format:

sample_post -> {user:[like, comment, tagging,], place:[]location, }

In this model, there's only one retrieval per user. As a result, no full scan is needed. In addition to the interaction aggregation, we do the data clustering on pages and places so the similar pages or places will be in together. Therefore, we can achieve query expansion and search speed improvement.

3.3   Social Search Algorithm (Topic Participation Algorithm)

Due to the observation on users' behavior on Facebook, we discover that two possibilities exist for the reason why a certain user give a comment reply or like on a certain post:

(1) Based on Friendship Like Conversations
Users' tend to interact more frequently base on their friendship focus the interaction on a single person. Lots of comments reply have nothing to do with the post content

but is a chat conversations. The other extreme case is the people in couple relationship tend to gives like to every single post published by the other.

(2) Based on the Interest of the Content
Users' tend to reply their opinion on specific topic focus not only the content they publish but also the interaction on various person but the similar post content. In this case, we consider this kind of interaction can reveal a person's interest and preference on specific topic.

By the above two cases, we propose a social search algorithm called Topic Participation Algorithm (TPA). TPA considers the degree of participation of users in a specific topic. The topic is generated by the user's query which is a set of keywords. In addition, there are two types of participation which is direct and indirect. The direct participation is a role of user who has one of the following activities:

A. The author of posts, photos, check-ins, or links.
B. The fan who poses like on a page.
C. The person who is tagged in photos or check-ins.

The degree of direct participation is the frequency count of these activities associated to a user. The indirect participation is a role of user who has one of the following activities:

A. The person who gives comment reply to a post.
B. The person who gives a like to a post.

The degree of indirect participation is the frequency count of these activities associated to a user. We now generate the TPA scoring formula:

After discussing the related information, we now focus on TPA algorithm which is performed as below:

    A. Aggregate all the data objects including posts, photos, check-ins, links, and pages which the content matches the user's query, and form a single node called topic.

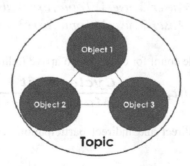

Fig. 2. TPA algorithm step 1

B. Connect each user who has direct participation to the topic node and count the frequency to arrive direct participation score.

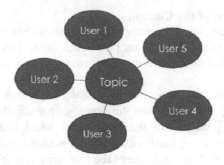

Fig. 3. TPA algorithm step 2

C. Connect each user who has indirect participation to the author of the object in the topic node where this participation occurred.

Fig. 4. TPA algorithm step 3

D. For each user, count the number of unique circle that the user belongs in this graph. We give an example cycle of user3 existed on the above figure:

$$Cycle1=\{user'3, user1, Topic, user4, user'3\}$$
$$Cycle2=\{user'3, user1, Topic, user3, user'3\}$$

E. Normalize the cycle count for each user to arrive indirect participation score.

$$Cyclecount_{user} = \frac{Cyclecount - Mincount}{Maxcount - Mincount}$$

F. Multiplying both direct and indirect participation score to get the final user score.

$$Score = Direct\ Interaction \cdot Indirect\ Interaction$$

4   Implementation

In this paper, we have already collected the data based on snowball sampling from Facebook API through user's authentication. These include 6,000 users, 150,000 photos, 1 million posts, 80,000 pages, and 36,000 places.

In order to submit and handle user query, real time processing is necessary. We build a Hadoop cluster which comprises 9 Mac mini servers to do the data indexing, clustering, and interaction aggregation based on map-reduce computation paradigm. In this section, we will demonstrate a fully-functional prototype to show the whole process and use coffee as keyword.

Fig. 5. The system interface

In figure 5, this is the first page when user's Facebook data has been fetched, and indexing has been accomplished. The search bar show in figure 5 can receive multiple keyword and support boolean query. By default, and operation has been set. Below the search-bar, the category list is introduced dividing our search result into these six categories which respectively are friend, post, photo, video, place, and page.

In figure 6, after receiving coffee keyword, TPA starts the analysis the relevance between user's friend and the keyword. The result will then show in friend section. The ranking scheme is from left to right and up to down which means the high relevance to low relevance. In our coffee example, the result tells us who likes coffee or has a strong connection to coffee in user's friend list. The score used in friend section will contribute to the remaining section.

In figure 7, the related posts published by friends in friend section will be gathered and showed here. The post includes checkin, status, photo, and video. The ranking of these posts is determined by keyword similarity and the social relationship score from TPA which is used in friend section. User can click each post and link to its original Facebook page to see the detail about comment replies and likes.

Fig. 6. The search result-Friends

Fig. 7. The search result-Posts

Fig. 8. The search result-Photos

In figure 8, we filter out the photos in the previous post section. The separation of post and photo can help user focus on the content in images because image can express more information such as facial emotion expression, background view, or others that text can not fully described.

In figure 9, videos will also be filtered out from posts. In addition to the function like photo section, we add a related user list showed in figure 9. We observed the same or similar video will repeatedly show up in user's timeline. We aggregate the user who has high score from TPA to these videos and list them in here. Our user can easily determine friends who have strong preference.

Fig. 9. The search result-Videos

Fig. 10. Related user list

In figure 10, all the location, in our coffee example, associated to coffee including coffee shops, cafes, or whatsoever visited by friends in friend section will be listed here with related check-ins. Location is another valuable information in Facebook. By using location information, our user can quickly realize where is the candidate place they can visit by looking into the detail of the check-in associated to each place from user's friends. This result can also be the source of place recommendation engine

Fig. 11. The search result-Check-in

In our last section, page section showed in figure 11 lists the page liked by friends in friend section. Sometimes page contains additional information like sales promotion, production introduction, and opinion from page fans. Our user can further use these pages to gain more detail by clicking the page icon in the result and linking to its original Facebook page.

Fig. 12. The search result-Pages

5 Conclusion and Future Works

In this paper, we have proposed the architecture of social search engine based on the data from Facebook. The topic participation algorithm (TPA) is conducted according to the user's query and the social interaction among the expected topic. The essence of this part is to discover the indirect participation of a user to a specific topic. A case study will be included to present how the search engine and the TPA work based on real data collected from Facebook. However, there is no explicit way to evaluate the quality of the search result since the data is generated within users' social network. Different user's social network has different structure and context. Due to this fact, our future work will focus on how to evaluate the social search process and further make the improvement when necessary.

References

1. Armbrust, M., et al.: A View of Cloud Computing. Commun. ACM 53(4), 50–58 (2010)
2. Baer, M.: The strength-of-weak-ties perspective on creativity: a comprehensive examination and extension. J. Appl. Psychol. 95(3), 592–601 (2010)
3. Jain, A.K., et al.: Data Clustering: A Review. ACM Comput. Surv. 31(3), 264–323 (1999)
4. Lewis, K., et al.: Tastes, ties, and time: A new social network dataset using Facebook.com. Soc. Networks (2008)
5. Lu, H.-H., et al.: A Novel Search Engine Based on Social Relationships in Online Social Networking Website (2012)
6. Manoj, M., Jacob, E.: Information retrieval on Internet using meta-search engines: A review, pp. 739–746 (October 2008)
7. Nizamani, S., et al.: CCM: A Text Classification Model by Clustering. In: 2011 Int. Conf. Adv. Soc. Networks Anal. Min., pp. 461–467 (2011)
8. Russell, M.A.: Mining the Social Web: Data Mining Facebook, Twitter, LinkedIn, Google+, GitHub, and More. O'Reilly Media (2013)
9. Scott, J.: Social Network Analysis: A Handbook. SAGE Publication, London (2000)

10. Ting, I.-H., et al.: Constructing a Cloud Computing Based Social Networks Data Warehousing and Analyzing System. In: 2011 Int. Conf. Adv. Soc. Networks Anal. Min., pp. 735–740 (2011)
11. Vieira, M.V., et al.: Ef cient Search Ranking in Social Networks. Search, 563–572
12. Wang, C.-S., et al.: Taiwan Academic Network Discussion via Social Networks Analysis Perspective. In: 2011 Int. Conf. Adv. Soc. Networks Anal. Min., pp. 685–689 (2011)
13. Xu, Z., et al.: Beyond Hadoop: Recent Directions in Data Computing for Internet Services. Int. J. Cloud Appl. Comput. 1(1), 45–61 (2011)
14. Yang, S.-H., et al.: Like Like Alike: Joint Friendship and Interest Propagation in Social Networks. In: Proceedings of the 20th International Conference on World Wide Web, pp. 537–546. ACM, New York (2011)
15. Zanardi, V., Capra, L.: Social Ranking: Uncovering Relevant Content Using Tag-based Recommender Systems. In: Proceedings of the 2008 ACM Conference on Recommender Systems, pp. 51–58. ACM, New York (2008)
16. Zinoviev, D., Duong, V.: Toward Understanding Friendship in Online Social Networks. CoRR. abs/0902.4 (2009)

Utility Knowledge Fusion in a Multi-site Environment

Guo-Cheng Lan[1], Tzung-Pei Hong[2,3,*], Yu-Chieh Tseng[3], and Shyue-Liang Wang[4]

[1] Department of Computer Science and Information Engineering,
National Cheng-Kung University, Tainan, 701, Taiwan
[2] Department of Computer Science and Information Engineering,
National University of Kaohsiung, Kaohsiung, 811, Taiwan
[3] Department of Computer Science and Engineering,
National Sun Yat-Sen University, Kaohsiung, 804, Taiwan
[4] Department of Information Management
National University of Kaohsiung, Kaohsiung 811, Taiwan
rrfoheiay@gmail.com, {tphong,slwang}@nuk.edu.tw,
jeff790925@hotmail.com

Abstract. According to multi-site relationship, this work presents a new issue named online multi-site utility mining, which considers not only quantities and profits of items in transactions but also online mining in a multi-site environment, to effectively address the distributed utility mining in multiple sites. In addition, an effective online framework, namely *TP-OMU* (Three-Phase Online Multi-site Utility mining algorithm), is proposed for coping with this problem, and the predicting strategy is designed to reduce the number of unpromising candidates by their utility upper-bounds in mining. Finally, the experimental results show *TP-OMU* has good efficiency in comparison with the traditional two-phase utility mining approach.

Keywords: Data mining; multi-sites; large-scale data; utility mining; online mining.

1 Introduction

In data mining, utility mining techniques, which considered both the quantities and profits of items in transactions to evaluate the actual utilities of the items, have been widely applied to various real applications, such as supermarket data, mobile data, etc. However, most of the existing studies related to utility mining [5][7][10] cannot be applied to effectively discover high utility patterns from multiple data sources in a multi-site environment under a user's querying condition. Intuitively, the problem of utility mining in such environment is more complex than that of traditional utility mining and association-rule mining in a single database environment. To efficiently provide information satisfying the querying conditions in real time, designing an effective online utility mining approach in a multi-site environment is thus an important issue.

---

* Corresponding author.

L.S.-L. Wang et al. (Eds.): MISNC 2014, CCIS 473, pp. 171–178, 2014.

This work thus considers multi-site relationship to present a new research issue named online multi-site utility mining, which allows users to quickly find the high utility patterns satisfying the querying conditions in a multi-site environment with large-scale and distributed data. In addition, a Three-Phase Online Multi-site Utility mining algorithm (abbreviated as *TP-OMU*) is developed to achieve the goal. For the proposed *TP-OMU* algorithm, an efficient and effective strategy is designed to tighten upper-bounds of utilities for candidate itemsets. Hence, many unpromising candidate itemsets can be pruned early before the data scan is executed. Finally, the experimental results show the proposed approach has good performance in execution efficiency.

The remaining parts of this paper are organized as follows. Some related studies are reviewed in Section 2, and the problem and some definitions are given in Section 3. The details of the proposed *TP-OMU* approach with consideration of multi-sites environment are described in Section 4. Finally, the experimental results and the conclusions are provided in Sections 5 and 6, respectively.

2   Review of Related Works

The main propose of data mining [1][2][6][9] is to extract useful information in various types of data, such as biomedical data, multimedia data, etc. To address this problem, Agrawal *et al.* first proposed several mining algorithms to find association rules from transaction databases [1][2][6][9]. In real applications, however, a transaction in a database usually involves quantities and profits of items. Due to this reason, association-rule mining techniques are insufficient to be used to cope with such data [1][2][6][9]. For example, diamond may not be a high-frequency product, but it may be a high-profit one when compared to food and drink in a transaction database. To address this, Yao *et al.* proposed a utility function [10], which considered not only the quantities of the items but also their individual profits in transactions, to find itemsets with high-utility from a database. However, utility mining lacked the downward-closure property in association-rule mining [10] so that it is much harder than the problem of association-rule mining. To effectively reduce search space in mining, Liu *et al.* proposed a two-phase (*TP*) utility mining algorithm to achieve this goal by adopting the downward-closure property, and this model with the property was named as transaction-utility upper-bound (*TUUB*) model [7]. The main principle of the model is that the summation of utilities of all the items in a transaction is regarded as the upper bound of any subset in that transaction. Afterward, most of existing approaches were based on the *TP* algorithm to cope with various applications of utility mining, such as on-shelf utility mining [5], and so on.

However, the traditional mining technology cannot be applied to extract useful rules or patterns from multiple data sources in a multi-site environment. Hidber presented [3] a research issue named online association rule mining (abbreviated as *OARM*) to quickly provide rules or patterns for users by using required sets of patterns without re-processing the entire database whenever user-specified thresholds are changed. Under the online association rule mining framework, users are able to dynamically adjust thresholds according to intermediate results. Wang *et al.* [8] then proposed flexible online association rule mining based on multidimensional pattern relations to effectively conduct the mining task by using a table called the multidimensional pattern relation. Although the studies [3][8] could be applied to find association rules, they did not consider the quantities and profits of items in transactions. To provide more useful information for users, it is quite important to develop an effective online framework for coping with the problem of utility mining in multi-site environment.

Based on the above reasons, this motivates our attempt to present a new research issue named online multi-site utility mining and to develop an effective approach for finding such patterns in a multi-site environment with the large-scale data problem.

3   Problem Statement and Definitions

In this study, $I = \{i_1, i_2, ..., i_n\}$ is a set of items that may appear in the transactions, and an itemset X is a subset of the items, where $X \subseteq I$. If $| X | = r$, the X is called an r-itemset. In addition, a transaction ($Trans$) consists of a set items purchased with their quantities. A database D is then composed of a set of transactions. That is, $D = \{Trans_1, Trans_2, ..., Trans_y, ..., Trans_z\}$, where $Trans_y$ is the y-th transaction in D.

Based on Yao et al.'s utility function, the utility u_{yi} of an item i in $Trans_y$ is the external utility s_i multiplied by the quantity q_{zi} of i in $Trans_y$, and the utility u_{yX} of an itemset X in $Trans_y$ is the summation of the utilities of all items in X in $Trans_y$. The actual utility ratio aur_X of X is the summation of the utilities of X in the transactions including X of D over the summation of the transaction utilities of all transactions D. Finally, let λ be the predefined minimum utility threshold. An itemset X is called a high utility itemset (abbreviated as HU) if $au_X \geq \lambda$.

Next, the main concept of the $TUUB$ model [7] is that the transaction utility of a transaction is used as the upper-bound of any subsets in that transaction, and the transaction-utility upper-bound $tuubr_X$ of an itemset X in D is the summation of the transaction utilities of the transactions including X in D over the summation of transaction utilities of all transactions in D. If $tuubr_X \geq \lambda$, the itemset X is called a high transaction-utility upper-bound itemset ($HTUUB$).

With the $TUUB$ model, the traditional utility mining can be divided into two phases, (1) finding high transaction-utility upper-bound itemsets and (2) finding high utility itemsets, and the procedures are explained below. In the first phase, all possible itemsets with high transaction-utility upper-bound satisfying the minimum utility threshold in a database were found. In the second phase, one additional data scan was performed to find the actual utility of each itemset from the set of high transaction-utility upper-bound itemsets ($HTUUBs$). Finally, all HUs with actual utilities larger than or equal to the minimum utility threshold are outputted. In our study, the set of transactions in a time period can be seen as a single database in traditional utility mining. Accordingly, all HUs within each time period in each store can be found by using the $TUUB$ model.

Based on the above definitions, assume there is a centralized company with multiple sub-stores. For example, assume there is a centralized company with the three stores, respectively named Taipei, Tainan and Kaohsiung, and also assume the process of mining high utility itemsets for the three stores has been done within three time periods, 2013/10, 2013/11 and 2013/12, as shown in Table 1. In Table 1, each record consists of seven features, including (1) record identification, (2) region, (3) branch (also called site), (4) time period, (5) total utility in a time period, (6) number of patterns, and (7) the set of the patterns with their actual utility ratio. In addition, assume there are four items sold in the three stores, respectively denoted as A to D.

Table 1. The information of high utility itemsets in each sub-store for this example

ID	Region	Branch	Time	Total_Utility	No_Itemsets	Pattern_sets (itemset, utility)
1	Northern	Taipei	2013/10	2000	1	(C, 5%)
2	Northern	Taipei	2013/11	3000	2	(A, 4%), (C, 2%)
3	Northern	Taipei	2013/12	2600	2	(A, 3%), (C, 5%)
4	Southern	Tainan	2013/10	4300	3	(A, 4%), (C, 4%), (AC, 3%)
5	Southern	Tainan	2013/11	5000	2	(A, 5%), (D, 2%)
6	Southern	Tainan	2013/12	5000	3	(A, 4%), (B, 4%), (AB, 3%)
7	Southern	Kaohsiung	2013/10	3750	3	(A, 10%), (B, 2%), (C, 4%)
8	Southern	Kaohsiung	2013/11	4500	3	(B, 3%), (C, 4%), (D, 2%)
9	Southern	Kaohsiung	2013/12	4500	2	(A, 4%), (B, 4%)

Formally, $S = \{s_1, s_2, ..., s_w\}$ is a set of sites (or called stores), where s_w denotes the w-th site in a company with multi-site, S. $P = \{p_1, p_2, ..., p_j, ..., p_n\}$ is a set of mutually disjoint time periods, where p_j denotes the j-th time period in the whole set of periods, P. A context attribute CX for a site s_w (a store) is the existing information of s_w. A content attribute CN for a site s_w (a store) is the mining information for s_w. An extended multidimensional pattern relation $empr_j\{ID_j, CX_{j1}, CX_{j2}, ..., CX_{jn}, CN_{j1}, CN_{j2}, ..., CN_{jn}\}$ is composed of several context attributes and content attributes, where ID_j, CX_{jn} and CN_{jn} are the j-th identification record, the n-th context attribute of the j-th record, and the n-th content attribute of the j-th record, respectively. For example, in Table 1, $empr_1 = \{1, \text{Northern, Taipei}, 2013/10, 2000, 1, (C, 5\%)\}$. An extended multidimensional pattern relation schema $EMPR$ is composed of a set of extended multidimensional pattern relations. That is, $EMPR = \{empr_1, empr_2, ..., empr_l, ..., empr_z\}$, where $empr_l$ is the l-th extended multidimensional pattern relation in $EMPR$.

$Q = \{sp_1, sp_2, ..., sp_z\}$ is a set of tuples, where sp_z denotes the specified time periods of the z-th site. For example, assume the user-specified query is "[Taipei: 2013/10; Tainan: 2013/10; Kaohsiung: 2013/10]", and the query represents that the user wants to know what the online multi-site high utility itemsets are within the union of the time periods in the three sites.

The online multi-site actual utility ratio $omaur_X$ of an itemset X is the summation of the actual utilities of X in the relations including X in the extended multidimensional pattern relation schema satisfying the user-specified query ($EMPR_q$) over the summation of the total utilities of all relations in $EMPR_q$. Take item A as an example. The relevant information of item A in the time periods in the three sites is showed in Table 2.

Then, the online multi-site actual utility ratio $omaur_{\{A\}}$ of item A can be calculated as $(2000*5\% + 4300*4\% + 3750*10\%) / (2000 + 4300 + 3750)$, which is 6.44%. Finally, Let λ_q be a predefined querying minimum utility threshold. X is called an online multi-site high utility itemset $(OMHU)$ if $maur_X \geqq \lambda_q$. For example, in Table 2, if $\lambda_q = 4\%$, $\{A\}$ is an $OMHU$ in $EMPR$ since $maur_{\{A\}} = 6.44\% \geqq \lambda_q$.

Table 2. The relevant information of item A in each sub-store for this example

Branch	Time	Total_Utility	Pattern_sets (itemset, utility)
Taipei	2013/10	2000	(A, 5%)
Tainan	2013/10	4300	(A, 4%)
Kaohsiung	2013/10	3750	(A, 10%)

4   The Proposed Mining Algorithm

In the section, the proposed TP-OMU algorithm consists of the three phases: (1) generation of possible online multi-site high utility itemsets, (2) reduction of unnecessary online multi-site high utility itemsets, and (3) discovery of online multi-site high utility itemsets. The detailed process of the TP-OMU is then stated below.

INPUT: An extended multidimensional pattern relation $EMPR$ based on an initial minimum utility threshold λ, a utility table with a set of all items, each with a profit value, a mining request q with a set of contexts CX_q and the querying minimum utility threshold λ_q.
OUTPUT: A set of online multi-site high utility itemsets satisfying the request q.

Phase 1: Generation of Possible Online Multi-Site High Utility Itemsets
STEP 1: Collect the records which satisfy mining request q in $EMPR$, and denote the records as $EMPR_q$.
STEP 2: Make a union set for the itemsets in $EMPR_q$, and denote the union set as US_q.

Phase 2: Reduction of Unnecessary Online Multi-Site High Utility Itemsets
STEP 3: For each itemset X in the union set US_q, do the following substeps:
  (a) Calculate the sum $u_X^{appearing}$ of all known utilities of X in US_q. That is,

$$u_X^{appearing} = \sum_{X \in HU_j \wedge p_j \in EMPRq} au_{jX} ,$$

  where au_{jX} is the actual utility of X within the j-th time period p_j.
  (b) Calculate the sum u_X^{ub} of the upper-bounds of all utilities of X. That is,

$$u_X^{ub} = \sum_{X \notin HU_j \wedge p_j \in EMPRq} (TTU_j * \lambda - 1),$$

where TTU_j is the sum of transaction utilities of all transactions within the j-th time period p_j, and λ is the original minimum utility threshold in p_j.

(c)   Calculate the utility upper-bound ratio of X. That is,

$$UUBR_X = \frac{u_X^{appearing} + u_X^{ub}}{\sum_{p_j \in EMPRq} TTU_j}.$$

(d)   If the utility upper-bound ratio $UUBR_X^{UB}$ of X is less than the λ_q, then discard X from the set US_q; otherwise, keep X in the set US_q.

Phase 3: Discovery of Online Multi-Site High Utility Itemsets
STEP 4: Initially set the set of online multi-site high utility itemsets $OMHU_q$ as empty.
STEP 5: For each candidate itemset X in the union set US_q, do the following substeps:
(a)   Calculate the sum of the actual values of the unknown utilities of X by scanning the data blocks of the matched records in which X is not a high utility itemset. That is,

$$au_X^{ub} = \sum_{X \notin HU_j \wedge p_j \in EMPRq} au_{jX},$$

where au_{jX} is the actual value of the unknown utility of X within p_j.
(b)   Calculate the online multi-site actual utility ratio of X in $EMPR_q$. That is,

$$OMAUR_X = \frac{u_X^{appearing} + au_X^{ub}}{\sum_{p_j \in EMPRq} TTU_j}.$$

(c)   If the online multi-site actual utility ratio $OMAUR_X$ of X is more than the threshold λ_q, put X in the set of $OMHU_q$; otherwise, omit X.
STEP 6: Output the set of online multi-site high utility itemsets satisfying q, $OMHU_q$.

5   Experimental Evaluation

In the section, a series of experiments was conducted to evaluate the performance of the proposed *TP-OMU* in effectiveness of online multi-site high utility itemsets, and showed the performance of the proposed *TP-OMU* and the traditional *TP* in terms of execution efficiency. The experiments were implemented in J2SDK 1.7.0 and executed on a PC with 3.3 GHz CPU and 4GB memory. In the experiments, the public *IBM* data generator [4] was used to produce the experimental data. Figure 1 showed the numbers of online multi-site high utility itemsets under various T when λ_q varied from 0.12% to 0.20%, and λ was set at 0.10%

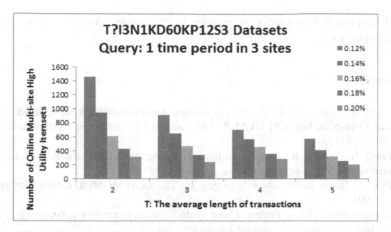

Fig. 1. Number of online multi-site high utility itemsets under various T

Figure 2 showed the performance of the proposed *TP-OMU*, traditional *TP* and the total process under various λ_q. As shown in the Figure 2, the execution time of our proposed *TP-OMU* is much fewer than the traditional *TP*.

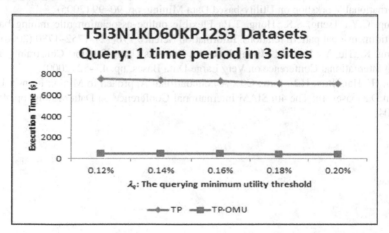

Fig. 2. Efficiency comparison of the two approaches under different thresholds λ_q

6   Conclusion

To effectively solve the large-scale data problem in a multi-site environment, this work has introduced a new research issue named online multi-site utility mining. In addition, the *TP-OMU* approach is developed to find online multi-site high utility itemsets in a multi-site environment. To our best knowledge, this is the first work on utility mining in a multi-site environment. In particular, this work also presents an effective strategy, which can effectively reduce the candidate utility itemsets in mining. The experimental results on several synthetic datasets reveal the proposed

TP-OMU approach has good performance in execution efficiency under different parameter settings.

References

1. Agrawal, R., Imielinksi, T., Swami, A.: Mining Association Rules between Sets of Items in Large Database. In: ACM SIGMOD International Conference on Management of Data, pp. 207–216 (1993)
2. Agrawal, R., Srikant, R.: Fast Algorithm for Mining Association Rules. In: International Conference on Very Large Data Bases, pp. 487–499 (1994)
3. Hidber, C.: Online association rule mining. In: The ACM SIGMOD Conference, pp. 145–156 (1999)
4. IBM Quest Data Mining Project, Quest synthetic data generation code, http://www.almaden.ibm.com/cs/quest/syndata.html
5. Lan, G.C., Hong, T.P., Tseng, V.S.: Discovery of High Utility Itemsets from On-Shelf Time Periods of Products. Expert Systems with Applications 38(5), 5851–5857 (2011)
6. Liu, B., Hsu, W., Ma, Y.: Mining Association Rules with Multiple Minimum Supports. In: International Conference on Knowledge Discovery and Data Mining, pp. 337–341 (1999)
7. Liu, Y., Liao, W.K., Choudhary, A.: A Fast High Utility Itemsets Mining Algorithm. In: International Workshop on Utility-based Data Mining, pp. 90–99 (2005)
8. Wang, C.Y., Tseng, S.S., Hong, T.P.: Flexible online association rule mining based on multidimensional pattern relations. Information Science 176(12), 1752–1780 (2006)
9. Wang, K., He, Y., Han, J.: Mining Frequent Itemsets Using Support Constraints. In: The 26th International Conference on Very Large Data Bases, pp. 43–52 (2000)
10. Yao, H., Hamilton, H.J., Butz, C.J.: A Foundational Approach to Mining Itemset Utilities from Databases. In: The 4th SIAM International Conference on Data Mining, pp. 482–486 (2004)

Importance of Attributions Including Academic Life Log Concerning to Win a Job

Eri Domoto[1], Antonio Oliveira Nzinga Rene[2], and Koji Okuhara[2]

[1] Department of Media Business
Faculty of Economics, Hiroshima University of Economics
5-37-1, Gion, Asaminami-Ku, Hiroshima, 731-0192 Japan,
er-domo@hue.ac.jp
[2] Department of Information and Physical Sciences
Graduate School of Information Science and Technology, Osaka University
1-5 Yamadaoka, Suita, Osaka 565-0871, Japan

Abstract. In this study, may reveal data and history of class selection of up to university graduation of students, clubs belong, such as from high school, or if it is being given what effect the career development of students which is the object. To reveal what the item is whether they affect the employment using the Decision Making Trial and Evaluation Laboratory method from student data.

Keywords: Decision Making Trial and Evaluation Laboratory method, Shapley value, Employment support.

1   Introduction

Analysis of employment support for students or their career development is an important issue for a University. One result expected from the university education is a student who can be benefit to the society. These indicators have been analyzed by macro surveys of universities in Japan. Top-down macro-information includes among others self-inspection and evaluation forms with e-learning initiatives using micro-information. In practice, the proposed methods focus the student ability only on a test paper which constitutes an example of measuring the formation of such characteristic. Rich test environment using multimedia, candidates enabling questions to interactive e-testing, store learning artifacts, such as reports and share e-portfolio, students meet evaluates learning outcomes of peer assessment, data mining stored large amounts of e-portfolio and e-learning in learning history [1-5]. These efforts, emerged in Japan and abroad. Expected advances by rich cloud environment as future transitions to such education from both sides of the micro and macro teaching information management becomes possible. But at this stage the education system is in transition, information on university education as macro data lies in fact that publishers do not know its status quo.

Referring to the student data analysis, in this research we made use of DEMATEL (Decision Making Trial and Evaluation Laboratory) methods [6-11] which indicates

L.S.-L. Wang et al. (Eds.): MISNC 2014, CCIS 473, pp. 179–188, 2014.
© Springer-Verlag Berlin Heidelberg 2014

the items that would affect the employment. Additionally, based on the results of the DEMATEL method, we also indicate a model to determining the Shapley value in game theoretic concepts.

The rest of the paper is structured as follows. In Section 2, we employ the DEMATEL approach for education data analysis. Section 3, describes the Shapley value in terms of linear programming (LP) model. The paper concludes in the following section.

2   Education Data Analysis by DEMATEL Method

In this section, we analyze a students' dataset in terms of how each factor affects their job, using the DEMATEL method. Table 1 is an example of the students' data. More details regarding the data under study are presented as follows.

- Sex : (1: male, 2: female)
- Department : (111: Economy, 112: Management, ⋯)
- Alma mater : (1: A high school, 2: B high school, ⋯)
- Entrance Examination : (1: General admissions, 2: Recommendation entrance examination, ⋯)
- Thesis advisor : ID number of each teacher
- GPA : Grade Point Average
- Attendance : Class attendance rate (%)
- Certificate : (0: None, 1~: Each number)
- Scholarship : Type number of each scholarship
- Club : Each club number
- Project : Each project number
- Employment as desired : (1: Yes, 0: No)

DEMATEL as a tool to support decision-makers is used for evaluating experimental attempts. It was developed in Battelle Institute of America to face worldwide complex problems. Its application area lies in the analysis of complex and uncertain factors within a problem. Its feature is trying to take full advantage from the institution information and as well people related to the problem under study.

Having this in mind, we used this approach in order to understand the overall impact of the relationship between the items related to a university education's information. First, the information regarding university education is arranged vertically and horizontally. Then, the items are represented into five criteria and through a paired comparison the influence of an item i over j is analyzed given a direct relation matrix $A: [a_{ij}]$. Based on this, the initial direct influence $X: [x_{ij}]$ which relatively expressed strength was calculated using formula (1).

Table 1. Examples of Student Data

No	Student ID number	1	2	3	4	5	6	7	8	9	10	11	12	13	14	15
1	Sex	1	1	2	1	1	1	1	2	1	2	1	1	2	1	1
2	Department	111	112	111	113	111	111	114	111	112	115	113	111	115	113	114
3	Alma mater	78	3	35	88	47	83	33	57	48	78	99	24	41	42	32
4	Entrance examination	6	10	4	9	6	7	9	8	2	3	1	10	2	8	2
5	Thesis advisor	95	9	69	89	72	48	44	100	19	94	82	50	78	46	19
6	GPA	3.08	3.08	3.25	3.23	3.61	3.15	3.78	3.54	2.98	3.69	3.34	3.09	2.96	3.28	3.69
7	Attendance(%)	68	65	95	81	74	75	97	91	73	62	65	88	72	76	90
8	Certificate	3	0	0	5	0	0	2	0	0	0	4	0	0	1	0
9	Scholarship	0	1	0	1	0	0	3	0	0	0	0	0	0	1	0
10	Club	20	38	0	7	1	30	29	13	17	4	0	43	0	40	24
11	Project	4	6	0	3	6	0	7	0	0	0	2	5	1	0	0
12	Employment as desired	1	0	0	1	0	0	0	1	0	0	0	0	1	0	1

Table 2. Direct Relation Matrix A

No	Item	1	2	3	4	5	6	7	8	9	10	11	12
1	Sex	0	0	0	0	0	0	0	0	0	0	0	1
2	Department	0	0	0	0	0	0	0	0	0	2	0	2
3	Alma mater	0	0	0	1	0	2	0	1	0	1	0	0
4	Entrance examination	0	2	1	0	0	1	0	1	0	0	3	0
5	Thesis advisor	0	0	0	0	0	1	1	1	0	0	0	2
6	GPA	0	0	2	0	0	0	2	0	4	0	0	3
7	Attendance	0	0	0	0	0	5	0	0	1	0	0	2
8	Certificate	0	0	0	0	0	0	0	0	0	0	0	1
9	Scholarship	0	0	0	0	0	0	0	0	0	0	0	1
10	Club	0	0	0	0	0	0	0	0	0	0	0	0
11	Project	0	0	0	0	0	0	0	0	0	0	0	1
12	Employment as desired	1	0	0	0	0	1	0	0	0	0	0	0

Table 3. Total Relation Matrix F

No	Item	1	2	3	4	5	6	7	8	9	10	11	12	D	D+R
1	Sex	0.009	0	0.002	0	0	0.01	0.002	0	0.004	0	0	0.095	0.122	0.385
2	Department	0.017	0	0.004	0	0	0.02	0.004	0	0.007	0.182	0	0.19	0.424	0.631
3	Alma mater	0.008	0.017	0.05	0.095	0	0.227	0.041	0.104	0.086	0.099	0.026	0.093	0.846	1.394
4	Entrance examination	0.01	0.184	0.116	0.011	0	0.135	0.024	0.102	0.051	0.044	0.276	0.115	1.068	1.208
5	Thesis advisor	0.025	0	0.033	0.003	0	0.178	0.123	0.094	0.076	0.003	0	0.271	0.806	0.806
6	GPA	0.037	0.004	0.215	0.02	0	0.175	0.214	0.021	0.447	0.02	0.005	0.406	1.564	3.002
7	Attendance	0.035	0.002	0.102	0.009	0	0.555	0.101	0.01	0.302	0.01	0.003	0.383	1.512	2.047
8	Certificate	0.009	0	0.002	0	0	0.01	0.002	0	0.004	0	0	0.095	0.122	0.455
9	Scholarship	0.009	0	0.002	0	0	0.01	0.002	0	0.004	0	0	0.095	0.122	1.148
10	Club	0	0	0	0	0	0	0	0	0	0	0	0	0	0.36
11	Project	0.009	0	0.002	0	0	0.01	0.002	0	0.004	0	0	0.095	0.122	0.432
12	Employment as desired	0.095	0	0.02	0.002	0	0.108	0.02	0.002	0.041	0.002	0	0.046	0.336	2.22

Table 4. Various Entropies

No	Item	$S^1_{(i)}$	$S^2_{(i)}$	$S^3_{(i)}$	$S^4_{(i)}$	$S^5_{(i)}$	$S^-_{(i)}$	$\Delta S_{(i)}$
1	Sex	0.845	1.000				0.887	0.113
2	Department	0.918	0.000	0.918	1.000	1.000	0.818	0.182
3	Alma mater	0.811	0.918	0.971			0.907	0.093
4	Entrance examination	1.000	0.000	1.000	0.918	0.811	0.800	0.200
5	Thesis advisor	0.000	0.000	0.000	0.918	0.971	0.507	0.493
6	GPA	1.000	0.811	0.811	0.000	0.811	0.782	0.218
7	Attendance	0.000	0.811	0.722	1.000	1.000	0.857	0.143
8	Certificate	0.881	0.971				0.911	0.089
9	Scholarship	0.946	0.811				0.910	0.090
10	Club	0.918	1.000	0.650			0.844	0.156
11	Project	0.863	0.954				0.912	0.088

The initial direct influence matrix X computed the sum of each line of the direct relation matrix A by breaking each element of A and maximizing them.

$$X = \frac{1}{\max\limits_{1 \leq j \leq n} \sum_{i=1}^{n} |a_{ij}|} \cdot A \tag{1}$$

The indirect influences were computed in (2) introducing an identity matrix I in the total relation matrix : $[f_{ij}]$ as presented next.

$$F = X + X^2 + X^3 + \cdots = X(I - X)^{-1} \tag{2}$$

Consider a direct impact matrix as shown in Table 2 referring to the data in Table 1 herein. Table 2, from which we based our analysis, shows an example where are evaluated 5 relative strengths of the causal relationship leading to "Attendance" from "GPA". Table 3 presents the Total Relation Matrix F obtained by the DEMATEL method. Row sum D of the Total Relation Matrix F represents the sum of the effects that certain items give to other items. $D + R$ corresponds to the sum of non-impact and impact, the item indicates whether I has a central role in the extent to which the issue structure.

Figure 1 shows the Causal relationship between the items through a directed graph. The positive direction of the X axis represents the cause of the positive direction and Y axis the center of the case.

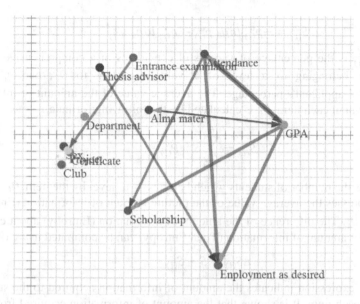

Fig. 1. Causal relationship between the items through directed graph

In addition, ij element of the square of the direct relation matrix A represents the number of paths to reach item j from item i. Similarly, ij element of the cube of the A represents the number of lengths of the path from i to j. The sum A, A^2, $A^3 \cdots$ represents the number of paths in order to represent the possibility of arrival, i.e., the presence of the path to obtain a reachability matrix R [12,13].

$$R = \frac{I}{I - A} \equiv I + A + A^2 + \cdots \tag{3}$$

$$\tilde{R} = R - I \tag{4}$$

Here, the importance over each item is introduced further. It is assumed that the score table which consists of u items $O_1, O_2, \cdots, O_n$ was given by the following E to N student $S_1, S_2, \cdots, S_N$.

$$E = e_j^l \tag{5}$$

Where, $e_j^l \in [0, 1]$ on $\mathbb{R}$. The success or failure is decided based on the overall points to everybody. That is, everybody is sifted out by suitable threshold. The value of θ is performed for Column of E by suitable thresholds. Next, equation (6) represents the amount of information (Entropy) S.

$$S = -p \, log_2 \, p - q \, log_2 \, q \tag{6}$$

Here, p is the ratio of the number of successful candidates; q is the ratio of the number of unsuccessful candidates. The amount of information (Entropy) about each item O_i is as follows.

$$S_{(i)}^\alpha = -p_{(i)}^\alpha \, log_2 \, p_{(i)}^\alpha - q_{(i)}^\alpha \, log_2 \, q_{(i)}^\alpha \tag{7}$$

Where, $p_{(i)}^\alpha$ is the percentage of successful candidates in a comprehensive score of the people belonging to α level of θ value regarding to O_i, and $q_{(i)}^\alpha$ is the unsuccessful applicant's percentage. The average value of the amount of information (Entropy) about each item O_i is calculated by the following formula.

$$S_{(i)}^- = \sum_{\alpha=1}^{\theta} S_{(i)}^\alpha P_{(i)}^\alpha \tag{8}$$

P^α denotes the number of students which belongs to α level by the item O_i.

$S_{(i)}^-$ expresses the average value of the picture amount of information (Entropy) in each item O_i. The difference of the amount of information (Entropy) about each item and the total amount of information (Total Entropy) can be computed using (9).

$$\Delta S_{(i)} = S - S_{(i)}^- \tag{9}$$

The above expression indicates the amount of information acquired to the success or failure by each item. It is shown that the amount of information acquired from O_i by the evaluation to the whole understanding is large and therefore, the value of $\Delta S_{(i)}$ is large. That is, $\Delta S_{(i)}$ expresses the size of importance in terms of the meaning with the item over whole understanding. The value of $\Delta S_{(i)}$ is defined as importance $I_1(i)$ of the item i.

Here, the amount of information (Entropy) is calculated based on the data of Table 1.

The success or failure is set with the item "employment as desired" on Table 1. In order to calculate Entropy in each item, in item 1 (Sex), it is considered as two-step evaluation (theta= 2). In item 2 (Department), it is considered as five-step evaluation (theta= 5) and the other items are set up as shown in Table 4. The value of item 1 (Sex) regarding to the number of 1 (male) is 11 and the number of 2 (female) is 4. Moreover, the value of item 1 (Sex) is 1 (male) and the value of item 12 (Employment as desired) of the number of 2 (Yes) is 3, and the number of 1 (No) is 8. Finally, the amount of information (Entropy) for male with respect to item 1 (Sex) is as follows.

$$S_{(1)}^0 = -\frac{3}{11} \, log_2 \, \frac{3}{11} - \frac{8}{11} \, log_2 \, \frac{8}{11} = 0.845351 \cdots \tag{10}$$

Similarly, the value of item 1 (Sex) is 2 (female), and also the value of item 12 (Employment as desired) of the number of 2 (Yes) is 2, and the number of 1 (No) is 2 in a similar manner. Then, the amount of information (Entropy) for female regarding to item 1 (Sex) is as follows.

$$S_{(1)}^1 = -\frac{2}{4} \, log_2 \, \frac{2}{4} - \frac{2}{4} \, log_2 \, \frac{2}{4} = 1 \tag{11}$$

Thus, the average value of the amount of information (Entropy) about the item 1 (Sex) is as follows.

$$S_{(1)}^- = -\frac{11}{15} * (0.845351 \cdots) + \frac{4}{11} * (1.0) = 0.886591 \cdots \tag{12}$$

Moreover, the acquired amount of information (Total Entropy) is as follows.

$$\Delta S_{(1)} = 1 - 0.886591 \cdots = 0.113409 \cdots \tag{13}$$

Calculating all the other items similarly we obtained the results exhibited in Table 5.

The reachability matrices in order to find the relationship between the direct and indirect gains when the cooperation occurs are defined by the formulae (14-15).

$$v(S) = \sum_{i \in S} v_i + \Delta v(S) \tag{14}$$

$$\Delta v(S) = \sum_{i \neq j \in S} \tilde{r}_{ij}(v_i + v_j) \tag{15}$$

3   Shapley Value as a Result of the DEMATEL Method

This section introduces few basic concepts of cooperative game theory and suggests a model to compute Shapley values in LP framework.

A game (N, V) has a vector imputation $z(N, v) = (z_1(N, v), z_2(N, v), \cdots, z_n(N, v))$ which satisfies the following properties:

• Individual rationality:

$$z_d(N, v) \geq v(\{d\}), d = 1, 2, \cdots, N$$

• Grand coalition rationality:

$$\sum_{d=1}^N z_d(N, v) = v(N)$$

Let $M_{r,s}, (r = 2, 3, 4, \cdots, R; s = 1, 2, \cdots, r - 1)$ be a set of weights with nonnegative values and at has least one positive value for each r. Consider, now, the following problem:

$$\text{Minimize} \sum_{S \subset R, S \neq R} M_{r,s}\left(v(S) - \sum_{d \in S} z_d(R, v)\right)^2 \tag{16}$$

$$\text{Subject to} \sum_{d \in R}(R, v) = v(R)$$

$z_d^+(R, v)$ in (16) can be obtained through the following expression gives the solution for (17) through the following expression:

$$z_d^+(R,v) = \frac{1}{r}\left\{v(R) + \sum_{i \in R}(c_{di} - c_{id'})\right\} \tag{17}$$

and

$$\Gamma(d^+, d'^-) = \{S \subset R \mid d \in S, d' \notin S\}.$$

The set of weights $M_{r,s}$ is obtained from (18) as follows.

$$M_{r,s} = \frac{1}{r-1}\{_{r-2}C_{s-1}\}^{-1} \tag{18}$$

The Shapley value as defined next is equivalent to the above mentioned vector imputation

$$\phi_d(R,v) = \sum_{\substack{S \subset R \\ d \in S}} \frac{(s-1)!\,(n-s)!}{n!}\{v(S) - v(S-d)\} \tag{19}$$

Shapley value $\phi_d(R,v)$ represents the mathematical expectation of the marginal contribution of player d when all orders of formation of the grand coalition are equiprobable. Here, s is the number of members of coalition S, n the amount of players in the game, $v(S)$ denotes the grand coalition or characteristic function and d the player under evaluation.

Thus,

$$z_d^+(R,v) = \phi_d(R,v) \tag{20}$$

For more results on this topic, we refer readers to [14-16] and the references therein.

Suppose now a sample d from a given dataset, and consider x_i and y_i as values of the sample. A residual error can be obtained by the inner product (21) as follows.

$$c_d = f_d(A, M, v) = (A^{\mathrm{T}} M v - A^{\mathrm{T}} M A_z)_d \tag{21}$$

where $(\ )_d$ denotes the selection of the d-th row value, thus the sum of all error functions $E = \sum_d |e_d|$ using the multiple linear regression model that minimize the sum of the absolute values of the residuals. This problem can be restated as an LP problem:

Minimize ϵ

Subject to $A^{\mathrm{T}} M v + s^+ - s^- = A^{\mathrm{T}} M A_z$

$$\sum_{d \in K} z_d(K, v) = v(K) \tag{22}$$

$$0 \le s^+ \le \epsilon, 0 \le s^- \le \epsilon$$

where

$$s^+ = [s_1^+, s_2^+, s_3^+, \cdots, s_n^+]^T$$

and

$$s^- = [s_1^-, s_2^-, s_3^-, \cdots, s_n^-]^T.$$

A is a matrix which observes the supperaditivity property; v is a column matrix whose elements are the real values $v(S)$, i.e., the characteristic function obtained from the coalitions among the players in the game. M = diag($M_{k,s}$) whose elements are the weight defined in (18).

$$M = \begin{bmatrix} M_{k,S} & 0 & \cdots & 0 \\ 0 & M_{k,s} & \cdots & 0 \\ \vdots & \vdots & \ddots & \vdots \\ 0 & 0 & \cdots & M_{k,s-1} \end{bmatrix} \tag{23}$$

4   Concluding Remarks

In this paper, historical university students' dataset from class selection, clubs attended, etc was analyzed using the DEMATEL approach and showed the main factors that can influence on the students' career formation after their graduation. Additionally, from the obtained results we also indicate a Shapley model which can be used in order to support decision-makers.

Acknowledgment. This work was supported by JSPS KAKENHI Grant Number 25350309.

References

1. Hasegawa, O., Yamakawa, H., Imai, J.: Proposal of e-Learning Operational Model for the Utilization on Elementary and Secondary Education. JSiSE Research Report 28(7), 73–80 (2014)
2. Sonoya, T.: Issues of the promotion policy for practical use of ICT at school: In terms of appropriate utilization. Bulletin of the Educational Research and Practice 23, 241–249 (2014)
3. Fijimoto, Y., Nonaka, Y., Sonoda, F.: Research on the Use of Digital Textbook and Other Teaching Materials by an English Teacher. Transactions of Japanese Society for Information and Systems in Education 31(1), 153–158 (2014)
4. Murai, R., Hayashi, T., Yaegashi, R.: Study of modeling for intellectual property education using ICT. IEICE Technical Report 113(482), 205–208 (2014)
5. Ban, H., Minagawa, J.: A Consideration on English Learning for Undergraduates Using Nintendo DS. IEICE Technical Report 113(229), 105–108 (2013)
6. Sato, T.: Analyzing the structure of management problems in the local government: using the DEMATEL method. Government Auditing Review (37), 87–97 (2008)

7. Kinoshita, S., Okamoto, S., Sasaki, H., Takeya, M.: Retrieval Method of Leading Documents Based on the Reference Relationships Using the DEMATEL Method. IEICE Technical Report. Educational Technology 99(500), 127–134 (1999)
8. Moro, Y., Hoshino, S.: Study on multiple effects of Women's Group Activities in Rural Communities by the DEMATEL method. Journal of Rural Planning Association 23, 151–156 (2004)
9. Kamaike, M.: Design Elements in the Passenger Car Development: The Classification and the Influence Analysis in Case of Recreational Vehicle. Japanese Society for the Science of Design 48(1), 29–38 (2001)
10. Naito, N., Nagase, H., Showji, K.: Development of a Support System to Create Learning Plans to Meet Students' Needs Based on the DEMATEL Method. In: Proceedings of the Annual Conference of JSET, vol. 14, pp. 143–144 (1998)
11. Matuo, H.: ISM, DEMATEL analysis software, http://www27.atpages.jp/hidmat/ism_dematel/
12. Toyota, N., Saegusa, T.: Importance Measure of Instructional Elements in Teaching Strategies. IEICE Technical Report. Educational Technology 96(148), 121–128 (1996)
13. Toyoda, T., Horii, H.: Application of Structural Modeling Analysis to Social Problems: For Falsification Issues at Tepco's BWR Plant. Sociotechnology Research Network 1, 16–24
14. Rene, A.O.N., Okuhara, K., Domoto, E.: Allocation of Weights by linear Solvable Process in a Decision Game. International Journal of Innovative Computing, Information and Control 8(3), 907–914 (2014)
15. Ruiz, L.M., Valenciano, F., Zarzuelo, J.M.: Some New Results on Least Square Values for TU Games. Sociedad de Estadistica e Investigacion Operativa 6(1), 139–158 (1998)
16. Namekata, T.: Probabilistic Interpretation of Nyu-value (a Solution for TU game). Bulletin of the Economic Review at Otaru University of Commerce 56(2-3), 33–40 (2005) (in Japanese)

Incorporating Human Sensors into Event Contexts
for Emergency Management

Chung-Hong Lee, Chih-Hung Wu, and Shih-Jan Lin

Department of Electrical Engineering,
National Kaohsiung University of Applied Sciences,
Kaohsiung 807, Taiwan
leechung@mail.ee.kuas.edu.tw,
{williamwu.tw,linshihjan.tw}@gmail.com

Abstract. In recent years, the concept of human-in-the-loop has been utilized to support environment sensing. Along with *IoT* (*Internet of Things*) and wearable computing technologies, connecting people and devices to the internet provides a significant advantage for real-time emergency management. For development of context-aware applications, it is important to utilize higher-level semantic information, such as human activity, social emotions, and human behaviors for event monitoring. Therefore, human users may become part of sensor networks by using mobile devices and social media to report local information around them. In this work, we mainly focus on the use of social messages spreading by human users to model the real-world events, in order to incorporate human sensors into event contexts for situational awareness. First, our algorithm computes the energy of each collected event messages, and then encapsulates ranked temporal, spatial and topical keywords into a structured node, which could reinforce the alert collected from physical nodes. The experimental results show that the proposed approach is able to extract essential entities of events for incorporating human sensors into event contexts for event prevention and risk management.

Keywords: Social data Entity extraction, Ontology Engineering, Knowledge management.

1 Introduction

Context awareness of emergent events is a critical issue for event modeling. As Schmidt [26] suggested that, the context is composed of two aspects: the physical environment (e.g., light, pressure, acceleration) and human factors (e.g., users, social environment, activities). In many applications, machine sensors have become the main stream for extracting context information from physical environments. However, the machine sensors are not as ubiquitously deployed in the environment as needed for many applications. Furthermore, in most cases sensing the human factors of context is much more important and difficult. To eliminate the gaps, social-messages (e.g. tweets) associated with specific events, produced by people (i.e.

L.S.-L. Wang et al. (Eds.): MISNC 2014, CCIS 473, pp. 189–198, 2014.
© Springer-Verlag Berlin Heidelberg 2014

human sensors), can act as a very useful resource for situational awareness. Social-messages (e.g. tweets) contain fruitful information which is valuable for understanding the development of real-world events, which can contribute to solutions for event awareness and crisis management. Fusing such diverse kinds of information presents a challenge to anyone seeking to establish a coherent picture of the developing situation and to enable sensible decision-making operations. For practical use of context, the general challenge is to identify the set of relevant features in terms of which a situation can be captured sufficiently. Motivated by this need, in this work we utilize social messages collected from Twitter for extracting essential event features, identified as key entities of an event model, to cope with emerging disasters events. It is expected that the discovered event features can be used to locate essential information to cope with novel disasters.

The system was developed using a novel event modeling method based on a data-cluster slicing method. By analyzing the publicly available social streams, many large accidents or disasters could be immediately detected for investigation [5; 13]. Although the applications related to event detection by using social datasets (e.g. tweets) have been largely developed in recent years [1; 2; 4; 7; 23], little work has been done to produce a consistent information source as a virtual sensor (i.e. human sensor) for obtaining local information by using real-time social data. The view is taken, therefore, in this work we first implement an event detection system which provides a faster way to detect events in their early stage using a new term weighting technique and social media streams. This enables an online extensible event model to be built for extracting essential event features, identified as key entities of an event model for emergency management.

2   Related Work

The integration of physical sensors and human sensors can enhance the accuracy and response time of context awareness for emergency management. Along with *IoT* (*Internet of Things*) and wearable computing technologies, connecting people and devices to the internet provides a significant advantage for real-time emergency management. For development of context-aware applications, it is important to utilize higher-level semantic information, such as human activity, social emotions, and human behaviors for event monitoring. Therefore, human users may become part of sensor networks by using personal portable devices and social media to report local information around them. In this work, we mainly focus on the use of social messages spreading by human users to model the real-world events, in order to incorporate human sensors into event contexts for situational awareness. Still, there are lots of challenges in fusing data obtained from physical and human sensor [24]. [27] proposed a platform called CROSS which joins signed volunteers who locate around the physical sensors as human sensors to observe and report real-time information of sensed events. On the other hand, the key issue for applying social data as human sensors is to verify the information collected from internet. [25] described a maximum likelihood estimation approach with an optimal solution which is obtained by solving an expectation maximization problem and can directly lead to an analytically founded quantification of the correctness of measurements as well as the reliability of participants.

Another approach is to utilize developed ontology for emergency response. Several ontology-based approaches have been developed to support crisis management and decision making [20]. For web-based applications, [3] focuses on developing an ontology structure of elements which provides a valuable guideline in creating new WB-NDMS Web sites as well as in benchmarking the sufficiency and comprehensiveness of existing Web sites for Web-based disaster management systems. Joshi [9] proposed an ontology-based approach, called Disaster Mitigation and Model (DMM), for disaster mitigation by using Web Ontology Language (OWL) which is envisioned to allow for seamless integration and management of heterogeneous, multi-lateral data from different local, state as well as federal agencies. In Ye's work, three categories of ontologies including static ontology, dynamic ontology, and social ontology [8], were developed to tackle challenging issues in different perspectives for crisis contagion management in financial institutions. Kang proposed a knowledge level model (KLM), which integrates top-level and domain-level ontologies, for systemic risk management in financial institutions [21]. Wang [18] proposed a conceptual modeling approach which was developed to detect real time quality problems and support quality information sharing between agents in the whole milk supply-chain. Still, lots of developed ontology applications mainly rely on people with domain expertise to construct the ontology manually.

We propose a method designed for automatic extraction of event entities, which combines techniques of online event modeling and entity extraction by using user generated content from social media. Considering the 5W1H (What, Who, When, Where, Why, How) of NOEM (News Ontology Event Model) [19] and constituent properties (atPlace, atTime, circa, illustrate, inSpace, involved, involvedAgent) of LODE (Linking Open Descriptions of Events) [17], the proposed system extracts the primary entities, including time, location and topic, which were mapped to the temporal, spatial and topic features, to provide true pictures of real-world events for emergency management.

3   Entity Extraction Using Data of Human Sensors

In this work, we develop a system framework for incorporating human sensors into event contexts for emergency management by extracting and reconstructing key entities from online social media streams (e.g. Twitter messages). As shown in Fig 1, the proposed system combines several technical modules, including Preprocessing, Event Modeling, and Event Entity Extraction modules. In order to encapsulate the key features (timeline, geographical center, cluster energy) and main concept of a critical event for social media streams, a modeling process is applied to normalize the clustered data sets with a dynamic weighting function and a self-adaptive clustering model. In our modeling process the system extracts a real-time event in the early stage by monitoring the evolvement of event energy in each thread from social media streams. Social media stream provides a good opportunity to facilitate human sensors due to its convergence of context information related to spatial, temporal, and other conceptual relationships.

Several research work have reported that the use of social datasets with text-stream mining techniques can obtain great results in the field of online event detection and extraction [6; 15; 16; 22]. However, most of them focus largely on grouping or finding hot topics of events without dealing with evolving entities of events. To learn the events in depth, a sophisticated processing work should be carried out for representing the intensity of incoming messages as detected hot topics. The stream can then be clustered in a near real-time manner to enhance system performance. For instance, Kleinberg [10] used an approach based on modeling the stream using an infinite-state automaton in which bursts appear naturally as state transitions to extract important topics in email streams.

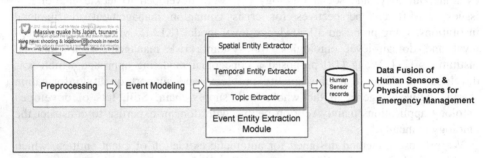

Fig. 1. System framework for incorporating human sensors for emergency management

The preprocessing and event modeling modules for online event detection have been successfully implemented in our previous work [11; 12; 13; 14]. In the preprocessing module, two major operations are accomplished by using the Twitter API and collected messages:

- Monitoring the temporal variability of a term and the intensity of online messages.
- Tracking the evolvement of an incident according to its density and life-cycle.

The pre-processing procedure was designed for representing the intensity of incoming messages. The streams can then be clustered near real-time for leveraging the efficiency and accuracy. The design of weighting process of real-time message should be constantly updated. In this work we apply our developed term weighting scheme BursT [11, 12] to deal with incoming event messages. The steps for term weighting and event-message clustering are carried out in the event modeling module. Fig 1 illustrates our system framework. The event entities obtained from the human sensors in this system are being used to connect sensing features of physical sensors investigated in the other project to form a hierarchically organized feature space for event contexts.

At the top level we propose to distinguish context related to human factors in the widest sense, and context related to the physical environment. In the physical environment, context information obtained from sensors is at a low level of abstraction, capturing isolated features of a situation. In this framework, we present a sensor-fusion approach which is expected to combine physical and human sensors for obtaining context information at a level of abstraction that relates more to situations than to specific physical conditions. In this paper, we only address the development process of event entity extraction module. Detailed description of our approach is described as follows.

3.1   Spatial Named Entity Extractor

Users of microblogs often use a limited vocabulary in the messages, including names of objects are used in singular and plural; a smaller number of words are used to produce many of the compound terms and phrases. For instance, in the event of Mar.11 2011 Japan Earthquake, we can find out that the terms tokyo and japan were used more intensively than other spatial words in the messages, and the importance of each word could be represented. The candidate of spatial entity ($E_{spatial}$) for a selected event can be obtained by calculating the product of message number and bursty score of each keyword. Then we can get the most possible candidate of $E_{spatial}$.

$$E_{spatial}^{evt_i} = \max_{w \in S} \left(\sum_{t \in TF} MS_{W_{evt_i}^t} \times BS_{W_{evt_i}^t} \right) \tag{1}$$

where S is collected terminology of spatial entity, e.g. GeoNames geographical database (http://www.geonames.org/), TF is the time frame of event evt_i, $MS_{W_{evt_i}^t}$ is the number of messages containing a specific geographical name and in the event evt_i at time t.

3.2   Temporal Feature Extractor

The temporal features were extracted from clustered events. The starting time could be set as the creating time point of first cluster, and vice versa the ending time is the last updating time point of the last cluster. As regards the definition of Event-Instance Relations in [9], the causal relations of events can be potentially extracted by distinguishing the relation of temporal interval as Fig 2. For instance, the atTime property of continuous sub-events, e.g. $e1$ and $e2$, of a selected event evt_i is the beginning time of $e1$, and the circa property can be defined as:

$$E_{circa}^{evt_i} = t'_{e2} - t_{e1} \tag{2}$$

Moreover, if $e1$ and $e2$ are independent, the temporal property will be decided by the lifecycle of each event.

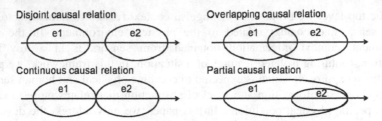

Fig. 2. Causal relations over time

3.3 Topic Extractor

The most possible topic words of event evt_i can be selected by weighting their number and bursty score mentioned in [11] as:

$$W_{evt_i}^{topic} = \max_{w \in D} (\Sigma_{t \in TF} MS_{w_{evt_i}^t} \times BS_{w_{evt_i}^t}) \qquad (3)$$

where D is the collection of domain specific terminology, e.g. natural disaster, and the candidate topic entity can be constructed as the combination of temporal feature, spatial entity and topic word as:

$$E_{topic}^{evt_i} = E_{circa}^{evt_i} + E_{spatial}^{evt_i} + W_{evt_i}^{topic} \qquad (4)$$

The topic extractor can be extended to different domain by applying different domain specific terminologies. As Fig 3, the temporal, spatial and candidate topic can be extracted by ranking their bursting score, and then these key entities would be composed as a new instance of some emergency ontology for further analysis.

311 - Japan - Tsunami (Earthquake)   Event Topic

Temporal Feature   Keyword Feature1   Keyword Feature2   Spatial Feature

eID	BS	StartTime	EndTime	KW1	KW1%	KW2	KW2%	L1	L1%	L2	L2%
36155789:129987529	2.630036446	Fri Mar 11 14:28:12	Fri Mar 11 14:42:49	tsunami	6.15%	earthquake	3.41%	japan	78.18%	tokyo	5.45%
36160938:129987617	3.661421143	Fri Mar 11 14:42:57	Fri Mar 11 14:47:38	tsunami	13.93%	earthquake	5.25%	japan	61.42%	philippines	4.72%
36162756:129987649	7.433504033	Fri Mar 11 14:48:11	Fri Mar 11 14:53:33	tsunami	14.21%	earthquake	4.43%	japan	59.41%	philippines	6.93%
36164546:129987682	12.08302991	Fri Mar 11 14:53:44	Fri Mar 11 14:59:58	tsunami	9.71%	earthquake	5.13%	japan	73.08%	tokyo	1.10%
36166848:129987722	11.8933249	Fri Mar 11 15:00:20	Fri Mar 11 15:03:11	tsunami	19.50%	earthquake	2.37%	japan	50%	taiwan	7.79%
36167933:129987739	6.509132447	Fri Mar 11 15:03:15	Fri Mar 11 15:04:15	tsunami	23.98%	earthquake	5.24%	japan	50.62%	indonesia	7.41%
36168613:129987750	22.18908585	Fri Mar 11 15:05:06	Fri Mar 11 15:07:46	tsunami	17.35%	earthquake	4.25%	japan	66.36%	taiwan	3.64%
36169576:129987766	31.39012775	Fri Mar 11 15:07:47	Fri Mar 11 15:10:25	tsunami	18.43%	earthquake	3.91%	japan	54.63%	taiwan	6.71%
36170540:129987782	21.50148077	Fri Mar 11 15:10:28	Fri Mar 11 15:12:24	tsunami	21.90%	earthquake	2.44%	japan	52.07%	philippines	4.73%
36171262:129987795	37.64577136	Fri Mar 11 15:12:30	Fri Mar 11 15:14:48	tsunami	30.80%	earthquake	2.68%	japan	48.75%	indonesia	6.45%
36172102:129987808	16.62280969	Fri Mar 11 15:14:49	Fri Mar 11 15:15:37	tsunami	31.55%	earthquake	2.49%	japan	46.32%	indonesia	7.37%
36172387:129987813	41.9865764	Fri Mar 11 15:15:38	Fri Mar 11 15:17:52	tsunami	33.41%	earthquake	2.61%	japan	51.60%	indonesia	4.80%
36173310:129987829	36.43095551	Fri Mar 11 15:18:11	Fri Mar 11 15:19:36	tsunami	34.18%	earthquake	2.90%	japan	47.69%	indonesia	6.15%
36173848:129987837	38.42318533	Fri Mar 11 15:19:36	Fri Mar 11 15:20:59	tsunami	30.04%	earthquake	2.91%	japan	62.50%	indonesia	7.14%
36174412:129987846	55.62022589	Fri Mar 11 15:21:02	Fri Mar 11 15:23:07	tsunami	33.93%	earthquake	2.33%	japan	47.46%	indonesia	8.14%

Fig. 3. Extraction of event entities for human sensor[14]

4   Experimental Result

We verified our framework for human sensor analysis by the event "May 20, 2013 - Oklahoma Moore - Tornado", and the experimental result is described below:

4.1   Case Study: "May 20, 2013 - Oklahoma Moore - Tornado" Event

In the case study, we investigate a real-world event regarding a massive, mile-wide tornado with winds up to 200 mph, which killed at least 51 people during 40 terrifying minutes of destruction across southern Oklahoma City and its suburbs.

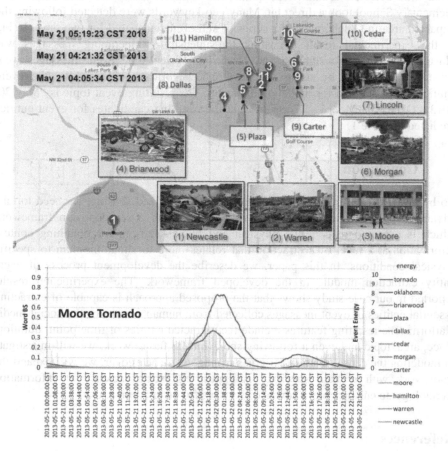

Fig. 4. The evolvement of event energy and bursty scores of extracted keywords on "May 20, 2013 - Oklahoma Moore - Tornado" event

Table 1. Resulting entities extracted from the event data sets

Term Name	Type	Candidate Feature
Event (Topic)	class	May 20, 2013 - Oklahoma Moore - Tornado
atPlace	property	Oklahoma, Moore, Briarwood, Newcastle
atTime	property	20/05/2013
circa	property	20/05/2013 to 21/05/2013
illustrate	property	tornado, briarwood, plaza

The storm struck Moore, Oklahoma, and adjacent areas. As shown in Fig 4, the event was detected around 3:00 p.m., but intensely discussed after Briarwood Elementary School took a direct hit. Many geo-keywords were detected following the impacted area in Oklahoma, and a series of sub-events appeared for two days. In Fig 4, the experimental result indicating the intensity of incoming messages was illustrated. The term weighting and clustering of real-time event messages were constantly computed by using our developed algorithms. Meanwhile, the event entities were extracted for modeling the event. The resulting event topic is "May 20, 2013 - Oklahoma Moore - Tornado" and the resulting candidates for event entities are listed in Table 1.

5   Conclusion

To be aware of real-time event situation, along with fixed sensing, we need to take advantage of social sensing. In this work, we present a sensor-fusion framework which is expected to combine physical and human sensors for obtaining context information at a level of abstraction that relates more to situations than to specific physical conditions. In this paper, we describe the development process of event entity extraction module in the developed framework. The experimental results reported in our case study shows that the proposed approach is capable of extracting essential entities of event messages collected from human sensors over a social-media platform. The resulting features include the extracted entities of time point, duration, place, area, topic and even a following sub-event. In future work, we will investigate to modeling the causal relation for crisis management. Moreover, we will expand the framework with domain specific ontologies, e.g. social science or information technology, to enhance more advanced services for other applications.

References

1. Aggarwal, C.C., Han, J., Wang, J., Yu, P.S.: A framework for clustering evolving data streams. In: 29th International Conference on Very Large Data Bases, vol. 29, pp. 81–92. VLDB Endowment Press, Berlin (2003)
2. Chi, Y., Song, X., Zhou, D., Hino, K., Tseng, B.L.: Evolutionary spectral clustering by incorporating temporal smoothness. In: 13th ACM SIGKDD International Conference on Knowledge Discovery and Data Mining, pp. 153–162. ACM Press, San Jose (2007)

3. Chou, C.H., Zahedi, F.M., Zhao, H.: Ontology for developing web sites for natural disaster management: methodology and implementation. IEEE Transactions on Systems, Man and Cybernetics, Part A: Systems and Humans 41, 50–62 (2011)
4. Feng, C., Martin, E., Weining, Q., Aoying, Z.: Density-based clustering over an evolving data stream with noise, Victoria, British Columbia, Canada (2006)
5. Gong, L., Zeng, J., Zhang, S.: Text stream clustering algorithm based on adaptive feature selection. Expert Systems with Applications 38, 1393–1399 (2011)
6. Heverin, T., Zach, L.: Microblogging for Crisis Communication: Examination of Twitter Use in Response to a 2009 Violent Crisis in Seattle-Tacoma, Washington Area. In: Seventh International ISCRAM Conference, Seattle, Washington (2010)
7. Chen, H.L., Chen, M.S., Lin, S.C.: A Framework for Clustering Concept-Drifting Categorical Data. IEEE Trans. on Knowl. and Data 21, 652–665 (2009)
8. Jurisica, I., Mylopoulos, J., Yu, E.: Ontologies for knowledge management: an information systems perspective. Knowledge and Information Systems 6, 380–401 (2004)
9. Kaneiwa, K., Iwazume, M., Fukuda, K.: An upper ontology for event classifications and relations. In: Orgun, M.A., Thornton, J. (eds.) AI 2007. LNCS (LNAI), vol. 4830, pp. 394–403. Springer, Heidelberg (2007)
10. Kleinberg, J.: Bursty and hierarchical structure in streams. In: 8th ACM SIGKDD International Conference on Knowledge Discovery and Data Mining, Edmonton, Alberta, Canada, vol. 7, pp. 91–101. ACM (2002)
11. Lee, C.H.: Mining spatio-temporal information on microblogging streams using a density-based online clustering method. Expert Systems with Applications 39, 9623–9641 (2012)
12. Lee, C.-H., Wu, C.-H., Chien, T.-F.: *BursT*: A Dynamic Term Weighting Scheme for Mining Microblogging Messages. In: Liu, D., Zhang, H., Polycarpou, M., Alippi, C., He, H. (eds.) ISNN 2011, Part III. LNCS, vol. 6677, pp. 548–557. Springer, Heidelberg (2011)
13. Lee, C.H., Wu, C.H., Yang, H.C., Wen, W.S.: Computing Event Relatedness Based on a Novel Evaluation of Social-Media Streams. In: Park, J.J., Leung, V.C.M., Wang, C.L., Shon, T. (eds.) Future Information Technology, Application, and Service. LNEE, vol. 164, pp. 697–707. Springer, Heidelberg (2012)
14. Lee, C.H., Wu, C.H., Yang, H.C., Wen, W.S., Chiang, C.Y.: Exploiting Online Social Data in Ontology Learning for Event Tracking and Emergency Response. In: 2013 IEEE/ACM International Conference on Advances in Social Networks Analysis and Mining, pp. 1167–1174. IEEE Press, Niagara Falls (2013)
15. Sakaki, T., Okazaki, M., Matsuo, Y.: Earthquake shakes Twitter users: real-time event detection by social sensors. In: 19th International Conference on World Wide Web, pp. 851–860. ACM Press, Raleigh (2010)
16. Sayyadi, H., Hurst, M., Maykov, A.: Event detection and tracking in social streams. In: 3rd AAAI International Conference on Weblogs and Social Media, ICWSM, pp. 311–314. AAAI Press, San Jose (2009)
17. Shaw, R., Troncy, R., Hardman, L.: LODE: Linking open descriptions of events. In: Gómez-Pérez, A., Yu, Y., Ding, Y. (eds.) ASWC 2009. LNCS, vol. 5926, pp. 153–167. Springer, Heidelberg (2009)
18. Wang, S., Yan, J., Xu, K., Liu, Y., Liu, L., Wang, H.: A conceptual modeling approach to quality management in the context of diary supply chain. In: 2nd International Conference on Information Science and Engineering (ICISE), pp. 13–16. IEEE Press, Hangzhou (2010)
19. Wang, W., Zhao, D.: Ontology-Based Event Modeling for Semantic Understanding of Chinese News Story. In: Zhou, M., Zhou, G., Zhao, D., Liu, Q., Zou, L. (eds.) NLPCC 2012. CCIS, vol. 333, pp. 58–68. Springer, Heidelberg (2012)

20. Ye, K., Wang, S., Yan, J., Wang, H., Miao, B.: Ontologies for crisis contagion management in financial institutions. Journal of Information Science 35, 548–562 (2009)
21. Ye, K., Yan, J., Wang, S., Wang, H., Miao, B.: Knowledge level modeling for systemic risk management in financial institutions. Expert Systems with Applications 38, 3528–3538 (2011)
22. Zhao, Q., Mitra, P., Chen, B.: Temporal and information flow based event detection from social text streams. In: The National Conference on Artificial Intelligence, pp. 1501–1506. AAAI Press, MIT Press, Menlo Park, Cambridge (2007)
23. Zhu, Y., Shasha, D.: StatStream: statistical monitoring of thousands of data streams in real time. In: 28th International Conference on Very Large Data Bases, pp. 358–369. VLDB Endowment Press, Hong Kong (2002)
24. Tsai, P.H., Lin, Y.C., Ou, Y.Z., Chu, E.T.H., Liu, J.W.S.: A Framework for Fusion of Human Sensor and Physical Sensor Data. IEEE Transactions on Systems, Man, and Cybernetics: Systems, 2168–2216 (2013)
25. Wang, D., Le, H., Kaplan, L., Abdelzaher, T.: On Truth Discovery in Social Sensing: A Maximum Likelihood Estimation Approach. In: 11th International Conference on Information Processing in Sensor Networks, pp. 233–244. ACM Press, New York (2012)
26. Schmidt, A., Beigl, M., Gellersen, H.W.: There is more to context than location. Computers & Graphics 23, 893–901 (1999)
27. Chu, E.T., Chen, Y.L., Lin, J.Y., Liu, W.S.: Crowdsourcing support system for disaster surveillance and response. In: 15th International Symposium on Wireless Personal Multimedia Communications, pp. 21–25. IEEE Press, Taipei (2012)

Mining the Most Influential Authors in Academic Publication Networks through Scholastic Actions Propagation

Shing H. Doong

ShuTe University
59 HenShan Rd., Yanchao District, Kaohsiung City, Taiwan
tungsh@stu.edu.tw

Abstract. The influence degree of academic authors can be judged in one way by citation-based indices such as the h-index [1]. Marketing researchers develop a theory of action propagation to measure consumers' influence in a market; by going viral, an action of marketing essential such as product adoption or brand awareness is spread to the majority of a market autonomously. Selecting the most influential consumers as marketing seeds is a hot topic in marketing research. In this study, we mine the most influential authors from the perspective of their ability to propagate scholastic actions such as attending a serial conference or publishing in a specific journal. The credit distribution model from Goyal et al. [2] is chosen as the propagation model with two academic publication networks (citation and coauthoring). Real data consisting of 10 years publication records from DBLP and ACM were used in the experiments. It is found that the citation network is more efficient than the coauthoring network to propagate scholastic actions. Top influential authors who can effectively affect fellows to attend a serial conference or publish in a specific journal are mined with the citation publication network.

Keywords: Influence diffusion, action propagation, viral marketing, academic publication network.

1   Introduction

The scientific importance of an academic journal is usually measured by its impact factor. Academic authors with publication in outlets of high impact factors are no doubt influential in one sense, because their articles may be cited more often than others. In order to balance quality and quantity of scientific research work, Hirsch proposed the h-index to measure the impact and productivity of a scholar [1]. In addition to the citation based index, we can also measure the influence degree of a scholar by investigating how he or she can impact others to take some scholastic actions such as attending a serial conference or publishing in a specific journal. Action based influence degree and its applications have long been a hot research issue in marketing theory.

L.S.-L. Wang et al. (Eds.): MISNC 2014, CCIS 473, pp. 199–212, 2014.

The dream of most marketers in today's social media era is for a marketing campaign to *go viral*. By going viral, the marketing message is propagated to the most intended population like a human virus without too much marketing effort. Viral marketing was previously based on the word-of-mouth communication between people, and is now founded on the rapid growth of social network sites like Facebook or Twitter. Finding the most influential customers as marketing seeds for a campaign so that a majority of the market eventually accepts the campaign has practical applications in marketing practices.

We assume that influence is conducted through social networks of people who take actions according to their own belief and influence exerted by neighbors that have taken the action earlier. Actions are defined broadly here such as adopting a product in marketing campaigns, joining a community in online social networks or attending a serial conference in academic publication network. Finding a small number of seeds so that most people of the network can be influenced to take the same action has been called the influence maximization problem in literature [3][4].

Influence maximization has many interesting applications in marketing theory and practices. On the other hand, the academic community still prefers citation-based criteria to judge the importance of a researcher. We argue that citation-based indices may represent the quality and impact level of academic papers, but they may not faithfully represent the influence degree of a researcher on his or her fellows in terms of scholastic actions. Scholastic actions include attending a serial conference such as the annual Hawaii International Conference on Systems Sciences (HICSSS) or publishing in a specific journal like the Communication of ACM. A specific reason for the argument is these indices are primarily based on the citation numbers of academic papers, and they do not take into account the presence of scholastic actions.

If we recognize that scholastic actions stated above are also an important ingredient of an academic life, it is interesting to study who are the most influential scholars in terms of their professional actions. This study is concerned with finding influential scholars that propagate scholastic actions effectively in academic networks. We adopt a data mining approach to investigate the issue. Due to the availability of empirical data, two types of academic networks are considered here: the citation publication network and the coauthoring publication network. We intend to discover which type of the network is more efficient to propagate scholastic actions, and mine the most influential researchers from this academic publication network.

This paper is organized as follows. In section 2, we discuss literature related to influence diffusion, classical models of influence propagation and the specific propagation model used in this study. In the following section, we describe how to collect the experimental data and to preprocess them. Statistics of these data are also presented. In section 4, we present experimental results and discuss implications of these results. We conclude the paper with a few remarks in section 5.

2   Literature Review

An old Chinese proverb says "A picture is worth a thousand words". In marketing practices, an action is worth a thousand verbal promises. Marketers are happier if they see that intended customers take the actions proposed in a marketing campaign. Word-of-mouth marketing practices have been attempted by major players in the industry. However, it was difficult to start such a campaign due to the knowledge lack of social networks among customers. In today's digital media era, virtual (online) social networks can often be inferred from the traces left from web surfing. Some social network sites even ask their users to officially acknowledge a link request before it can be established.

With such a bundle of online social networks available to practical applications, researchers from all areas including marketing, physics, mathematics, information sciences and etc. have established a new field of scientific research: how to effectively use the available network data to maximize the propagation of intended actions. In this section, we review literature regarding influence diffusion via social networks, classical influence propagation models and the specific propagation model used in this study.

2.1   Influence Diffusion

Can influence be diffused through social networks? An old doctrine in sociology is the concept of homophily which explains the tendency of becoming friends between two persons with similar characteristics. Two people may form the link of a social network because they like the same actions such as joining a special discussion board. In this case, influence of actions is not propagated through the network; on the other hand, the social tie is founded on the same action taken. Confounding is another explanation for correlations between actions taken and socially connected actors. Anagnostopoulos et al. defined confounding as external influences from the environment to affect both the tie formation and actions taken [5]. For example, if two people live in the same city (geographical confounding factor), they are more likely to form a social tie and take the same scenic pictures (actions taken).

A major contribution from Anagnostopoulos et al. [5] is to distinguish genuine social influence (through network) from social correlation founded on homophily or confounding. Statistical tests based on randomization were proposed to test the genuine degree of social influence. Based on the proposed procedure, the researchers found that though "there is significant social correlation in tagging pictures on the Flickr system, this correlation cannot be attributed to social influence" [5].

On the other hand, Watts and Dodds [6] have questioned the validity of an influential hypothesis that has been assumed in marketing theory and practices. The hypothesis assumes that by activating a small number of influential nodes, an exceptional number of their peers can be activated later on. Through computer simulations of interpersonal influence processes, they found that a critical mass of early adopters must be activated in order to reach a high spread level of the actions. Watts and Dodds called this the "big seed" marketing.

In this study, we assume that scholastic actions can be influenced through academic publication network. For example, when I know an author whose articles have been cited by me (the citation network) attended a conference, I may follow his trace to attend the same conference. Similarly, the influence of actions may be propagated through the coauthoring network. Whether taking the same scholastic actions is founded on homophily or confounding is out of scope of this study.

2.2 Two Classical Models for Influence Propagation and Maximization

Influence propagation models have been studied in [3], which used Markov random fields to model the expected lift of profits in undirected networks. After Kempe et al. published their seminal paper in [4], the issue of influence propagation and maximization has gained the attention of many researchers. Kempe et al. assumed that a social network, directed or undirected, is known in advance. An initial set of seeds is selected and activated for some marketing campaign. These seeds activate their social neighbors through certain random mechanisms under two assumptions: (1) once a node is activated, it stays active for the rest of the influence propagation process; and (2) the more active friends an inactive node has, the more likely it will be activated. The goal of influence maximization is to choose seeds of a preset size so that most nodes can be activated when the influence propagation process stops. In the following, we describe the two basic influence propagation models introduced in [4].

Linear Threshold Model (LTM). Let $N(v)$ contain all network neighbors of a node v, and each $w \in N(v)$ exerts a weight b_{vw} to activate v. The weights are fixed properties of a network with the sum from all neighbors of a node no more than 1. Influence propagation proceeds in discrete steps of time. At the start of each simulation run, each node v chooses a random threshold θ_v uniformly from $U(0,1)$ to indicate his satisfactory level of active neighbors in order to be activated. At $t=0$, only seed nodes are active. Let A_t denote nodes activated up to time t. Since an active node cannot be turned inactive, the set of active nodes A_t increases as time progresses. For the next time step $t+1$, an inactive node is activated once the total weight from its active neighbors in A_t is larger than his chosen threshold θ_v. To obtain a stable expected spread value, 10000 simulation runs are normally conducted with each run starting with different thresholds.

Independent Cascade Model (ICM). In the independent cascade model, each node v has a probability of p_{vw} to successfully activate its neighbor $w \in N(v)$. The influence propagation process advances in discrete steps of time. At time t, each newly activated node $v \in A_t$ has a one shot chance to activate each inactive neighbor w with the probability p_{vw}. When the activation is successful, the neighbor becomes a newly activated node in A_{t+1}; otherwise, the node v does not have any further chance to activate w. The simulation process continues until no new nodes can be activated. In ICM, influence probabilities p_{vw} are fixed properties of the network. In each simulation run, these probabilities are compared with numbers randomly generated from $U(0,1)$ to decide whether a trial of activation is successful or not. Again, 10000 runs of simulation are conducted to calculate the expected spread of the set of seeds.

It is time consuming to conduct 10000 runs of simulation in either LTM or ICM to calculate the expected spread of a seed set. Kempe et al. [4] showed that finding the optimal seeds is an NP-hard problem. However, because the objective function is monotone and submodular, a simple greedy algorithm provides a guarantee level (around 67%) of the optimal performance. The simple greedy algorithm is to add the seed one by one. At each stage of the algorithm, a node that makes the most contribution to the objective function is added to the seed set. Even though effective heuristics such as CELF in [7] have been proposed to mitigate the tedious simulation steps, the simple greedy algorithm still does not scale well to large networks commonly found in today's social network sites. Recent research in this field has focused on designing good heuristics to attack really larger social networks [8][9].

2.3   The Credit Distribution Model

In addition to the scaling problem, another critical issue associated with LTM and ICM is the learning procedure for network properties, i.e. the influence weights b_{vw} in LTM or influence probabilities p_{vw} in ICM. Not until recently, most influence maximization studies have assumed that the weights or probabilities are pre-assigned for a network. Saito et al. proposed an expectation-maximization based solution to learn influence probabilities in ICM [10]. On the other hand, Goyal et al. used the concept of partial credits to learn the weights in LTM [11].

Based on the successful implementation of partial credits algorithm, Goyal et al. proposed a new objective function in [2] to find optimal seeds in influence propagation. We have selected the credit distribution model of [2] for influence maximization in this study because of two reasons: (1) Recent studies have shown that network properties such as weights and probabilities should be learned from real data; data based approach for influence propagation should be considered when real data are available [12][13][14]. (2) The model should be effective and efficient based on real data, i.e., it should learn influence probabilities efficiently and infer influence propagation effectively.

Using the credit of influencing a node based on observed actions, Goyal et al. [2] considered the following objective function in influence propagation problems:

$$\sigma_{cd}(S) = \sum_{u \in V} \kappa_{S,u} \tag{1}$$

In equation (1), $\kappa_{S,u}$ is the total credits given to the set S for influencing u to take some actions. The total influencing credits for the set is defined as the sum of credits for influencing each individual node u. Goyal et al. showed that influence maximization with the credit distribution model is NP-hard. However, like the ICM or LTM, this objective function is monotone and submodular, thus the simple greedy algorithm provides a guaranteed level (around 67%) of the optimal performance.

Efficient algorithm has been designed to compute incremental influence credits $\sigma_{cd}(S + x) - \sigma_{cd}(S)$, which is needed in the simple greedy algorithm. In their experiments, Goyal et al. showed that based on real action log data, the credit distribution model predicts actions more accurately than ICM with probabilities learned from [10] or LTM with weights learned from [11]. For detailed algorithm, we refer the reader to [2].

2.4   Academic Publication Network

In order to use the credit distribution model, empirical data should be prepared accordingly. Two types of data are needed for the credit distribution model: social networks and action logs. As explained in [12], one of these two types of data is often directly available while the other one has to be inferred or synthesized by some other means. A social network consists of a set V of nodes and a set E of links (ties, edges) between nodes. Links may be directed or undirected. Action logs contain data in the form of $(u, a, t_u(a))$ where u is a member of V, a is a type of action and $t_u(a)$ the time when u performs the action a.

Though the original influence propagation models in [4] seem to imply a unique type of actions such as adopting a special product/brand, data based influence propagation approaches [2][10][13] tend to consider different types of actions, for examples attending different serial conferences. Thus, in terms of our study goal, attending HICSS and ICIS (International Conference on Information Systems) will be considered two types of actions in the data based approach. If user u does not perform action a, its time of action $t_u(a)$ may be considered infinite. In practice, the log set will not contain an entry for this triplet.

To fulfill our study goal, we have selected publication data from DBLP and ACM for the experiments. Detailed information about the data will be described in the next section. Each publication record includes name of author(s), year and outlet (conference name or journal name) of publication, and references in the publication among other useful information. Thus, an entry in the action log may look like ('John Doe', 'HICSS', 2011). For our purpose of scholastic actions propagation, it is easy to construct action logs out of these records.

Unfortunately, social networks of researchers must be synthesized from the publication records. We consider two types of social networks in the study: citation publication network and coauthoring publication network. It is easy to synthesize the coauthoring network. When authors u and v publish in the same (conference or journal) paper in year yr, an undirected link (u, v, yr) is established. For data based approach in influence propagation [13][14], it is important to verify that a network link is established before both users take the same action. That is, yr is less than or equal to both $t_u(a)$ and $t_v(a)$. This is to make sure that influence is genuinely propagated through established links. Otherwise, we might end up with the homophily situation, i.e., links are established because of similar actions taken by neighbors. The homophily case can happen when both $t_u(a)$ and $t_v(a)$ are less than or equal to yr.

It is trickier to synthesize the citation network. Each publication record contains references cited by that publication. Thus, a directed link is established from every author of a cited article to every author of the citing article. The establishing time of the link is the year when the citing article was published. Similar to the coauthoring network, we need to make sure that the link establishment time is earlier than action time of both users taking the same actions.

3   Data for a Mining Project

Large data sets of publication records are available online today. However, due to varying styles of name presentation, the same author may have different types of spelling in different publication outlets. On the other hand, the same spelling and presentation of the author name may point to different authors in reality. Entity recognition and disambiguation is an important research issue in digital library [15]. To save the efforts of downloading data from DBLP, we choose to use the processed data from ArnetMiner for our experiments [16]. We explain how to preprocess data downloaded from http://www.arnetminer.org for our research purpose.

3.1   Data Description and Statistics

We downloaded the DBLP-Citation-Network V5 data set from ArnetMiner website. This corpus includes 1,572,277 papers and 2,084,019 citation relationships. The citations come from both DBLP and ACM databases. Each publication record in the data set contains the following fields that are relevant to this study: authors with each name separated by a comma, year of publication, publication venue (conference name or journal name), index id of this record, the id of references cited by this paper. Since nodes of our social networks are academic authors, names must be separated from commas if a paper has multiple authors. An example of the publication records is shown below followed by the record template [16].

> #*Spatial Data Structures.
> #@Hanan Samet
> #year1995
> #confModern Database Systems
> #citation2743
> #index25
> #arnetid27
> #%165
> #!An overview is presented of the use of spatial data structures in spatial data bases. The focus is on hierarchical data structures, including a number of variants of quadtrees, which sort the data with respect to the space occupied by it....

> #* --- paper title
> #@ --- authors
> #year ---- year
> #conf --- publication venue
> #citation --- citation number (both -1 and 0 means none)
> #index ---- index id of this paper

#arnetid ---- pid in arnet database
#% ---- the id of references of this paper (multiple lines for multiple references)
#! --- Abstract

For the citation network, we used citation ids to find authors of a cited article. Directed links are established from the cited authors to the citing authors. In this case, we assume that the cited authors start to have influences on the citing authors beginning from the publication year of the citing article. With both the coauthoring network and citation network, it may happen that a pair of nodes has several links established in different years. If this happens, we choose the earliest year to represent the link year for the nodes.

We narrow down the data set to include papers published between 2001 and 2010. With this specification of the base data set, there are still 242,002 articles spanning over 7917 venues (the same serial conferences in different years are counted as different venues, e.g., HICSS 2004 and HICSS 2005 are different venues.) The base data set contains 244,108 different authors (nodes), which make our social networks prohibitively large to be handled in a reasonable time. The number of venues is also too big for the current study. We further process these two issues in the following.

3.2   Data Preprocessing

The base data set is scanned to count the publication frequency of each author. The number of authors who have published more than 10, 20, 30, 40 and 50 articles in the base data set is respectively 11273, 3360, 1322, 592 and 317. Similarly, the number of serial venues that have appeared in more than 2 and 4 years is respectively 1044 and 581. We denote these sets of venues as C_2 and C_4 respectively. Thus, the HICSS venue will be counted as a single record in C_2 and C_4. In order to balance the processing speed and the interestingness of results, we consider the subsets of authors who have published more than 20, 30 and 40 articles in the base data set. These author sets are denoted as V_{20}, V_{30} and V_{40}. After rescanning the base data set with authors in V_{20}, V_{30} and V_{40} only, the number of articles has shrunk to 78010, 46672 and 28794 respectively. The shrunk data sets of articles are denoted as D_{20}, D_{30} and D_{40} respectively. Using the shrunk article corpus and the method described above including the aggregation of multiple links with different establishment years, we form the citation network and the coauthoring network with V_{20}, V_{30} and V_{40} respectively as nodes. Statistics of these networks are reported in Tables 1 and 2.

Table 1. Statistics of the citation network

Node set	Number of nodes	Number of edges
V_{20}	3360	119375
V_{30}	1322	42437
V_{40}	592	16179

Table 2. Statistics of the coauthoring network

Node set	Number of nodes	Number of edges
V_{20}	3360	13944
V_{30}	1322	4819
V_{40}	592	1849

The action logs are formed by using the shrunk data set of articles D_{20}, D_{30} and D_{40} with venues C_2 and C_4. The log set formed by D_{xx} and C_y will by recorded as Log_{xxy}. For example, the log set Log_{202} is formed with D_{20} and C_2, and this log set works with networks with nodes V_{20}. Sizes of these log sets are reported in Table 3.

Table 3. Size of log sets composed of article sets and venue sets

	C_2	C_4
D_{20}	46988	37229
D_{30}	23828	18850
D_{40}	12856	10156

Note that the same log file is used for both the citation and coauthoring networks with the appropriate node set. We intend to use the same set of action logs to investigate the efficiency of action propagation in different networks.

4   Experiments and Results

The first purpose of experiments is to find out which academic publication network (citation or coauthoring) is more efficient to propagate scholastic actions. For this purpose, the set of venues (actions) will be divided into a training set and a test set. The training set consists of training actions and associated action traces. Thus, if the HICSS venue belongs to the training set, all log records with HICSS as the venue such as (u, 'HICSS', yr) will be in the training set. On the other hand, if venue ICIS belongs to the test set, then all log records like (v, 'ICIS', yr) are included in the test set. Log records of the same venue are never divided into training and test sets. The training set is used to train the credit distribution model, while the test set is used to validate the trained model. For the C_2 based log files, the ratio of the number of training actions over the number of test actions is around 2.25, and for the C_4 based log files, this ratio is around 3.3.

The trained model is validated with the test set where nodes of an action without pre-activated parents in the social network are considered seeds of the action. The ground truth for the spread of this action is obtained from the log file: all those nodes that take the action with influencing parents or not belong to the final spread of this action. We assess the spread from two perspectives: (1) the spread rate (spr) from all actions in the test set; (2) the improvement rate (imp) over the seeds. Let s be the number of seeds, p be the number of predicted nodes and g be the number of ground truth nodes, then the spread rate and improvement rate are defined respectively as:

$$spr = \frac{p}{g} \tag{2}$$

$$imp = \frac{p-s}{g} \tag{3}$$

It is desirable to have both rates as high as possible. The higher improvement rate indicates a higher propagation efficiency level over the network. We then mine the top 50 influential authors from the network of a higher efficiency level.

4.1   Prediction Assessments

The seed number (s), predicted number of spread (p), ground truth number (g), spread rate (spr) and improvement rate (imp) for the citation network are reported in Table 4. Results of prediction assessment for the coauthoring network are reported in Table 5.

Table 4. Assessments on test set of different log files with the citation network

	Log_{202}	Log_{204}	Log_{302}	Log_{304}	Log_{402}	Log_{404}
s	6403	6576	3185	3244	1720	1735
p	9158.4	9895.158	4431.45	4705.16	2333.25	2439.426
g	11448	12082	5718	6040	3050	3214
spr	0.8	0.819	0.775	0.779	0.765	0.759
imp	0.241	0.274	0.218	0.242	0.201	0.219

Table 5. Assessments on test set of different log files with the coauthoring network

	Log_{202}	Log_{204}	Log_{302}	Log_{304}	Log_{402}	Log_{404}
s	8842	9347	4466	4696	2441	2541
p	9685.008	10390.52	4808.838	5121.92	2610.8	2754.398
g	11448	12082	5718	6040	3050	3214
spr	0.846	0.86	0.841	0.848	0.856	0.857
imp	0.074	0.087	0.06	0.07	0.056	0.066

One way to interpret the results is the coauthoring network has generally yielded a higher spread rate. However, if we compare these two networks log file by log file, we see that the spread improvement over the citation network is not very significant. On the other hand, the citation network has provided a higher improvement rate than the coauthoring network and the advantage is significant, e.g. 0.241 over 0.074 for the Log_{202} data set. An explanatory reason for the above phenomena is the coauthoring network uses a substantially large number of seeds. We declare the citation network is more efficient to propagate scholastic actions because of its higher improvement rate.

4.2   The Top 50 Influential Authors of the Citation Network

Using the citation network as the social network to propagate scholastic actions, we mine the top 50 influential authors from the whole log file. This time, the entire log file is used to estimate the influence credit distribution model, and the simple greedy algorithm is used to find the 50 most influential persons of the citation network. For the top 5 authors in the list, it appears that author publication threshold value is more important than the venue threshold value. That is, Log_{202} and Log_{204} have produced the same top 5 authors with different orders. Similar phenomenon is observed for the Log_{30y} and Log_{40y} files. In Table 6 we list the top 5 authors for the C_2 venue set. Three authors (Mahmut T. Kandemir, Hector Garcia-Molina and Luca Benini) have appeared in all three publication threshold values.

Table 6. The top 5 influential authors to propagate scholastic actions in the citation network

Log_{202}	Log_{302}	Log_{402}
Mahmut T. Kandemir	Luca Benini	Mahmut T. Kandemir
Hector Garcia-Molina	Hector Garcia-Molina	Luca Benini
Ian T. Foster	Jiawei Han	Jiawei Han
Mani B. Srivastava	Mahmut T. Kandemir	Hector Garcia-Molina
Luca Benini	Deborah Estrin	Deborah Estrin

4.3   Characteristics of the Top Influential Authors

Since objective function (equation (1)) of the credit distribution model is monotone and submodular, the simple greedy algorithm is used to mine top influential authors. According to the algorithm, node making the most contribution to the objective function is added to the seed set one by one. Take Log_{402} as an example, Mahmut T. Kandemir is the author that makes the most contribution to the function of equation (1) among all authors. Then, Luca Benini is the author that maximizes the incremental contribution $\sigma_{cd}(\text{Kandemir} + x) - \sigma_{cd}(\text{Kandemir})$. The other authors can be interpreted similarly.

An interesting question arises: what characteristics do these influential authors possess? Do they have the most ties? Previous research [4] has shown that node degree is not the characteristic to maximize information propagation; we also observed this phenomenon in the result. For example, Mahmut T. Kandemir has an out degree of 39, i.e., influencing 39 other authors through the citation network, but this just puts him in the 133rd place of the list of authors ordered by a decreasing out degree. Thus, he would not make it to the top 50 influential authors should we use the out degree to select authors. What Mahmut T. Kandemir possesses uniquely is he has published in 12 venues (number 1) in 2001 (the beginning year). The other four authors also have similar characteristics. For example, Hector Garcia-Molina published in 8 venues (number 4) in 2001. In terms of the out degree ranking, Hector Garcia-Molina is number 1 with an out degree of 182. Citation-based indices will probably choose Hector Garcia-Molina as the number 1 influential author, but he ranked number 4 in our list.

The other three authors (Luca Benini, Jiawei Han, Deborah Estrin) all ranked within the 50th place in both ranking systems (number of publication venues in 2001 and the out degree). Thus, to become an influential author in the credit distribution model, it seems that an author needs to publish early in more diversified venues. This will broaden his or her influencing power to propagate scholastic actions.

4.4   Discussion

It is commonly believed that coauthoring relationship is more powerful to propagate scholastic actions among researchers. For example, a researcher may be influenced by his or her coauthors to attend a specific conference or publish in a special journal. This belief seems well grounded if we consider the situation that junior researchers often follow the footprint of their senior fellows in scholastic actions. From our empirical study of the citation data obtained from DBLP and ACM, we observed that the coauthoring network is not as efficient as the citation network to propagate scholastic actions. That is, a researcher seems more likely to follow the footprint of the fellows whose work has been cited by the researcher. The coauthoring network has provided a higher spread rate only because in that network, more researchers are considered seeds for most actions. This could be attributed to the relatively sparse network structure of the coauthoring network. Comparing the results in Tables 1 and 2, we see that the citation network can provide 10-fold links than the coauthoring network, thus providing more chances for scholastic actions to propagate through this network.

5   Conclusion

Whether a researcher is influential in the academic publication network can be judged in different perspectives. Traditionally we use citation-based indices such as the h-index to measure the quality and quantity of scientific output of a researcher. But, a high h-index may not imply the researcher is influential enough to affect fellows to take some scholastic actions. In this study, we define scholastic actions as those activities of attending serial conferences or publishing in a special journal. As scholastic actions are becoming more important in a researcher's academic life, it is interesting to find influential authors who can motivate other fellows to take some scholastic actions.

Action propagation has been studied and analyzed by marketing researchers for some time now. A key mission in viral marketing is to find influential seeds (preferably a small seed set due to the marketing expenses) to start a marketing campaign and hope that the campaign will be carried well into the mass majority via these seeds and their followers. Though finding optimal seeds is computationally expensive (NP-hard), several alternative models and heuristics have been proposed to mitigate this issue.

In this study, we apply influence maximization theory to find the most influential authors who can effectively propagate scholastic actions to their fellows. It is found that the citation network can propagate scholastic actions more efficiently than the coauthoring network. However, it is not clear whether this is caused by a sparser network structure of the coauthoring network. Future research may use more empirical data of all sorts to confirm or refute the above finding.

Using the citation network as the vehicle to propagate influences on scholastic actions, it is found that the most influential authors may not possess very high out degrees. Out degrees in a citation network are major components used in most citation-based indices for scholars' assessment. In order to affect more fellows to take similar scholastic actions, it is better to make diversified actions earlier, i.e., to publish in more venues at an early stage.

Besides finding the most influential authors in academic publication network, the theory of influence maximization can be applied to other areas as well, for example, to find the influential users of an appraisal site. Similar to the current study, we expect that either the social network structure or the actions log has to be inferred from other sources before the theory is applicable.

Acknowledgments. This work is supported in part by grants from the ministry of science and technology (Taiwan) under the contract numbers MOST-103-2410-H-366-001 and MOST-103-2632-E-366-001. The author appreciates valuable comments from two anonymous reviewers to improve quality of the paper.

References

1. Hirsch, J.E.: An Index to Quantify an Individual's Scientific Research Output. PNAS 102(46), 16569–16572 (2005)
2. Goyal, A., Bonchi, F., Lakshmanan, L.V.S.: A Data-Based Approach to Social Influence Maximization. Proceedings of the VLDB Endowment 5(1), 73–84 (2012)
3. Richardson, M., Domingos, P.: Mining Knowledge-Sharing Sites for Viral Marketing. In: SIGKDD 2002. ACM (2002)
4. Kempe, D., Kleinberg, J., Tardos, E.: Maximizing the Spread of Influence through a Social Network. In: SIGKDD 2003. ACM (2003)
5. Anagnostopoulos, A., Kumar, R., Mahdian, M.: Influence and Correlation in Social Networks. In: KDD 2008. ACM (2008)
6. Watts, D.J., Dodds, P.S.: Influentials, Networks, and Public Opinion Formation. J. Consumer Research 34, 441–458 (2007)
7. Leskovec, J., Krause, A., Guestrin, C., Faloutsos, C., VanBriesen, J., Glance, N.S.: Cost-effective Outbreak Detection in Networks. In: KDD 2007. ACM (2007)
8. Wang, C., Chen, W., Wang, Y.: Scalable Influence Maximization for Independent Cascade Model in Large-scale Social Networks. Data Min. Knowl. Disc. 25(3), 545–576 (2012)
9. Chen, W., Yuan, Y., Zhang, L.: Scalable Influence Maximization in Social Networks under the Linear Threshold Model. In: 2010 IEEE Int. Conf. on Data Mining, pp. 88–97. IEEE Press, New York (2010)
10. Saito, K., Nakano, R., Kimura, M.: Prediction of Information Diffusion Probabilities for Independent Cascade Model. In: Lovrek, I., Howlett, R.J., Jain, L.C. (eds.) KES 2008, Part III. LNCS (LNAI), vol. 5179, pp. 67–75. Springer, Heidelberg (2008)

11. Goyal, A., Bonchi, F., Lakshmanan, L.V.S.: Learning Influence Probabilities in Social Networks. In: Proceedings of WSDM 2010, New York City (2010)
12. Chen, W., Lakshmanan, L.V.S., Castillo, C.: Information and Influence Propagation in Social Networks. Morgan & Claypool Publishers, San Francisco (2014)
13. Bonchi, F.: Influence Propagation in Social Networks: a Data Mining Perspective. In: WI-LAT, pp. 573–582 (2011)
14. Mathioudakis, M., Bonchi, F., Castillo, C., Gionis, A., Ukkonen, A.: Sparcification of Influence Networks. In: KDD 2011. ACM (2011)
15. Tang, J., Fong, A.C.M., Wang, B., Zhang, J.: A Unified Probabilistic Framework for Name Disambiguation in Digital Library. IEEE Trans. Knowl. Data Engi. 24(6), 975–987 (2012)
16. Tang, J., Zhang, J., Yao, L., Li, J., Zhang, L., Su, Z.: ArnetMiner: Extraction and Mining of Academic Social Networks. In: KDD 2008. ACM (2008)

A New Method of Identifying Individuals' Roles in Mobile Telecom Subscriber Data for Improved Group Recommendations

Saravanan Mohan[1] and Manisha Subramanian[2]

[1] Ericsson Research India, Chennai, India
[2] Ericsson India Global Services, Chennai, India
{m.saravanan,manisha.subramanian}@ericsson.com

Abstract. The presently available methods are highly capable of offering personalized recommendations to individuals and also group recommendations to the subscriber hub. Operators who generate recommendations expect the information to be dissipated to a large number of subscribers in a specific space and time. In this paper, our interest is to combine the concepts of recommendations and information dissipation to identify the unique role players who can spread information rapidly with more specific to mobile telecom communities. It generates two improvements: reduces the extra cost incurred by the service provider in advertising to the entire community and also establishes a spatio-temporal mechanism to understand the sphere of influence of users in a dynamic environment. This is done by considering some essential factors to assign weights to each user in terms of the amount of influence they can exert within the group. Recent studies indicate that the feedback of a campaign is higher when recommended by peers rather than by operators. Our proposed method has another advantage that it can easily be extended to social network communities for similar purposes. We have also looked into the roles played by the individuals for different recommendations and the significance of the different role players' impact on the community with the help of domain experts. Finally, we have evaluated our proposed method on mobile telecom data and compared with customer ranking algorithm statistically through paired t-test method.

Keywords: Mobile Telecom Data, Group Recommendation, Individual Role Player, Mobile Network Communities, Information Diffusion.

1 Introduction

As per the Handbook of Recommender Systems [1], Recommender Systems are software tools and techniques providing suggestions for items to be of use to a user. For example, if we consider mobile telecom networks, we can generate recommendations to individual users based on their interests that can be determined from their communication patterns. The interests that can be determined from

L.S.-L. Wang et al. (Eds.): MISNC 2014, CCIS 473, pp. 213–227, 2014.

communications include the browsing history, usage of different applications and schemes used by a particular user.

Group Recommender Systems are an extension of recommender systems and they generate recommendations to specific groups of users instead of single users [1]. A group recommender for a mobile subscriber could generate recommendations to groups of users, where the groups are generated by clustering people of similar interests (location, plan type, demographic profile, etc.) together. The generation of groups for a group recommendation system is a separate problem which does not depend on the recommendation system itself. It involves the usage of community mining algorithms to form groups. A group or community refers to a set of people who are connected to each other by one or more social relationships. Community mining is closely related to the ideas of graph partitioning in graph theory and computer science, and hierarchical clustering in sociology [2].

For effective dissipation of information by influencers within a community, three criteria must be satisfied[3]:

1. Existence of social influence within a community
2. Presence of influencer who exert disproportionate influence on their peers
3. Ability of service providers to identify such influencers

The diffusion of information within a network starts with the dissipation of information by individual nodes. This may be in the form of voice calls, SMSs and, recently through, emails as far as a mobile Telecom network is concerned. Social influence is defined as "A social phenomenon that individuals can undergo or exert, also called imitation, translating the fact that actions of a user can induce his connections to behave in a similar way" [4]. The more influential or connected a node is, the faster the information spreads. Identifying the most influential spreaders in a network is critical for ensuring efficient diffusion of information. For instance, a campaign can be optimized by targeting influential individuals who can trigger large cascades of further adoptions [4]. However, no individual can be a universal influencer, and members tend to be influential in only a limited number of peer groups. To calculate the influence of each user, the entire network must be divided into smaller sub-networks or communities. One can then say that every community has a few nodes that have the ability to influence others in the community: these nodes are said to be influential nodes [5] and the community in which these nodes can exert their influence is known as their sphere of influence. By targeting these influential users, we can ensure that the entire community is covered. Moreover communities are dynamic in nature [6]. In a mobile telecom network, the same users do not remain active at all times. According to [7], a typical day can be divided into 3 time periods based on the level of activity. Further, a user may use one of several different methods to connect with her peers (voice calls, SMS, emails, etc.). The role of influential users is identified according to the time period and level of activity in the network.

Users in a network may be classified as information disseminators and information receptors [8]. The measure of the influence exerted by an information disseminator depends on the level of consumption of information by the receptors. This means that we need to maintain scores for both disseminators and receptors. For this, we developed a new algorithm based on the HITS algorithm [8, 4, 6] to determine the score for the disseminators and receptors in a community. The new algorithm which we will call MOBILE-HITS (M-HITS) algorithm will take into account 3 new creative factors (*Recurrence, Relevance* and *Recency*) to determine the users who possess the maximum amount of influence within the community at the moment when recommendations have to be dispatched. The basic flow of the new method which we have addressed in this paper will be determined in three steps as follows:

Step 1: Extract spatio-temporal data from the original dataset

Step2: Apply community mining algorithms to the spatio-temporal data to obtain reasonably sized communities

Step3: Introduce new M-HITS algorithm by defining R^3 factors to find the *Role Players*

Hence, in this paper, we state that it is sufficient for a group recommender system to communicate only with the role players of a group and these influential members will be responsible for the diffusion of information throughout the group. We illustrated our proposed algorithm through specific applications using mobile telecom subscriber data. Also we evaluate our algorithm by observing the statistical significance between the results obtained from traditional and our proposed method.

2   Related Work

Group recommendations are much difficult to generate than personalized recommendations for individuals. This is because when we consider a group of people, it is not always possible to generate a recommendation which satisfies every member in the group. Clustering or community mining algorithms are used to determine the groups for the group recommendation systems.

There are a number of hierarchical clustering algorithms [2] which define a similarity measure quantifying some (usually topological) type of similarity between node pairs. Commonly used measures include the cosine similarity, the Jaccard index, and the Hamming distance between rows of the adjacency matrix. Then one groups similar nodes into communities according to this measure. The CNM or FastQ algorithm [11] uses modularity to guide a greedy hierarchical agglomeration process. The Louvain method [12], also known as the *Fast Unfolding algorithm*, is a greedy optimization method that attempts to optimize the modularity of a partition of the network whose nodes are the communities.

Influential nodes are the important nodes in a network whose behavior largely affects the behavior of their peers. If we were to represent a community as a graph where users represent the nodes and the social relationships that exist between the users represent the edges, these influential users will represent the set of nodes which are essential for the graph to be connected. If we remove these influential nodes, the graph will split into a number of smaller graphs or even unconnected nodes. If these influential nodes were to leave the network, then there may be a lack of communication within the community and the service provider will also suffer huge losses.

The HITS algorithm (Hyperlink-Induced Topic Search) [9] made use of the link structure of the web in order to discover and rank pages relevant for a particular topic in Page Ranking. The algorithm categorized pages into two types: *authority* and *hub*. Page *i* is called an authority for the given query if it contains valuable information on the subject. There is a second category of pages relevant to the process of finding the authoritative pages, called hubs. Their role is to advertise the authoritative pages. There is a mutually reinforcing relationship between authorities and hubs: a good authority should be pointed to by many good hubs while a good hub should point to many authorities. HITS algorithm makes use of this mutual relationship between them and gets the page ranks by iterative computation.

A few studies identified certain problems with the traditional HITS algorithm [10]:

1. The HITS algorithm does not consider the content of the pages at all, so it easily generates topic drift phenomena.
2. When HITS algorithm calculated the authority weights of the pages, it only used the old web pages and ignored the new ones.

Many improved versions of the HITS algorithm were proposed to overcome the problems identified. One such improvement [10] argued that a rational user will choose a page relevant to the search topic according to the page abstract. Therefore, in order to reflect the user's judgment, a weight factor of user's click-behavior is introduced in the HITS algorithm. The improved algorithm also proved that pages which were updated recently are accessed more number of times than pages which contain stale information, because users always tend to prefer the latest information on a topic.

Given the fact that the connections between mobile users are just like the links between Web pages, the basic idea of HITS algorithm can be used in mobile communication for customer behavior analysis. The HITS algorithm was successfully adapted to the telecom domain in [5] and it was called Customer Ranking algorithm. Here, the hub refers to the calling party (*caller*) and the authority represents the called party (*callee*). A good hub is a user who calls many other users and a good authority refers to a user who is called by many users. The operations were modified to take into consideration the cost incurred by the user in calling another. A new category of user called *connector* was introduced. Connectors act as a bridge between the hubs and authorities.

Game-theoretic approaches have been used to solve the problem of influential user identification for quite some time [13]. The advantage of such approaches is that a combination of nodes can be considered while analyzing a network instead of considering individual nodes only. One such method that could be used to benchmark our results is the *Shapley value* [13]. The Shapley value represents the influential score for any node in the network.

In this paper, we have proposed a new algorithm called M-HITS algorithm, which will be based on the improved HITS algorithm given in [10]. The algorithm will take into consideration some new factors which will help us in deriving the new role players *Performer, Spectator* and *Linker* relevant to the mobile telecom subscribers.

2.1   Drawbacks in the Existing Systems

Group recommender systems choose different strategies to satisfy all the members of a particular group. This includes maximizing the satisfaction or minimizing the misery of all members collectively. However, the studies have not taken into consideration the collective impact of the structure of the group and the diffusion of information within the group. Individuals within a group communicate with each other and the probability that a member is influenced by other peers within the group is very high because of the social relationships that exist between them.

Further, the problem of identifying influential users within a community is dynamic in nature. It is important to identify influential users based on the latest information that can be obtained about their behaviour. The problem with the existing methods is that none of them take into account the amount of influence the user possesses at the current instant of time. A user who is influential at one instant of time may not be as influential at another point of time. By considering the overlapping of communities, a user who is influential in one community may not necessarily be influential in other overlapping communities. We also consider the probability of a user being influential is higher if he or she has already been an influential user earlier. Similarly, a user who has been influential recently has more precedence than a subscriber who has been influential in the past. This could be considered as one of the important perspectives of our proposed algorithm.

In a mobile telecom scenario, based on the complex characteristics of communication patterns, the existing algorithms cannot be directly applied. So, there is a need for introducing a new algorithm which addresses some of the drawbacks mentioned by considering the time, location, frequency and history of communications.

3   Proposed System

Evidence suggests that people tend to rely more on recommendations from their friends than on recommendations from similar but anonymous individuals [14]. In this work, we will utilize this information by advertising recommendations to a few users and letting them propagate the information to the entire group instead of asking the service provider to advertise to everyone. By utilizing the concept of social marketing intelligence [15], the probability of the acceptance of the recommendations is increased.

We will consider a Mobile Telecom Network (MTN) to introduce our method and perform experiments based on the MTN data. An MTN will consist of millions of nodes and billions of edges. Analyzing an MTN as a whole is not only impossible but also unnecessary because though the MTN is very large, each subscriber is connected to a very small portion of the MTN only. This is the reason why we need to break down the entire network into smaller sub-networks or communities and analyze each community separately.

3.1   Community Detection Using Spatio-Temporal Information

A simple way of organizing the network into smaller communities is by considering the geographical location. But such communities will still be large in size. Community detection algorithms can be applied on geo-specific data to obtain logical communities in which users share behavioral aspects [5]. However, in a mobile network, users do not remain active all the time. In a single day, there may be periods of time when the activity on the network is high and it keeps varying depending on the time. For example, in a rural area, there are more calls during the afternoon whereas in urban areas, there is more activity on the network at night [16]. A typical day in an urban area can be split into three time durations based on the amount of activity [7]:

- 9 am to 5 pm: This period constitutes the typical working hours and hence, it is considered as the period of high activity.
- 5 pm to 1 am: This period is considered as the time for recreational activity after work, so it is the period of medium activity.
- 1 am to 9 am: This period is obviously the period of low activity, since most people are asleep and only a little activity occurs in this period.

To efficiently identify communities, we must not only consider geo-specific data but also temporal data which is compiled according to the three time slots. Now, if community detection algorithms are applied, we can group users more accurately into logical communities with a spatio-temporal tag. The next step is to identify the influential users in each community.

3.2   Influential User Identification

Users in a network may be classified as information disseminators and information receptors [8]. In this paper, we will introduce three new types of role players for the benefit of customer-oriented recommendations. We will call the information disseminators as *performers* and the information receptors as *spectators*. In an MTN, the calling party can be considered as a performer, whereas the party who is called is considered as a spectator. Then there are *linkers* who act as a bridge between the performers and spectators [5]. All three categories of users are important, even if the performers are the people who bring in most of the revenue to the telecom operator. It is important to note that a single user can simultaneously be a spectator, performer and also a linker. Also, a single user can exist in more than one community because of the temporal and/or geographical splitting of data: a user can move from one location to another and/or be connected to a different set of users based on the time of the day. The user may possess different amounts of influence in different communities.

3.3   Definitions

A *performer* is a user who is very active socially within a community and aggressively promotes campaigns and disseminates information to his peers.

A *spectator* is a passive influential user and is indirectly responsible for generation of revenue. Passive influential means that while the spectator might not initiate social links, once established, they will ensure that the ties are maintained.

Linkers establish connectivity between the performers and spectators. They connect users not only within a community, but also between communities.

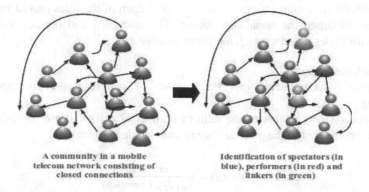

A community in a mobile telecom network consisting of closed connections

Identification of spectators (in blue), performers (in red) and linkers (in green)

Fig. 1. Identification of performers, spectators and linkers in a community

In addition to introducing three new role players, our proposed algorithm will consider certain new creative R^3 factors to determine the different influential scores for each user exclusively for MTN:

Relevance will represent the number of times the user has called or received a call from another user in the same community. Every user will have two values of relevance: $relevance_{spec}$ for incoming calls and $relevance_{perf}$ for outgoing calls.

The *recency* factor will store the timestamp of the last call made/received by the user. When the user makes/receives a call again, the difference between the current time and the time the previous call was made/received is calculated, so that we consider only the users who are currently active in the community rather than people who are very infrequent but who possess a certain degree of influence. A user can make as well as receive calls. So, there will be two recency values for every user. They will be denoted as $recency_{perf}$ and $recency_{spec}$.

The third factor, *recurrence*, will store the number of times a particular user has been judged to be influential by the M-Hits algorithm. The initial value for this factor will be zero. Again, an influential user can be active, passive or a linker, or a combination of the three. So three values of *recurrence* are represented as $recurrence_{perf}$, $recurrence_{spec}$ and $recurrence_{Link}$

These factors are selected with the idea of identifying the users who are influential at the time of generating recommendations. Since, the same subscriber need not be influential all the time; this algorithm must be used at the instant of generating recommendations, to get the best results.

If we were to represent a community C as a graph where users represent the vertices then the social relationships that exist between the users represent the edges.

The initial values of all attributes of a vertex v of a community C are assigned as follows:

$$recurrence_{spec}(v)=0$$
$$recurrence_{perf}(v)=0$$
$$recurrence_{link}(v)=0$$
$$p(v)=1$$
$$s(v)=1$$

The M-HITS algorithm assigns a rank weight to each of the roles played by a user. This weight is called the *recurrence* factor. The relevance and recency values are calculated for nodes u and v which represent an edge e.

Relevance Calculation

The *relevance* factor determines the number of people a user is capable of influencing.

$relevance_{perf}(u,v)$ represents the ratio of number of calls that user u has made to user v when compared to the total number of calls made by user u.

$$relevance_{perf}(u,v) = \frac{num(u,v)}{\sum_{p:(u,p)\in E} num(u,p)} \; . \tag{1}$$

Here, $num(i,j)$ represents the cumulative number of all calls made by user i to user j. E represents the list of all edges in the community.

$Relevance_{spec}(u,v)$ represents the ratio of number of calls that user v has received from user u when compared to the total number of calls received by user v.

$$relevance_{spec}(u,v) = \frac{num(u,v)}{\sum_{q:(q,v)\in E} num(q,v)} \; . \tag{2}$$

Recency Calculation

Since the time factor plays a crucial role in determining the currently active users, recency will help us to store information about how recently a user has been active. This is done by comparing the timestamps of the two latest calls made/received by the user.

$$recency_{perf}(u)= \sqrt{\frac{\alpha}{i+1}} \; . \tag{3}$$

$$recency_{spec}(u)= \sqrt{\frac{\alpha}{j+1}} \; . \tag{4}$$

where *i=difference between timestamp of current call and last call made by u*
j=difference between timestamp of current call and last call received by u

$recency_{perf}(u)$ represents the recency value of u with respect to the last call made by u, whereas $recency_{spec}(u)$ represents the recency value of u with respect to the last call received by u.

The values of i and j can be represented in days, hours, minutes or seconds depending on the application. If i and j are represented as hours, then α can take the value of 24.

3.4   Proposed Algorithm

We introduce a new algorithm: M-HITS. M-HITS finds the individual scores for all the users and distinguishes the role players. We can then use these scores to determine the top-k influential users. $s(u)$, $p(u)$ and $l(u)$ represent the spectator, performer and linker score of user u.

Algorithm: Calculate performer, spectator and linker weights.

```
procedure: M-HITS(C(V,E))  /*where C is a community */
  /* D₁ is the knowledge base containing vertices,
relevance, rank and recency values, spectator and
performer weights of each vertex */
for all u ∈ V
```

$$s(u) = \sum_{q:(q,u)\in E}\{cost(q,u)p(q)relevance_{spec}(q,u)\} \times recency_{spec}(u) + recurrence_{spec}(u) \,.$$

$$p(u) = \sum_{p:(u,p)\in E}\{cost(u,p)s(p)relevance_{perf}(u,p)\} \times recency_{perf}(u) + recurrence_{perf}(u) \,.$$

$$l(u) = \frac{1}{n}\left(\sum_{v:(v,u)\in E} connection(u,v) \times recency_{spec}(v)\right) \,.$$

```
end for
```

$cost(i,j)$ represents the cumulative cost of all calls made by user i to user j

$connection(u,v)$ is 1 if $num(u,v)$ is not 0. The linker weight also involves the $recency_{spec}$ values, so that it specifies the value of active connections maintained by a subscriber, rather than only the number of connections.

As per our claim in related work section, the top priority in the algorithm is given to the *recency* factor. This is because time plays a very critical role in determining the users who have been active in mobile telecom network. The next priority is for the *relevance* factor. A subscriber who interacts with a large number of peers frequently has an increased probability of influencing them, than a subscriber who rarely interacts with his peers. Lastly, we consider the *recurrence* factor to give a slight edge to subscribers who have remained consistently active within the community without significantly affecting the chances of other subscribers becoming influential.

The M-HITS algorithm should be applied on each community obtained by applying a suitable community detection algorithm on spatio-temporal data.

4   Experiments

Experiments have been performed in order to verify how our proposed algorithm addresses the drawbacks mentioned.

4.1   Mobile Telecom Subscribers Dataset

Call Detail Records (CDRs) from an MTN was considered as our dataset. A CDR contains a unique transaction ID, the phone numbers of the calling party and the called party, the time and duration of call, type of call (voice, SMS, etc.) the cost incurred and also information specifying the location of the parties involved. A CDR is generated for every call made.

We considered a Mobile Dataset of 1.2 million subscribers which consists of CDRs from 8 different locations of West African operator. The data has been preprocessed according to geographical and temporal specifications. Initially, community detection has been performed to obtain even more logical communities. The actual phone numbers are mapped to some unique integer values (IDs) to avoid breach of privacy. Of the 1037 communities detected by applying community mining algorithms on CDRs from one particular location, we randomly selected a community which represents subscribers during the period of high activity (9am – 5pm). The M-HITS algorithm was applied to this community and the *recency*, *relevance* and *recurrence* values are determined to distinguish the k-role players of the performer, spectator and linker based on the weights derived from each user.

4.2   Results

Table 1 presents the IDs of the top 10 performers which are obtained as a result of our algorithm. These are the users who have been dissipating information in the most recent timeframe. The recency and relevance values are calculated and using these values, the performer weights are determined. The users' corresponding recurrence values will be considered the next time the algorithm is run.

Table 1. Calculation of Performer Weight

ID	recency$_{perf}$	relevance$_{perf}$	recurrence$_{perf}$	Performer Weight
25	4.317	0.180556	0.500000	0.779461
64	4.284	0.080136	0.220218	0.343302
53	3.684	0.061750	0.145926	0.227487
136	4.234	0.053593	0.145559	0.226914
119	1.359	0.135865	0.118441	0.184640
304	2.414	0.075305	0.116610	0.181786
232	1.633	0.089939	0.094212	0.146870
215	2.000	0.068775	0.088234	0.137551
66	4.149	0.030699	0.081704	0.127370
35	3.420	0.031740	0.069632	0.108551

Table 2 presents the IDs of the top 10 spectators which are obtained as a result of our algorithm. These are the users who have been consuming the dissipated information most recently. Users with IDs 64 and 66 also appear in the list of the top 10 performers. This means that they are actively involved in receiving information as well as dissipating it.

Table 2. Calculation of Spectator Weight

ID	recency$_{spec}$	relevance$_{spec}$	recurrence$_{spec}$	Spectator Weight
120	3.435	0.198880	0.500000	0.683151
24	3.343	0.193277	0.472901	0.646126
106	3.991	0.140336	0.409925	0.560082
66	4.377	0.119148	0.381694	0.521509
64	3.816	0.104575	0.292072	0.399059
74	3.12	0.095238	0.217480	0.297143
103	3.651	0.074930	0.200226	0.273569
152	3.782	0.067227	0.186088	0.254252
15	4.242	0.057423	0.178283	0.243588
274	1.359	0.176471	0.175527	0.239824

Table 3 presents the IDs of the top 10 linkers who have been consuming the dissipated information most recently. These users not only receive information, but are also responsible for connecting many more spectators to the performers. User with ID 25 is simultaneously one of the top 10 performers as well as a good linker.

Table 3. Calculation of Linker Weight

ID	Linker Weight	Recurrence$_{link}$
48	0.160024	0.500000
144	0.075528	0.235988
125	0.072698	0.227148
3	0.064319	0.200967
93	0.061255	0.191393
32	0.061160	0.191095
25	0.060938	0.190402
79	0.058299	0.182158
72	0.052069	0.162690
107	0.043539	0.136038

By combining the three factors and prioritizing them according to importance, the role players are identified effectively.

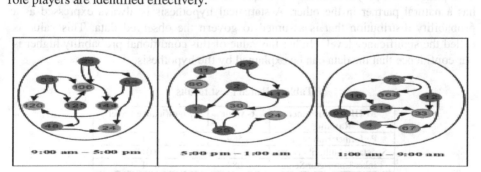

9:00 am – 5:00 pm   5:00 pm – 1:00 am   1:00 am – 9:00 am

Fig. 2. Identification of performers (red), spectators (blue) and linkers (green) within a community at different time-periods

Fig. 2 depicts the top 3 performers, spectators and linkers of a particular community of a specific location across different time-periods of activity. User with ID 25 is an active performer in the first two time slots. Similarly, user 67 is active in the second and third time periods, but acts as a performer in the second time period and as spectator in the third time slot.

Table 4. Results of Shapley Value Analysis vs. M-HITS Algorithm

ID obtained by Shapley Value Analysis	Top Linker ID obtained by M-HITS
48	48
3	144
79	125
32	3
144	93
93	32
72	25
127	79
104	72
125	107

To benchmark our results, we have applied a game-theoretic method (Shapley value) to the dataset. From Table 4, we find that the top 10 users obtained as a result of the Shapley value analysis do not vary very much from the top 10 linkers that we have determined. This proves that the usage of recency is validated. However, if we apply Shapley value to determine top spectators and performers, the results will vary. This is because our algorithm uses the mutual relationship between spectators and performers which is one of the implicit factors of mobile telecom data.

By comparing the results of IDs obtained by the customer ranking algorithm (based on HITS) and the M-HITS algorithms, we can see that the top-k (k being 10 in our example) influential users obtained by both the algorithms vary drastically.

A paired t-test [17] compares two samples in cases where each value in one sample has a natural partner in the other. A statistical hypothesis is always expressed as a probability distribution that is assumed to govern the observed data. This value is called the significance level. Higher the value of this conditional probability higher is our confidence that the data can be explained by the hypothesis.

Table 5. Paired t-test results

M-HITS/ Customer Ranking	t- value	Significance Level
Performer/Hub Score	3.569	0.01
Spectator/Authority Score	4.172	0.01
Linker/Connector Score	4.088	0.01

We applied the paired t-test on the performer, spectator and linker weights obtained from our algorithm and the Customer Ranking algorithm. From Table 5, it is clear that our algorithm has a better significance in identifying the role players for group recommendations than the Customer ranking algorithm. From the available dataset, we found that the IDs of the M-HITS algorithms seem to be more influential in their respective communities for various schemes proposed by the operators.

5   Inference

From the results, it is clear that all three role players of users (*performers, spectators* and *linkers*) are important for the service providers. The results specify that a user cannot be influential just because he/she generates a lot of revenue to the service provider. This emphasizes the fact that by including the creative factors (*recency, relevance* and *recurrence*), more valuable time-specific information is taken into account rather than just the cost factor. This will help us to generate recommendations which are more relevant to the subscribers in the community. A spectator or a linker may not be a direct source of revenue, but may be the cause for a number of other people to stay active on the network. A service provider must also give importance to people who make maximum use of their services in a consistent manner and are responsible for spreading their influence across a wide circle, thus inducing their peers to behave in a similar manner. Thus the different roles of influential users help us to determine loyal customers and they are the key to social marketing intelligence [15].

6   Applications

The identification of users as performers, spectators and linkers means that we would be able to target the users based on their behavior within the network. For example, the performers are loyal customers who promote campaigns aggressively. Hence they can be used as beta testers for new campaigns. Spectators, on the other hand, must be made into more active users within the network. Hence customer retention techniques (such as discounts at local restaurants or free movie tickets at select locations) can be applied to increase their activity. Linkers are used when we want to spread information rapidly, not only within a community but also between communities. By combining these findings with the behavioral analysis of mobile telecom users in urban and rural areas, service providers can effectively increase their customer base. For example, in a rural community, the communities are smaller in size but connectivity within a community is relatively higher. Urban communities are generally larger in, but connectivity within a single community is decreased [16]. This means that there are many performers within a rural community and many linkers within an urban community. Appropriate measures can be taken to increase connectivity in an urban area and the size of rural communities.

The algorithm can also be adapted for social data analysis by modifying the parameters used appropriately. For example, calls can be replaced by tweets, and we can find out the top-k influencers who are dissipating information on trending topics.

7   Conclusion

Our method ranks the influential users by the amount of influence they possess at the current instant of time. In this paper, we considered the *recency* factor which determines which users are currently active in their network. The next factor is *relevance*, which determines the number of people a user is capable of influencing. The last factor *recurrence* is important because a person who has been consistently influential among her peer group is likely to exert a larger amount of influence since her peers will have greater confidence in her opinions. This clearly means that if a group recommender system sends recommendations to the most influential members of the group, the information will be dissipated to all the members of the group, thus minimizing the expenses on the service provider. Also, by specifying different role players for the subscribers (performer, spectator, linker), the service provider can also advertise campaigns suitable for a specific user group. Thus our method serves a dual purpose of finding the influential users capable of dissipating information to the entire group and also identifying various role players of users for recommendations based on their roles. Our suggested method can very well be introduced into other verticals like transportation, health, hospitality etc. Moreover, it is also suitable for blooming social networks for influencing the groups for specific recommendations.

References

[1] Ricci, F., Rokach, L., Shapira, B., Kantor, P.B.: Recommender Systems Handbook, 1st edn. Springer (2011)
[2] Johnson, S.C.: Hierarchical Clustering Schemes. Psychometrika 32, 241–254 (1967)
[3] Iyengar, R., Van den Bulte, C., Valente, T.W.: Opinion leadership and social contagion in new product diffusion. Marketing Science 30(2), 195–212 (2001)
[4] Guille, A., Hacid, H., Favre, C., Zighed, D.A.: Information Diffusion in Online Social Networks: A Survey. SIGMOD Record 42(2) (June 2013)
[5] Brown, J., Broderick, A.J., Lee, N.: Word of mouth communication within online communities: conceptualizing the online social network. Journal of Interactive Marketing (2007)
[6] Rajiv, A., Aparna, V., Natarajan, J., Saravanan, M., Prasad, G.: Analysis of the behaviour of Influential Users in a Telecom Network. In: ACM SNAKDD (July 2010)
[7] Polepally, A., Saravanan, M.: Behavior Analysis of Telecom Data using Social Network Analysis. In: International Workshop on Behavior Informatics 2010, BI 2010-PAKDD, Held in Conjunction with The 14th [PSY]Pacific-Asia Conference on Knowledge Discovery and Data Mining (2010)
[8] Choudhury, M.D.: Discovery of Information Disseminators and Receptors on Online Social Media. In: HT 2010, June 13-16 (2010)
[9] Kleinberg, J.M.: Authoritative Sources in a Hyperlinked Environment. In: ACM-SIAM Symposium on Discrete Algorithms (1998)
[10] He, Y., Qiu, M., Jin, M., Xiong, T.: Improvement on HITS. Algorithm Appl. Math. Inf. Sci. 6-3S(3), 1075–1086 (2012), ACM 978-1-4503-0041-4/10/06

[11] Clauset, A., Newman, M.E.J., Moore, C.: Finding community structure in very large networks. Phys. Rev. E 70, 066111 (2004)

[12] Blondel, V.D., Guillaume, J., Lambiotte, R., Lefebvre, E.: Fast unfolding of communities in large networks. Journal of Statistical Mechanics: Theory and Experiment 1742-5468, P10008 (2008)

[13] Saravanan, M., Vijay Raajaa, G.S.: A Graph-Based Churn Prediction Model for Mobile Telecom Networks. In: Zhou, S., Zhang, S., Karypis, G. (eds.) ADMA 2012. LNCS, vol. 7713, pp. 367–382. Springer, Heidelberg (2012)

[14] Sinha, R.R., Swearingen, K.: Comparing recommendations made by online systems and friends. In: DELOS Workshop: Personalization and Recommender Systems in Digital Libraries (2001)

[15] Sharma, C., Herzog, J., Melfi, V.: Mobile Advertising. John Wiley & Sons (2008)

[16] Saravanan, M., Bharanidharan, S.: Analyzing the dynamics of Mobile phone customer behavior for improved marketing. In: International Conference on Business Intelligence, Analytics, and Knowledge Management (BIAKM 2013), April 17 (2013)

[17] Zimmerman, D.W.: A Note on Interpretation of the Paired-samples t Test. Journal of Educational and Behavioral Statistics 22(3), 349–360 (1997), doi:10.3102/10769986022003349

Novel Visualization Features
of Temporal Data Using PEVNET

Amer Rasheed and Uffe Kock Wiil

The Maersk Mc-Kinney Moeller Institute
University of Southern Denmark
Campusvej 55, 5230 Odense M, Denmark
{amras,ukwiil}@mmmi.sdu.dk

Abstract. The information visualization of networks has been a tricky task during the last decade. It is difficult to understand such large amounts of statistical data. A number of solutions have been proposed to tackle this bulk of information. By examining some dynamics of criminal networks and by making use of some novel interactive features, we have found that the prevailing challenges to information visualization can be eliminated to a large extent. The current study will help understand interesting patterns, which are extracted by way of monitoring the temporal data of a criminal activity. We have appended six more features to the PEVNET framework. These are 'Node color feature', 'Link size feature', 'Link details on demand feature', 'Detecting collaborating sub-cluster feature', 'Sub-cluster detection feature', and 'Temporal pattern feature'. A novel clustering algorithm has been proposed. We have proposed a unique way of visualizing the clustering of data, with which the analyst gets a sound visualization of the data.

Keywords: Information visualization, Dynamics of criminal networks, Temporal data, Clustering.

1   Introduction

It is not unusual to say that technology brings challenges with it [1]. There is a dire need to cope with these challenges. Huge information sources are among the big question marks. There are diverse views regarding information visualization. One view focuses on the need of proper filtering, slicing, and integration of different visualization techniques [2]. There is a requirement for novel multi-dimensional and hierarchical information visualization techniques along with dynamic hierarchy computation [3]. A sound visualization provides a showcase of clusters in easily understandable semantics [4]. Previous static networks were based on graphs. With the advancement, continuous snapshots of the network are required. Visualization of temporal data is not an easy task with graphical representation [5]. It is presentable and it can be shown in the shape of graphs, but there is much difficulty in detecting or representing the repetition of an activity over time. It cannot be easily shown with graphs. It requires novel visualization techniques.

L.S.-L. Wang et al. (Eds.): MISNC 2014, CCIS 473, pp. 228–241, 2014.

No criminal attack is possible without proper planning and coordination. According to Lindelauf [6], 'secrecy and efficiency' are the main properties of criminal networks. Much research has been carried out over assessing the coordination between criminal network entities. There has been a marked brute force and sophistication in these attacks as well. They can only be intercepted with effective pattern detection. Detecting patterns in the criminal activity is also a tedious task. There are patterns in terrorist attacks; for instance, the patterns in the attack could be recruiting members, acquiring subsequent visas one after another in the target country, etc.

In this paper, we have added more features to PEVNET [8]. We have addressed the issue of assessing the coordination of criminal networks by finding the patterns using the clustering of data. PEVNET focuses on the visualization of the networks with emphasis on temporal data. A Chicago Narcotics data set was used to show the effectiveness of the framework. We have used a case study from the Chicago Police Department (CPD) which is recognized as one of the oldest serving police forces after the New York Police Department (NYPD); CPD is the second largest police department in the United States.

The organization of the remainder of this paper is as follows. In the next section, challenges from the field of criminal network visualization are described. Section 3 describes related work. In Section 4, the proposed visualization features in PEVNET [8] are elaborated. Section 5 discusses the novel clustering algorithm. Section 6 describes the framework for detecting patterns in temporal data. A case study, based on Chicago Narcotics data along with crime types, is presented in Section 7. Section 8 provides a discussion. Finally, Section 9 concludes the paper and describes future work.

2   Challenges

There are a numerous challenges confronting existing visualization tools and technologies. For instance the existing system lacks a manual analysis and integration. For this purpose, patterns of the activities of those criminals, their footage [9], their links, and the orientation of the master minds of the attacks are required to be traced. There is a need to understand the pattern in the information supplied to the investigators. Mixing of complex data with the network information is another problem. There is a lack of inter and intra-cluster composite communication and the requirement of advanced filtration techniques. The matter becomes even worse due to scarcity of internal and external referencing techniques. Due to the dynamic nature of the networks, there is a lack of predicting [10] and pattern recognition techniques. There is a great deal of requirements for extracting information from statistical tabular data. More efforts are needed for refinement in locating the central person and the key clusters in the network. There is also a requirement of training visual analysis aptitude among the investigators as they still believe in a primitive node analysis technique on a white board. There is a dire need to elaborate the display complexity of multiple hierarchies.

3   Related Work

Clustering of data is an effective technique to better comprehend large data sets. It is achieved by grouping similar nodes together and viewing them individually as per requirement. Xingan and Martin [11] identified that data mining and visualization techniques are vital for assessing the patterns. Their proposed self organizing maps technique [11] used ScatterCounter [12] in the experiment. The authors performed validation by using advanced tools, nearest neighbor [13], Discriminant Analysis [16], and Viscovery SOMine. Viscovery SOMine was used for clustering objects with multi-dimensional attribute onto the 2-D plane. Self-organizing maps [12], proposed by Li and Juhola were used for assessing the structure of the network. These maps have the ability to cluster similar objects together. It is similar to NETSCAL proposed by Hutchinson [14]. Kreuseler et al. [3] used a famous technique called Dendrogram for this purpose. The theme is that, as the hierarchy of the group is descended, the similarity increases. Hierarchical cluster analysis addresses nested grouping of entities. Hierarchical clustering contains a sequence of partitions [4]. The recursive pattern technique [17] is based upon clustering of data. It is based on setting back and forth sequences i.e., from right to left and again from left to right and so on [4].

Linking by painting in the scatterplots makes it possible to see how clusters are related. Thus, linking integrates the partial information contained in the individual views into a coherent image of data. For focusing of some specific point of interest, the authors use the fish eye view [19], or cone tree structure [20].

Klerks [25]describes the identification of the position of power. In other words, it can be identifying leaders of a group or master minds. Table Lens [28]has a number of operations for controlling of the focal area. The Table Lens is much favorable for finding some relationships in data. It carries an ability to visualize it for better investigation by isolating large data sets. Table Lens uses different types of graphical presentations that are text, color, shading, length, and position.

Timebox [27] is used as a sound mechanism for specifying queries on temporal data. It is actually rectangular regions that are placed on time data and the results are manipulated. The boundaries of the rectangular region indicate the parameters of the query. It has an ability to pose a number of queries at the same time. There can be multiple attributes. When the queries are put forth, the data sets are displayed on the upper left corner of the window. The attributes, which can be dynamic ones also, are displayed and the values can be displayed as a line graph. The input query is made through the lower left corner. The query can be resized, moved and modified, after it is created. The details of the individual time boxes are on the lower hand corner of the window on the interface of the utility. The data envelope facility provides the maximum range of values that can possibly be queried. It increases the user cognitive ability by suggesting the suggested query areas by highlighting them. Finally, the percentage of the selected data is also displayed.

MeDISYS [31] is a European project for detecting public health patterns out of 1200 news websites in 35 different languages. As soon as there is activity of a disease report, it immediately sends out a warning. Lifelines [30] support routines and patterns in the data sets. It helps in detecting anomalous behaviors and provides easy access to details. The work has been used to eliminate anomalies in the parent's history. The features of advanced visualization, such as zooming, details on demand, color coding, and filtering, are included in Lifelines.

Chen [22] proposed pattern recognition by way of employing data mining techniques. Graph theory [15] is a mathematical study of structures consisting of nodes with links. Certain patterns can be traced with Graph theory [15]. The technique reveals patterns in the data that need fruitful structural representation [15].

The inference-based predictive technique is among the initial steps in the upbringing of the concept of prediction by Peterson et al. [23]. CrimeNet Explorer [24] focuses on structural analysis. It is based upon relational analysis and positional analysis. In the relational analysis, the focus is over identifying central member to find the central member. Whereas in positional analysis, the focus is on the fact of how similarly two network members connect to each other. Crimenet [24] focuses on identifying sub-clusters, identify the relationship patterns and finally locating the key person. Crime pattern theory [26] focuses on finding patterns in the criminal activity in metropolitan areas. "Crime Ridges" is proposed by Sang et al. [32]. The authors carried out the visualization of data that resembles mountain peaks termed as "Crime Ridges" by authors. The authors studied the directionality in the criminal activity by way of visualizing the patterns in the accumulation of crimes near major city centers. Kosara et al. [7] is of the view that there cannot be an attack without planning. There can be some patterns, specific range of frequencies or repetition of some special words that are seen online before that attack. Occurrences of certain entities bring a good reason for the identification of a warning.

For detecting patterns, data mining techniques were employed to detect any unusual exposure of certain entities that occurred again and again. Algorithmic and statistical methods are used for finding patterns in the temporal data sets [30].

In this paper, we have made a contribution by implementing six visualization features using PEVNET [8]. Among those features, four are visualization features, two are temporal pattern features. Since, according to author's best knowledge, there was no existing technique to detect hidden clusters. Furthermore, there is no considerable research found in detecting criminal activity based on temporal patterns. The proposed features help analysts in approaching the hidden criminal clusters and to trace their activity by making use of temporal features that are proposed in this study. The proposed features have been implemented by finding the gaps and challenges as discussed in section 2 above. A novel and unique clustering algorithm has been proposed in this work. The algorithm makes clustering of given data set for instance crime types. We have found certain patterns of reported crimes by employing our proposed framework PEVNET in closely observing the crime activity during the year 2011 in the Chicago Narcotics data set.

4   Proposed Visualization Features

Network visualization features, visualization features based on temporal data, and composite features are three types of features that are proposed in PEVNET [8]. All the features of PEVNET are showed in Table 1. The features are as follow:

4.1   Network Visualization Features

These features address details on demand, drag and drop, and focusing issues.

4.1.1 Node link color feature
4.1.2 Locating central person
4.1.3 Node size feature
4.1.4 Node link details on demand feature
4.1.5 Visual filtering
4.1.6 Network of clusters
4.1.7 Detecting collaborating sub-cluster feature

4.2   Visualization Features Based on Temporal Data

4.2.1 Sub-cluster detection feature
4.2.2 Synchronization feature
4.2.3 Encircle feature
4.2.4 Temporal pattern feature

4.3   Composite Feature

4.3.1 Expand collapse feature
4.3.2 Visualizing similar node feature

Six visualization features have been implemented in the current study. The first four features are network visualization features and are denoted as 4.1.1, 4.1.3, 4.1.4, and 4.1.7 in Table 1. The other two features are visualization features based on temporal data, denoted as 4.2.1 and 4.2.4. The features are as follows:

Node Color Feature. Based on the number of times certain crime is committed by the same person, the color will change from lighter colors to darker colors. The feature helps network analysts in locating the most wanted culprit. These apply to top five persons which are processed based on centrality. This feature, that exist in the literature, is re-examined and implemented in PEVNET.

Link Size Feature. The links become thicker with the increase in weight. The weights of the links are shown in the 'Criminal Record' window. The feature helps network analysts in gathering group data information of the links for visual analysis.

Node Details on Demand Feature. Details on the demand feature display the attribute information of the node, which is accessed. For instance, the node label, node weight, in-degree, and out-degree of each node are displayed. It is done with a mouse click on the node. All the details appear in a window called 'Criminal Record' window on the PEVNET desktop as shown in Fig. 1. The offence list tab indicates the number of crimes in which the node, which is accessed, is involved.

Detecting Collaborating Sub-Cluster Feature. There may be hidden sub-clusters inside networks, which are not visible to the analyst. The nodes in these sub-clusters can have links with external clusters, which can give further information. The linkage of nodes with those clusters can be detected with the proposed framework. By selecting the crime type information from the menu, the analyst can detect sub-clusters inside the network; for instance, if there is a crime type for instance '2092' detected, there is good possibility that the suspects may have been involved in some other related crime. There is also a probability that the same suspects may have been involved with some other collaborating network. The proposed 'Detecting Collaborating Sub-Cluster feature' addresses this issue and extracts any hidden crime or criminal network.

Sub-Cluster Detection Feature. Sometimes the clusters get mixed up with other clusters and are difficult to point out. Keeping in mind the dynamic nature of criminal networks, which stands for the creation and breaking up of groups, this feature has been proposed. The framework circumvents those clusters by color clues, which are provided as legends along with the crime types provided in the form of a list of values.

Temporal Pattern Feature. In the temporal part, it is used as a feature for observing the dynamics in the temporal data. Temporal patterns can be detected by changing the time calendar.

5   Novel Clustering Algorithm

We have proposed a novel clustering algorithm as shown below. It displays the cluster of the selected crime type as shown in Figs. 1 and 2. The nodes with the selected crime type are displayed in such a way that the cluster seems to pop up from the original network snapshot as shown in Fig. 2. Based on crime types, the algorithm first creates a list of nodes. After that, sorting is carried out with respect to the node's weight and crime type. Finally, the algorithm displays the selected nodes by ascertaining their x and y co-ordinates. The clustering of a crime type 'Hallucinogens' is performed. There are some sub-clusters of crime type 'Hallucinogens' in the given network as shown in Fig. 1. After implementing the clustering algorithm, the given network is manipulated and the results are generated. There are total of two sub-clusters of crime type 'Hallucinogens' that are detected as shown in Fig. 2. The original network can be seen in the background.

```
Create a list of nodes based on crime types named as
selected nodes;
      Sort selected nodes based on their weight
      Set radius to the value 250
      Loop through each node in selected nodes
      Display 1st node in the center of all
For displaying other nodes
  Calculate x and y coordinates
```

```
Calculate angle(•) = (360/total number of nodes less than
1))*( /180)*(index of current node-1);
    Assign  x  position  to  current  node:=center  screen
position + (Cos(•) * radius);
    Assign  y  position  to  current  node:=center  screen
position + (Sin(•) * radius);
```

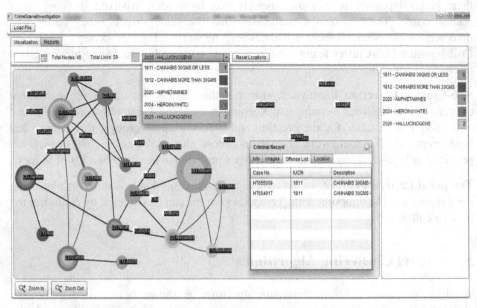

Fig. 1. Clustering using proposed novel clustering algorithm require selection of Crime Type from the list of values on the PEVNET desktop

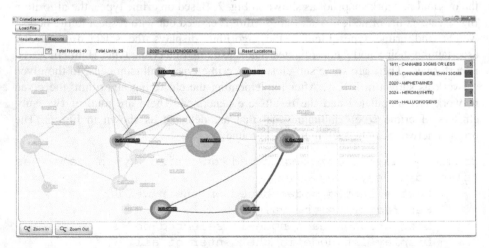

Fig. 2. On selecting the Crime type say 'Hllucinogens', the sub-clusters with the selected crime types are displayed. On close observation, the original criminal network is also evident in the background.

6   The Framework for Detecting Patterns in Temporal Data

Previously, four visualization features were incorporated in the PEVNET [8]. Briefly describing those features, the user was able to easily detect the central person since the central node was the thickest among the nodes. 'Node size feature' and 'Link details on demand' features were implemented. With the increase in the nodes' weights, their sizes became thicker. Top five persons were shown, in PEVNET, based on their centrality values. It was beneficial information for the analysts since they are in search of most wanted persons. Similarly, the user was able to detect the link activity between two nodes by clicking any link. A network of clusters was also incorporated in PEVNET.

The analyst worked in the interface of the FLEX[1] tool. The framework was implemented in Visual Studio.Net[2] using the C# programming language. The motivation behind using the .Net framework was that it supports visualization constructs. Strong database communication to handle large amounts of data is also a big plus for the .Net framework. C# is a powerful server side language. We also used Flex along with the SQL Server Express Edition 2012[3] as a transit database platform for the data flux. Flex was used due to its strong support for design patterns, the requirements of which are identical in various fields.

7   Case Study

The Chicago Narcotics data sets by CPD were selected with the purpose of testing the proposed framework. CPD is one of the prime law enforcement agencies in Illinois [26]. There are twenty three police districts. These districts are further divided into sub-sections called beats and wards. There are nearly forty thousand data records. They are data sets of the whole year 2011 with daily dates and time information against the numbers of reported crime types [8]. These crime types are denoted by IUCR-Illinois Uniform Crime Reporting, such as '2095' stands for 'Attempt possession narcotics', '2170' for 'Possession of drug equipment', and '2092' for 'Solicit narcotics on public way', and so on.   The various crime types are 'Cannabis 30gms or less'[4], 'Amphetamines'[5], 'Heroin (white)'[6], 'Crack'[7], 'Hallucinogens'[8], 'PCP'[9], 'Barbiturates'[10], 'Found suspect narcotics', 'Solicit narcotics on public way'[11],

---

[1] http://www.adobe.com/products/flex.html
[2] http://www.visualstudio.com/
[3] http://www.microsoft.com/en-us/download/details.aspx?id=29062
[4] http://www.iwu.edu/ccs/Illinois_Drug_Laws.htm
[5] http://www.cesar.umd.edu/cesar/drugs/amphetamines.pdf
[6] http://en.wikipedia.org/wiki/Heroin(white)
[7] http://en.wikipedia.org/wiki/Crack_cocaine
[8] http://en.wikipedia.org/wiki/ Hallucinogens
[9] http://en.wikipedia.org/wiki/Phencyclidine
[10] http://www.nlm.nih.gov/medlineplus/ency/article/000951.htm
[11] http://www.nlm.nih.gov/medlineplus/ency/article/000951.htm

'Cocaine'[12],and 'Forfeit property'. Colored circles indicate criminal nodes. These nodes have committed crimes either in a single capacity or with other criminal nodes as shown in Figs. 1 and 2. In the case of a single person, it is represented by a single circle called a singleton. It is the basic structure of the criminal network visualization. If there are two or three nodes, then they are represented by dyads or triads and, so on [8]. There are a number of criminal nodes that are collided to form clusters. The links among them are shown by a line connecting each of them in the shape of a cluster.

Due to the bulk of data, we have selected a small bunch of data in order to have a better understanding of the actual contribution. We have selected 'Hallucinogens'; these are drugs which stimulate the central nervous system. Fig.2 indicates the snapshot of PEVNET. The 'Hallucinogens' cluster indicates that a sub-cluster of three criminal nodes 'Chloe', 'Emily', and Daniel were found to have committed 'Hallucinogens'. Another sub-cluster of criminals, 'William', 'Madison', 'Olivia', and 'Alexander' were found to have committed the same crime. With the help of the proposed novel features 'Detecting collaborating sub-cluster feature' and 'Sub-cluster detection features, i.e. 4.1.7 and 4.2.1 features respectively in Table 1, we distinguished between these two sub-clusters of 'Hallucinogens'. The difference was that the first sub-cluster was found to have committed crimes with the crime types or IUCR #'1811', '2024', and '2025'as shown in the legend window on the right hand side of both Figs. 2 . They were found to have committed the crimes of 'Cannabis 30 grams or less', 'Heroin (White)', and 'Hallucinogens' whereas the members of second sub-cluster were found guilty of having committed the crime with crime types or IUCR #'1811' and '2025'. In simple words, they were found to have committed the crimes of 'Cannabis 30 grams or less' and 'Hallucinogens' only. So the usefulness of PEVNET was depicted. It became easy to distinguish between the two sub-clusters which was not possible before. So visualization by making use of sub-clusters detection feature, we have found that one sub-cluster was involved in another crime type 'Heroin (White), in addition to other crime types.

8   Discussion

The diversity in information makes it difficult for the analyst is to select his desired target application at hand. The core aim of the current study has been to alleviate the burden of managing these issues. The nature of the problem varies as discussed in Section 2. In Table 1, a comparison of existing criminal network visualization features with proposed features of PEVNET has been made. The existing features, which are available in the current literature, are shown in the vertical column on the left side of the table whereas sub-section numbers of the proposed PEVNET features, described in

---

[12] http://www.drugs.com/cocaine.html

Table 1. Table shows of existing features (first vertical column) against the visualization features of PEVNET (feature numbers of section 4). Last two rows indicate the information about the existence of feature and the features which are re-examined/implemented

Existing Features (Found in the literature review)	4.1.1	4.1.2	4.1.3	4.1.4	4.1.5	4.1.6	4.1.7	4.2.1	4.2.2	4.2.3	4.2.4	4.3.1	4.3.2
CrimeNet [15] Clustering sub-group feature	X	X	X			X							
Dynalink [34] Cluster of crime types feature	X	X	X										
Hierarchical clustering [19] Clustering based on similarity						X							
Multi-dimensional scaling [20] Arranging objects based on similarity						X							
Radviz [35] Visually grouping similar objects						X							
Theme space and galaxies [16] Extracting patterns from the text							X						
Timebox [28] Time series graph feature								X					
TempoViz [29] Time-slider features								X					
Radviz [35] Spiral view for temporal patterns								X					
Collapse Expand Composite feature [33]												X	
Prune Growth [22] Composite feature												X	
CrimeNet [15] The link thickness feature		X											
CFA [8] Node-link thickness feature		X	X										
Timebox [28] Details on demand feature in the window panel.				X									
NodeXL [30] Node centrality feature		X											
CrimeNet [15] Node label and Centrality information		X		X									
TempoVis [29] Time-based filtration feature					X								
NodeXL [30] Visual dynamic filtering feature					X								
KeyPathwayMiner [21] Boolean operation usage					X								
Spiral Display [31] Visualizing sub-clusters using Spiral technique.								X					
NodeXL [30] Clustering feature						X							
No Evidence found in the existing literature								X	X	X	X		X
Implemented / Re-examined in this study	X			X	X		X	X			X		

Section IV, are shown in the horizontal row on the top. The comparison of features was performed by marking 'X' in the Table 1. For instance two existing features: 'Details on demand feature' in Timebox [28] and 'Node label and centrality information' in CrimeNet [15] are equivalent to 'Node details on demand feature'.

Similarly the existing features such as 'Time series graph feature' [28], 'Spiral view for temporal patterns' [35], and 'Time-based filtration feature' [29] correspond to 'Sub-cluster detection' feature. 'Node link color feature' can be observed upon close observation of Fig. 1. The node color can be seen changing from lighter colors to darker ones. The fact that the criminal node has been accessed more is evident by the greater color variation from light to darker ones. Again, upon examining Fig. 1, the 'Link size feature' can be observed between the two nodes 'Chloe' and 'Emily'. By evaluating the data sets provided, or by using the 'Link Details on demand feature', it can be found that the nodes 'Chloe' and 'Emily' have more than one interaction between them. That is why the link has been shown thick unlike other nodes which have single interaction. So, these two nodes can be a good consideration for the network analysts. It could be a case where one of these two nodes were found and other nodes can be traced easily by examining the past history. It could also be possible that they have a common taste that that is based on sports, religion, family issues, study, etc. So with this feature, they can be traced easily by undergoing some network analysis techniques. Furthermore, the feature is useful in detecting missing links in the network. According to the author's best knowledge, such implementation has not been found in the literature. The 'Criminal record' window shows the 'offence list' of the selected link as shown in Fig. 1. The selected node was 'William'. The selected node was first of all depicting the highest centrality. Secondly, upon clicking any of the link with the mouse, the information of the nodes, connecting those links, appeared in the 'Criminal record' window. The 'offence list' showed all of the crimes with their 'Case No.', 'IUCR', and 'Description'. Then, the weight of the selected node, i.e., '6' in this case, was being shown on the PEVNET desktop. The proposed 'link details on demand visualization feature' helps the user in watching the parameters on a run-time basis unlike the feedback provided in Operand Foci [5]. The user just has to click with the mouse over the node or link and the information is at his/her disposal. The 'Detecting collaborating sub-cluster feature' can be understood by examining Fig. 1 again. The theme of this feature is to find out hidden sub-clusters. It is achieved by way of depicting the crime information with the legends on the top right corner of the PEVNET desktop. For instance, the node 'Chloe' was shown to have been involved in 'Cannabis 30 GMS OR LESS', a crime with the code '1811' with another node 'Emily'. The legend of 'Cannabis 30 GMS OR LESS' was shown with a dull green color in the legend of the PEVNET desktop and inside one of the node colors. But upon closer observation, it was also seen involved with two other nodes, 'Emily' and 'Daniel' over the crime type 'Hallucinogens' with the code '2025'. The legend of 'Hallucinogens' was shown with a bright green color in the legend of the PEVNET desktop and inside one of the node colors. Getting a step ahead, the similar node 'Chloe' was involved with three other nodes, 'Mikael', 'Ethan', and again 'Daniel' over the crime type 'Heroin (White)' with the code '2024'. The legend of 'Heroin (White)' was shown with a brown color in the legend and inside one of the node colors. This novel feature can be a good source of information in detecting the hidden sub-clusters, which could not otherwise be easily available with statistical data or even by using advanced query optimization techniques in relational databases. The 'Sub-cluster detection feature' can be seen again by observing Fig. 1. It is based on the proposed 'Novel clustering algorithm' in Section 5 above. Upon selecting some crime type, the cluster can be seen, as shown in Fig. 2. The information based on this feature, also indicated that the selected cluster 'Hallucinogens' with the code '2025' carried two co-offending sub-clusters. The first one comprised of 'Chloe', 'Emily' and 'Daniel' and the second sub-cluster comprised of 'Williams, 'Alexander,' Olivia, and 'Madison'. Another crucial piece of information was that 'Williams', the main master mind and the most central person, was not part and parcel of the first sub-cluster. So with 'Sub-cluster detection feature', we can demarcate the sub-clusters easily unlike existing clustering techniques. For making a choice of certain clusters, with some crime types, the legends can also be used. The 'Temporal pattern feature' can be executed by shuffling the date calendar. Upon shuffling the date calendar, the information of the selected date, month, and year can be seen. Dynamics can be observed in the shape of new links or sub-clusters; for instance, there was a new link established in the above data shuffling, i.e., between 'Chloe' and 'Mia'. Similarly, emerging patterns can be traced with this 'Temporal pattern feature'.

Besides the features that are implemented, there were some features that were proposed but have still not been implemented. Those features are among the visualization features based on temporal data and composite features. The 'Synchronization feature' and 'Encircle feature' are features based on temporal data whereas the 'Expand collapse feature' and 'Visualizing similar node feature' are composite features. With our proposed 'Visualizing similar node feature', the analyst can see the nodes that have the same crime type but in another geographical location. If a person is involved in a crime type, say 'Possession of drug equipment', in Illinois and the criminal expert wants to see the collaboration of that suspect in another geographical location, say New York, then he/she can trace it with the help of the proposed feature. The 'Encircle feature' integrates multiple views of different instances of time; for instance, the cluster detected in the year 2004 is again detected in the subsequent year. Thus, with the proposed 'Encircle feature', the past clusters can be traced [8]. Similar is the functionality of the 'Synchronization feature'. Finally, the 'Expand collapse feature', as obvious from the feature name, is concerned more with composites. The proposed features present a good solution to 'Visual cluttering[13]', i.e., too many affects in one visual scene. The issue is addressed with filtering and clustering of data. The proposed features are analogous to 'Link Analysis' and is quite beneficial to the network analysts in a variety of ways. It is the term used for finding patterns and trends in the network data sets. It is concerned with extracting, discovering, and linking together sequences from the data set [10, 11].

9   Conclusion and Future Works

Additional features to PEVNET [8] have been presented in this study. These features are six in number. Out of six, four features are visualization features while the remaining two visualization features are based on temporal data. A comparison has been made in Table 1 to show the correspondence of existing features with the proposed features. The question arises how the features are novel? The novelty of the features lies in their implementation. According to authors best knowledge these features, with the exception of 'node color feature' [33], are not found in the existing literature. Moreover, we have implemented the features in a unique way for instance the clustering of crime type 'Hallucinogens' is a novel implementation. It is done in such a way that the entire network is seen in the background whereas the selection of analyst is visible. Similarly, the analyst can detect the suspicious criminal activity by using 'detecting collaborating sub-cluster feature' and other proposed temporal features. Last but not the least is our proposed novel clustering algorithm.

In the future, implementation of the remaining features proposed in PEVNET [8] will be carried out. Those include the 'Visualizing similar nodes feature', 'Composite feature', 'Synchronization feature', and 'Encircle feature'. The quantification of the temporal part may be the immediate extension to the proposed framework. Validation of the PEVNET features [8] will be performed by a qualitative survey and usability evaluation of network visualization features, which is in the pipeline. The features will then be compared with the existing tools and techniques. It would then become a comprehensive decision making tool for analysts.

---

[13] http://en.wikipedia.org/wiki/Clutter

References

1. Gina, S.: Privacy Protections for Personal Information Online (2006)
2. Huai-hsin Chi, E., John, T.R.: An operator interaction framework for visualization systems. In: Proceedings of the IEEE Symposium on Information Visualization. IEEE (1998)
3. Matthias, K., Norma, L., Heid, S.: A scalable framework for information visualization. In: IEEE Symposium on Information Visualization, InfoVis 2000. IEEE (2000)
4. Jaspeet, K., Dharmender, K.: Pixel-Oriented Technique: A technique to visualize multidimensional data. International Journal of Computing & Business Research (2012) ISSN: 2229–6166
5. Smriti, B.: Prediction Promotes Privacy in Dynamic Social Networks (2010)
6. Lindelauf, R., Borm, P., Hamers, K.: The influence of secrecy on the communication structure of covert networks. Social Networks 31, 126–137 (2008)
7. Robert, K., Helwig, H., Donna, L.G.: An interaction view on information visualization, State-of-the-Art Report. In: Proceedings of EUROGRAPHICS (2003)
8. Amer, R., Uffe, K.W.: PEVNET: A Novel Framework for Visualization of Criminal Networks (submitted for publication, 2014)
9. Stefan, F.: Secunia-stay secure, The Evolving Threats From Digital Footprints, http://www.secunia.com
10. Rhodes, C.J.: Inference approaches to constructing covert social network topologies. In: Memon, N., David Farley, J., Hicks, D.L., Rosenorn, T. (eds.) Mathematical Methods in Counterterrorism, pp. 127–140. Springer, Vienna (2009)
11. Xingan, L., Martti, J.: Country crime analysis using the self-organizing map, with special regard to demographic factors. SpringerLink (2013)
12. http://www.uta.fi/sis/cis/research_groups/darg/publications/scatterCounter_eng.pdf
13. http://en.wikipedia.org/wiki/Nearest_neighbour_algorithm
14. Wesley, H.: NETSCAL: A network scaling algorithm for nonsymmetric proximity data. Psychometrika 54(1), 25–51 (1989)
15. Roger, S.W., Francis, T.D., Donald, W.D.: Network structures in proximity data. The Psychology of Learning and Motivation 24, 249–284 (1989)
16. http://www.uk.sagepub.com/burns/website%20material/Chapter%2025%20-%20Discriminant%20Analysis.pdf
17. Daniel, A.K., Hans-Peter, K., Mihael, A.: Recursive Pattern: A Technique for Visualizing Very Large Amounts of Data, Institute for Computer Science, University of Munich, Leopoldstr. 11B, D-80802 Munich (1995)
18. John, V.C., Joseph, A.K.: Interactive Visualization of Serial Periodic Data (1998)
19. George, W.F.: Generalized fisheye views. In: Proceedings of the ACM Conference on Human Factors in Computer Systems, SIGCHI Bulletin, pp. 16–23. ACM Press (1986)
20. Robertson, G., Mackinlay, J.D., Stuart, K.C.: Cone trees: animated 3D visualizations of hierarchical information. In: Proceedings of the SIGCHI Conference on Human Factors in Computing Systems (1991)
21. Huai-hsin Chi, E., John, T.R.: An operator interaction framework for visualization systems. In: Proceedings of the IEEE Symposium on Information Visualization. IEEE (1998)
22. Hsinchun, C., Jenny, S., Roslin, V.H., Linda, R., Homa, A., Harsh, G., Chris, B., Kevin, R., Andy, W.C.: COPLINK: managing law enforcement data and knowledge. Communications of the ACM 46(1), 28–34 (2003)
23. Rasmus, R.P., Uffe, K.W.: Crimefighter investigator: A novel tool for criminal network investigation. In: European Intelligence and Security Informatics Conference (EISIC), pp. 197–202 (September 2011)

24. Xu, J.J., Hsinchun, C.: CrimeNet Explorer: A Framework for Criminal Network Knowledge Discovery University of Arizona. ACM Transactions on Information Systems (TOIS) 23(2), 201–226 (2005)
25. Peter, K.: The network paradigm applied to criminal organizations: Theoretical nitpicking or a relevant doctrine for investigators? Recent developments in the Netherlands. Connections 24(3), 53–65 (2001)
26. Romana, R., Stuart, K.C.: The table lens: merging graphical and symbolic representations in an interactive focus+ context visualization for tabular information. In: Proceedings of the SIGCHI Conference on Human Factors in Computing Systems. ACM
27. Harry, H., Ben, S.: Visual specification of queries for finding patterns in time-series data. In: Proceedings of Discovery Science (2001)
28. Stuart, K.C., Mackinlay, J.D., Ben, S. (eds.): Readings in information visualization: using vision to think. Morgan Kaufmann (1999)
29. Alan, D., Geoffrey, E.: Starting simple-adding value to static visualization through simple interaction. School of Computing, Staffordshire University, UK
30. Ahn, J.-W., Taieb-Maimon, M., Sopan, A., Plaisant, C., Shneiderman, B.: Temporal Visualization of Social Network Dynamics: Prototypes for Nation of Neighbors. In: Salerno, J., Yang, S.J., Nau, D., Chai, S.-K. (eds.) SBP 2011. LNCS, vol. 6589, pp. 309–316. Springer, Heidelberg (2011)
31. John, S.: COPLINK: Database Integration and Access for a Law Enforcement Intranet, National Criminal Justice Reference Service, NCJRS (2001)
32. Andrew, J.P., Herbert, H.T., Patricia, L.B.: Dynalink: A Framework for Dynamic Criminal Network Visualization. In: European Intelligence and Security Informatics Conference (2012)
33. Josch, P., Alcaraz, N., Junge, A., Baumbach, J.: KeyPathwayMiner 4.0: Condition-specific pathway analysis by combining multiple OMICs studies and networks (2013)
34. Dennis, L., Bertini, E., Hertzog, P., Bados, P.: Visual Analysis of Corporate Network Intelligence: Abstracting and Reasoning on Yesterdays for Acting Today
35. James, A.W., James, J.T., Kelly, P., David, L., Marc, P., Anne, S., Vern, C.: Visualizing the non-visual: Spatial analysis and interaction with information from text documents. In: Proc. Information Visualization Symposium (InfoVis 1995), pp. 51–58. IEEE, IEEE CS (1995)
36. Marc, A.S., Ben, S., Natasa, M., Eduarda, M.R., Vladimir, B., Cody, D., Tony, C., Adam, P., Eric, G.: Analyzing Social Media Networks with NodeXL: Insights from a Connected World. Morgan Kaufmann (2010)
37. John, V.C., Joseph, A.K.: Interactive Visualization of Serial Periodic Data. In: Proceedings of UIST 1998, the 11th Annual Sympoium on User Interface Software and Technology, San Francisco (November 1998)

A Vehicle-Based Central Registration Video Transmission Network on Social Mobile Communication Network

Lu Shuaibing[*], Fang Zhiyi, Ge Bingyu, and Qin Guihe

College of Computer Science and Technology
Jilin University
Changchun, 130012, P.R. China
lushuaibing11@sina.com

Abstract. In order to meet the increasing functional requirements of vehicular multimedia network video transmission, and improve the process of mobile video communication quality and the transmission bandwidth during the E-Commerce times. This paper proposed a center registration video transmission network on social mobile network which can solve the communication problems on such as mobile video conferencing, mobile office and so on. Through analyzing the design philosophy of vehicle-based central registration video transmission network node, at the same time discuss the realization of processing the video data during the social mobile communication under the center registration mechanism in the administration of media oriented system transport network. We use MOST150 video transmission node build a MOST150 video transmission network and via the output result which shows on the screen to prove the video transmission principle and the central registration mechanism of the MOST150 network. Experimental results show that the vehicle-based central registration video transmission network network based on center registration mechanism is propitious to the transmission of video data on social mobile communication network.

Keywords: video transmission network, media oriented system transport(MOST), social mobile network, mobile video office.

1   Introduction

With booming development of automotive E-commerce industry, the vehicle information process continue to comprehensively and thoroughly, requirement for the mobile office or communication by transmitting the video data through mobile social mobile network. The vehicle information and system increasingly that is developing from a single vehicle multimedia information equipment which is traditional high wiring harness and low bandwidth, gradually develop towards intelligent network with low wiring harness and high bandwidth for the users. Media Oriented System Transport (MOST) is a network technology which formulated by the MOST

---

[*] Corresponding author.

L.S.-L. Wang et al. (Eds.): MISNC 2014, CCIS 473, pp. 242–253, 2014.
© Springer-Verlag Berlin Heidelberg 2014

organization to communication protocol based multimedia information transmission. It is a kind of ring topology network for the transmission of oriented media, using optical fiber line or twisted pair as the physical transmission medium[1].It has the virtue that high transmission rate, low cost, high ability on anti-jamming and flexibility. The generation of the MOST network technology solves the high bandwidth requirements for the multimedia data transmission process and plug-and-play issues in network. Realizing the inside multimedia equipment interconnect with each other and at the same time it can meet the requirements of the vehicle multimedia network flexibility and replaceability for users.

At present, the domestic research on vehicle network mainly focuses on the synchronization transmission of audio or video data, most of them focuses on the study of the audio data-transmission in the network, ignoring the importance of the video data during the social mobile communication network. In this paper, according to the MOST network protocol specification, proposed a vehicle-based central registration video transmission network during the communication of social mobile network, realize the information query and storage of the nodes devices on the network at the same time to ensure the high quality of video stream which makes the e-commerce and mobile video conferencing become more fluent and clear.

2   Vehicle-Based Media Oriented System Transport Network Video Transmission

Vehicle Multimedia Network Transmission Protocol

Media Oriented System Transport (MOST) is a formulated by the MOST organization to communication protocol based on vehicular multimedia information transmission network technology[1]. MOST150 is the third generation on media oriented system transport network, the transmission speed can reach 150Mbps.It is a function oriented system which is able to transmit three different data type simultaneously including streaming data, packet data and control data. A frame in MOST network is divided into three part including one for the transmission of video or audio stream data on the synchronization region, one for the transmission of packet data in asynchronous regions and one for the transmission of control information area. The overall model of the vehicle-based video transmission network on social mobile communication network is the Fig1.

Synchronous and asynchronous regional border demarcation is achieved through the boundary identifier[1]. For the development of the MOST network application layer protocol and above with the function block for the core component, the function block incorporates all the methods and properties that control the MOST equipment. By accessing the function blocks of MOST nodes realizing the communication between devices above application layer. Therefore, for audio or video data processing and transmitting during the nodes' communication in MOST system network management is very important.

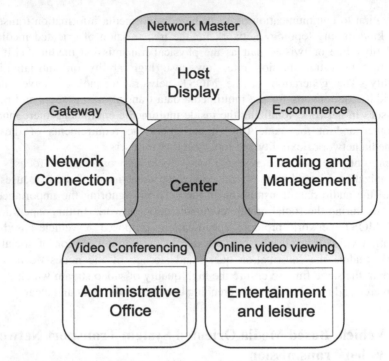

Fig. 1. Overall model of the vehicle-based video transmission network

Media Oriented System Transport (MOST150) Video Transportation Theory
The video transmission is divided into three parts in MOST150 network system, that is video compression, Creating and connecting sockets and the transmission of video streaming data in synchronization channel. MOST150 node is composed by a single video codec chip, an external host controller(EHC) and intelligent network interface controller(INIC).It adopt a strong anti-electrical magnetic interference optical fiber for transmission, Components of a MOST150 video node as shown in the figure1.

During the transmission of video data, source video data is first sent through the DVI interface into MOST150 video source node and then process it. In order to ensure the video conference quality during the electronics conference office, we use the video codec chip compress the input digital video signal into the H.264 data format[7]. H.264 has taken a more pragmatic, focused on approaching to addressing the problems and needs of current and emerging multimedia applications, the purpose is to ensure the quality of video transmission, on the basis of the number of bits occupied by reduced to a minimum, the compression process chip is located on the video input or output companion evaluation platform which is based on the H.264 video compression model implementation which is shown in figure2.

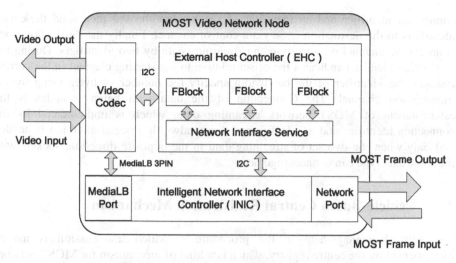

Fig. 2. Inside components of a MOST150 video node

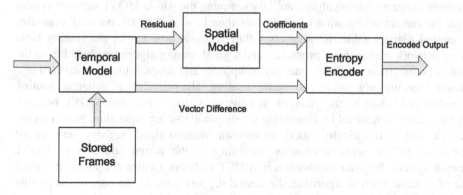

Fig. 3. Video encoder block diagram

After compression processing the video data transmit through the Media-LB 3PIN interface into the intelligent network controller(INIC), Media-LB can transport streaming data, control data and fast packet data simultaneously an by that supports all types of multimedia signals that need to be transported in modern information systems. Then process the video stream data into a format defined the MOST protocol frame format that can be transmitted in the network. Finally, transmit the data frames into the MOST network, network transmission process need to use the socket, Socket represents the MOST protocol routing of various data, which describes in detail a port data transfer type, transfer direction and data transmission bandwidth[5].Before INIC established the path between two ports in the channel two video boards MOST INIC ports should be open, and then these two ports to create a socket separately, if they have a socket, Intelligent Network controller(INIC) will return a confirmation message and an identification, to establish a connection to the data stream. In the process of establishing a channel for the video transmission, it will generate a connection identifier and an identifier of the bandwidth, and then send these two identifiers to the destination node via a control channel. Finally the destination node connects the channel of corresponding streaming data by two identifiers. During the connection identifier indicates the offset relative to the starting channel in the current channel, the identifier of the bandwidth specify the number of bytes using by the transmission channel. The Connection of the data path between nodes is the establishment of MOST network streaming data, which is implemented by the connection identifier and an identifier of the bandwidth. Special attention is needed that only when the two set of streaming data in the opposite direction, sockets with the same data type can connect together.

3   Vehicle-Based Central Registration Mechanism

The network management in the procedure of video transmission is mainly implemented by the centre registry, which is a kind of mechanism for MOST network query management is realized by setting up the central registry table through simulated dynamic routing algorithm[4].By scanning the whole MOST network system to get the current information of all devices then be stored by the network-master in the central registry table. Routing algorithm is about how to get the routing table during network management procedure and a good routing algorithm should have the nature of correctness, stability, integrity, simplicity and adaptability. It is divided into dynamic routing algorithm and static routing algorithm. For dynamic routing algorithm can adapt to the changes in network status better, the MOST network central registry is realized by simulating the dynamic routing algorithm. Nodes on the network can automatically build their own decentralized registry and adjust themselves to the network changes according to the actual situation of MOST network system. Register mechanism in MOST Network Center adopts the designed idea of dynamic routing algorithm, the central registry table as the core of the polling mechanism, which is similar with the routing table. Main realized procedure as follows: The first tapped button is determined as the network-master, whose logical address is set as 0x100 and stored it in the central registry table. Then network-master init the network by means of broadcasting. All nodes, including the network-master receive the massage then inform the network-master by sending a status message of the IDs of each function block. At the same time they initialize the logical node address by dynamically allocation according to the following equation upon initialization of the network.CA means that the logical address of the node that need to be calculated, Pos is the meaning of the physical location of the current network.

$$CA \, / \, 0x100 \, \rightarrow \, Pos \triangleright \tag{1}$$

During the initialization of MOST network system, the initialization of the device uses the address 0xFFFFF to respond the demand. If the off time of a device is longer than T(T is between 2 and 4 seconds),the logical node address is aborted[7].If the network-master receives a message from a slave-node whose address is 0xFFFF,there is a system conflict during the system operation. The network-master interrupts all services and broadcast messages to all nodes in the MOST system. It compares the newly registered function blocks to the one stored in its central registry table and broadcasts the system status throughout the system via the message. The system contains four basic states which are OK, Not-ok, New and Invalid. Any other states are based on the four basic states. When the system state is OK, while there is no changes in the central registry, at this moment network master sends messages Configuration. Status(OK) to all nodes in the ring. then the complete central registry can be queried, slave-nodes are able to build connections with the other nodes. While the message Configuration. Status(OK) is published, one node should send a delay message for another new node. When the system state is Not OK, while the current central registry does not match stored central registry or the system has a conflict, the network master sends messages Configuration. Status(Not-ok) to all nodes to stop all connections in the ring. The system keeps in Not-ok state before a OK message is send[8]. When the system state is New: the network structure will change if new nodes have been added to the network and then the network-master sends a message Configuration. Status(New) to all nodes in the ring. The message is available only in the OK state. After adding new nodes, Network-master will recheck all nodes function blocks to make sure it is able to recognize the new nodes and their addresses. If a new node is added in the middle of the ring, there is no system conflict if the new node has a free and unused logical address, otherwise there would be a conflict which should be solved by conflict resolution. Process of network-master state transition as shown below:

M1. There is no matching nodes on the network and no reproducible node-address nor new device configured with system configuration.

M2. The network-master detected the system configuration successfully.

M3. The function block is not available.

M4. The new function block is available in the system, and the properties of the function blocks could be queried by using callback functions.

M5. Opening the system configuration scanning.

M6. Ending state after scanning the system configuration.

M7. System conflict judgment.

M8. After system conflict judgment.

The corresponding form as follows:

Table 1. States set of network-master

Network Master States	ID	Events	Event- ID
M1	0X00	Network Starting-up	t1
M2	0X01	Address not available	t2
M3	0X02	Initialize the address	t3
M4	0X03	Event is triggered	t4
M5	0X10	Send NotOK(OK)	t5(t8)
M6	0X20	Start(End) scanning the system configuration	t6(t7)
M7	0X30	Reset address(Dot reset) address if there is a conflict	t9(t10)
M8	0X31	Finish reset the address	t11

The master node state diagram is shown below:

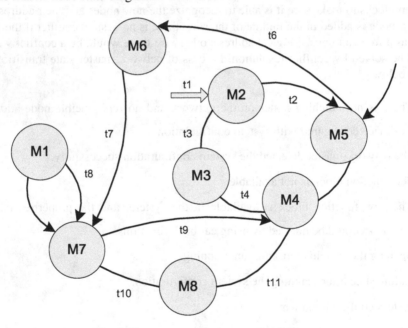

Fig. 4. The master node state transition diagram

The state of slave nodes due to the different events and produced change, one state to another determined state or be converted to a variety of different state between two different states, the specific conversion as shown in the slave-node state transition diagram. Process of slave-node state transition as shown below:

States of slave-node:

S1. There is no matching nodes on the network and no reproducible node-address nor new device configured with system configuration.

S2. The slave-node detected the system configuration successfully.

S3. FBlocks of the slave-node are not available and the invalid function block could be realized by using callback functions.

S4. There is new function block joining the system.

S5. The new function block is not available.

Table 2. States set of network-slave

Network Master States	ID	Events	Event- ID
S1	0X00	Network Starting-up	u1
S2	0X01	Address not available	u2
S3	0X02	Connection timeout	u3
S4	0X03	If (State == NotOK)	u4
S5	0X10	Start to connect the initial procedure	u5
S6	0X20	Check the received messages	u6

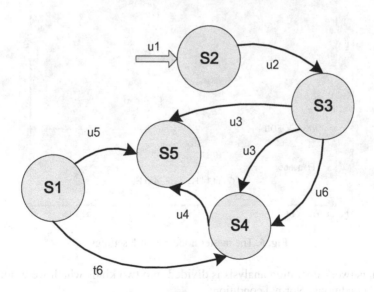

Fig. 5. The slave-node state transition diagram

4   Analysis of Experimental Results

In order to verify the correctness of the automotive multimedia network digital video transmission based on the principle of central registration system, and to show process of the central registration system, we do the following experiment. Laboratory equipment are MOST150 video boards and OptoLyzer Suite V1.4.5 provided by the SMSC company as the hardware-based, OptoLyzer Suite V1.4.5 is a tool for automotive multimedia network node test which was developed by SMSC. Experiment was divided into two parts. Firstly we do the network test of two nodes, then analysis center registration mechanism through the output. We will be an experimental process MOST150 video board as one of the video nodes, optolyzer as another video node network connection. We observe video workflow nodes in network management through data analysis center mechanism implemented in the network management process by the OptoLyzer Suite Viewer output. Since any node in the network can be used as the master node , so first we set the OptoLyzer as the network master. After the physical node is connected, each node enters the initialization state as figure5 shown:

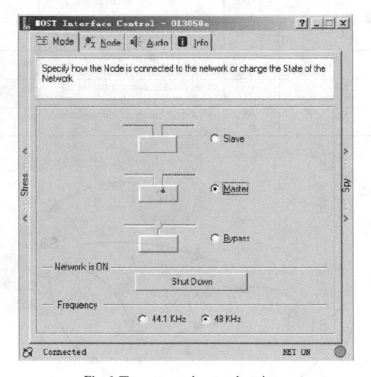

Fig. 6. The master node network settings

MOST network operation analysis is divided into two kinds which were normal and abnormal conditions. Normal condition:

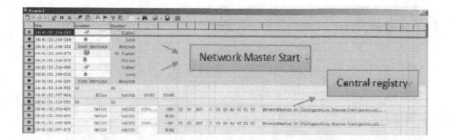

Fig. 7. Central registration of Network Master

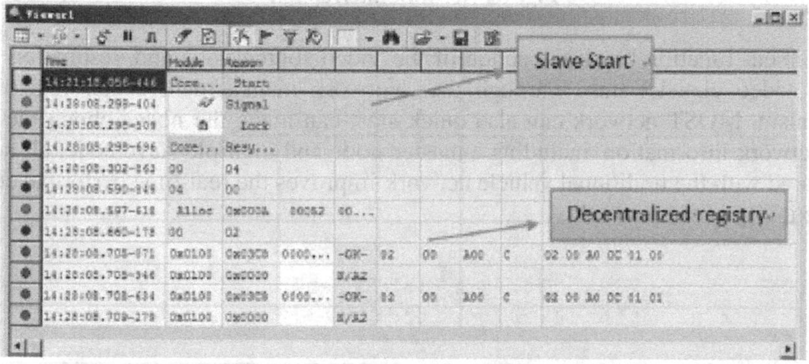

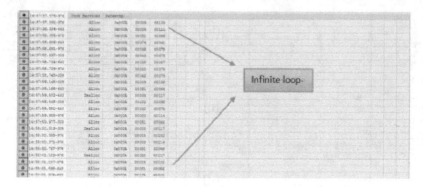

Fig. 8. Decentralized registry of Slave Node

Abnormal conditions: (1)When a node has been set to become Network Master, if the other nodes are forced into a Master, the registry will occur error. (2)When one of the nodes is abnormal, network interruption.

Fig. 9. In_nite loop

The serial output information displayed by the above graph we can see that the MOST network startup process, and the network after the start, in the connection

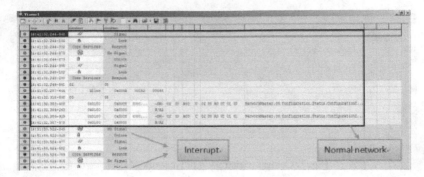

Fig. 10. Network interruption

management function block the action of the video source node and destination node set up video channel link. Through then network management center registration mechanism, MOST network can also quick start can make the nodes obtain the real-time network information, including a master node and multiple slave node, all nodes. Compared with the traditional vehicle network improves the real-time performance and the flexibility of the network.

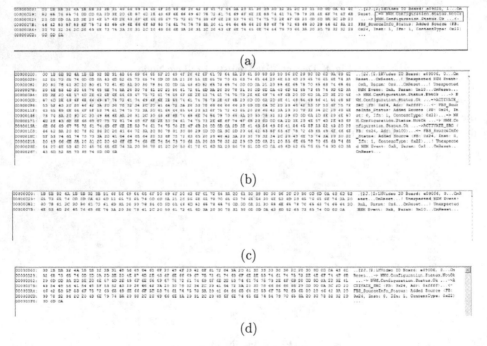

(a)

(b)

(c)

(d)

Fig. 11. (a)Network node initialization state (b)Network connection status changes to OK from NotOK.(c)MOST network transmission information content of the video frame;(d)The realization process of register mechanism in MOST Network Center.

5   Conclusion

This paper studied a vehicle-based central registration video transmission network on social mobile communication network, which is mainly to solve the multimedia information transmission bandwidth constrained problem during the mobile network information transmission, such as conference call, e-commerce and so on. At the same time, through the network management system realized the central register of centralized control and information integration of the network nodes. Central registration mechanism is the core of video transmission network on social mobile communication, network video node using the central registration mechanism can't always query the property of its attention, improve the flexibility and bandwidth of the vehicle network. Lay the foundation for the development of vehicle network video transmission vehicle multimedia network and make the video communications in the social mobile network unimpeded.

Acknowledgements. The authors of this paper would like to thank the re-viewers and editors for their helpful comments and suggestions.

References

[1]   MOST Cooperation, MOST Book 2[M], MOST Cooperation, pp. 30–35 (2011)

[2]   Jack, K.: Video Demystified, Newnes is an imprint of Elsevier, pp. 165–170 (2005)

[3]   Richardson, I.E.G.: H.264 and MPEG-4 Video Compression, pp. 29–35 (2003)

[4]   Tanenbaum, A.S., Wetherall, D.J.: Computer Networks, pp. 281–292 (2012)

[5]   MOST Cooperation, MOSTSpecification 3VOE2[M], MOST Cooperation, pp. 206–212 (2010)

[6]   MOST Cooperation, MOST Netservice Layer2 V03 00 xx-5 [M], MOST Cooperation, pp. 81–90 (2010)

[7]   MOST Cooperation, MOST UM VOIC Evaluation Platform V01 00[M], MOST Cooperation, pp. 19–25 (2011)

[8]   MOST Cooperation, MOST Dynamic Specification 3V0-1[M], MOST Cooperation, pp. 19–25 (2011)

[9]   Nan, Y., Qin, G.-H., Dong, J.-N., Gao, Y.: Vehicle-mounted MOST Network Management System Based on FSM Model. Computer Engineering 37(18) (2011)

[10]   Qin, G.-H., Nan, Y., Chen, Y.-H., He, W.: Connection Management Strategy in Media Oriented Systems Transport Network. Journal of Jilin University (Engineering and Technology Edition) 42(4) (2012)

[11]   Zhang, Y.-L., Qin, G.-H., Dong, J.-N., Hao, J.-Y.: Design of MOST audio network based on INIC. Computer Engineering and Design 32(4) (2011)

[12]   Qin, G.-H., Li, B.-L.: Design of In-Vehicle MOST Audio Network. Micro Computer Information 23(4-2) (2007)

Exercise Support Robotic System by Using Motion Detection

Eri Sato-Shimokawara, Yihsin Ho, Toru Yamaguchi, and Norio Tagawa

Faculty of System Design, Tokyo Metropolitan University, Tokyo, Japan
{eri,hoyihsin,yamachan,tagawa}@tmu.ac.jp

Abstract. In this research, we propose an exercise support robotic system. The system collects human daily life activity data by motion detection as a part of big data, and analyzes it for providing better service. The Normal exercise tools such as pedometer need to be carried while exercising. However, sometimes people forget to take such exercise sensor, when it happened many times, people will feel hard to continue exercise regularly. Thus, the proposed system focuses on user's behavior when exercise by Kinect, and based on it to recommends user a suitable exercise pattern through an interaction robot. Firstly, this paper discusses the user's exercise habits and their weariness in exercise, secondly, describes the proposed system approach, the method of detecting human motion, and maintaining exercise motivation. Finally, the experiment and the results are shown.

Keywords: Exercise Support, Robotic System, Motion Detection, Human-computer Interaction.

1 Introduction

Nowadays, human's daily life become convenience than a hundred years age, which changes peoples life-style. It also brings chronic conditions. One of the most common is lack of exercise, which often leads health problems. Having exercise habit is the best way to improve the situation. However, we usually feel hard to continue an exercise before making exercise a habit regularly. In accordance with National Health and Nutrition Survey Japan by MHLW (Ministry of Health, Labour and Welfare), from 2003 to 2010, less than 35% Japanese people (Male 34.8%, Female 28.5%) take exercise regularly [1], which shows a hidden health problem of Japanese people. For this reason, many exercise tools and applications are developed for supporting people with their exercise.

The most normal exercise tool is the pedometer. However, traditional pedometers only calculate user's calories by counting his/her steps. Currently, the newer pedometer, also known as activity meter, can also calculate the amount of activity by using records GPS data. Those data can also be continually accumulated to concretize in PC or mobile device for analyzing the user's exercise situation. However, these tools or applications depend on users' motivation. But people sometimes forget things. When users forget to take a sensor device or bother to synchronize the sensor data with PC or Internet, it will become hard for user to make exercise a habit, what

L.S.-L. Wang et al. (Eds.): MISNC 2014, CCIS 473, pp. 254–267, 2014.

constrains user to continue exercise. Therefore this paper proposes the system, which recommends an exercise not only according to user's amount of activity and life style, but also considers user's weariness to help user make exercise a habit. We also adapt the attachment theory to robot. When robot analyzes collected data to provide service, it changes its expression as a pet to let the user feel different emotions and have emotion resonance. Moreover, in this research, the robot is not only a communication tool but also plays a rule as social robotics, which as a social media robot that be applied as information media. The robot analyzes the collected data, which user's daily life activity data, as big data for providing service. These data are able to construct social network, which help extend users' social circle.

This paper describes related researches in Section 2, clarifies the problems from preliminary experiment in Section 3, describes the proposed exercise support robotic system in Section 4, shows the experiments of proposed system in Section 5, and summarizes the research and discusses the future works.

2   Related Research

There are researches describe about exercise support system. Shibano et al. [2] proposed an exercise system by using a robot. They compared exercise system with or without a robot by salivary amylase concentration. The experiment of this method has 3 situations: 1. The subjects do exercise without robots. 2. The subjects do exercise with a virtual robot. 3. The subjects do exercise with a real robot. This result shows that exercise with real robot reduces users' stress effectively. Moreover, the exercise system, which using robot to select and recommend an exercise type from 3 exercise patterns according to user's life style and health patterns, is also presented [3]. This system showed recommend an exercise according to users life style is important. However, in this research, the robot recommends an exercise based on an exercise schedule (ex. Let a user do stretching 8 o'clock). In this case, because the system doesn't consider user's weariness in exercise, the robot probably recommends an exercise when he/she does not want to. Moreover, our previous work [4] presents a method based on amount of exercise for user modeling. This research cluster user's activity data using data mining technique to find the group the user's exercise style belong to, as large as the users' data increase, as correct as the result will be. The concept help adapt to construct social network in Big data. Thus, in this paper, the authors detect user's weariness by movement in exercise to recommend a suitable exercise, communicate with the user by robot, and collect users' data through recoding user's activity and the communication situation.

3   Pre-experiment for Exercise Support System

For clarifying the reason of why are people hard to continue exercise, and finding the way to let user continue exercise. We design the following experiments that consist of exercise games and questionnaires.

3.1   Method for Pre-experiment

In accordance with Exercise Guide 2006 [5], every person needs to take 3 METs or more exercise in 1 hour per a day, and at least 23 Ex in a week for maintaining health condition. METs (Metabolic Equivalent) stands for exercise intensity. Ex (Exercise) stands for amount of exercise in an hour, for example: doing a 3 METs exercise in a hour means taking 3 Ex. METs is utilized for comparing the level of exercise strength and the energy by a motion or exercise. In general, "calories" is used to calculate an exercise, which depends on human's weight that varies from person to person. However, METs don't depend on users weight, same physical activity has same METs, thus it is more applicable to use for evaluate everyone's amount of exercise. Table 1 is an example list of METs.

For increasing the attraction of doing exercise, the authors choose to use exercise video games as the interface for these experiments. We utilize 2 sensors: Actimarker and Kinect to detect user's motions in exercise. Actimarker EW4800, a sensor by Panasonic, for calculating human's amount of exercise using movement is carried on user's waist. Kinect, a motion sensing input device made by Microsoft take human motion by human skeleton. We adapt it as the main sensor for recognizing user's motions. We adapt the concept of virtual keyboard and mouse to the experiment, what replace the input device from keyboard and mouse to Kinect sensor.

Table 1. The list of METs for daily exercise

Motion	METs
Sleeping	0.9
Sitting (not talking or reading)	1.0
Walking (67m/min.)	3.0
Walking (120m/min.)	6.3
Running (8km/h)	8.0
Running (12.1km/h)	12.5

In this paper, we selected 2 kinds of games from FLASH GAME, which are Street Fighter II [6] and Skipping Rope Dance [7]. In accordance with our experiments and METs table [5], the average amount of exercise of Street Fighter II is 4.8Ex and Skipping Rope Dance is 8.1 Ex. For executing our experiment, we convert the recognized human motion to mouse and key commands, for example, when a user raised his/her right hand, the system converts this gesture to right arrow key "→". The same setting can also be adapted to several games. However, because the experiments include Street Fighter II and Skipping Rope Dance, which have different key commands, thus we design different motions for these 2 games. Street Fighter II has 13 motions, and Skipping Rope Dance has 6 motions.

3.2   Experiment and Results

There are 6 persons (2 male, 4 female), who are between 23-27 years old, to participate this experiment. Firstly, subjects answered the questionnaire about their situation of doing exercises in daily life. The result is shown in Table 2. 2 subjects take exercise almost every day (1 male, 1 female). 2 female subjects take exercise sometimes. 2 subjects seldom exercise.

Table 2. The subjects' exercise situation in daily life

Frequency	Male (2 subjects)	Female (4 subjects)
Every day (over 15 min./day)	1	1
Sometimes	0	2
Seldom	1	1

We convert the recognized human motion to key commands, and design different actions for these 2 games. 6 participants are asked to do every motion 50 times for experiment the recognition accuracy. Through the experiment as show in Table 3 and 4, we know that the average recognition rates of Street Fighter II and Skipping Rope Dance are 73% and 94%, which can execute these games smoothly. The scenes of participants exercise using these games are shown in Fig. 1.

Table 3. The designed motions contrast to original key command of Street Fighter II

Body part	Motion	Recognition rate (%)
Right hand	Right punch	84
Left hand	Left punch	80
Both hands	Go to horizontal side	70
Right foot	Right Kick	74
Left foot	Left Kick	70
Whole body	Jump (over 5cm)	76
Whole body	Squat	70
Whole body	Move to right side	72
Whole body	Move to left side	72
Whole body	Shoryuken	70
Whole body	Hadouken	70
Whole body	Hurricane Kick (right)	72
Whole body	Hurricane Kick (left)	70
Total recognition rate		73

Table 4. The designed motions contrast to original key command of Skipping Rope Dance

Body part	Motion	Recognition rate (%)
Right hand	Go to horizontal side	94
Right hand	Raise	96
Left hand	Go to horizontal side	92
Left hand	Raise	90
Whole body	Move	100
Whole body	Jump	92
Total recognition rate		94

Fig. 1. The scenes of doing exercise using video games: Street Fighter II (Left) and Skipping Rope Dance (Right)

For examining whether the exercise video games can help maintain amount of exercise or not, we picked a subject who seldom exercises, and ask her to play these games 1-2 hours everyday for one week. As the result shown in Fig. 2, the total amount of exercise that over 3 METs is 25Ex. This result cleared the goal, which plays our exercise video game is able to get the value of Ex that is recommended by MHLW (Health, Labour and Welfare Ministry).

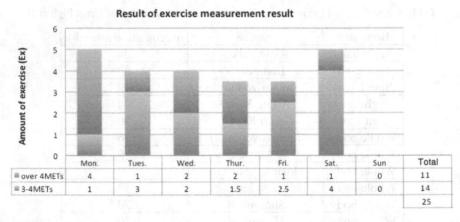

Result of exercise measurement result

	Mon.	Tues.	Wed.	Thur.	Fri.	Sat.	Sun	Total
over 4METs	4	1	2	2	1	1	0	11
3-4METs	1	3	2	1.5	2.5	4	0	14
								25

Fig. 2. The result of a subject played the exercise video game 1–2 hours everyday in a week. This result shows that the subject cleared the goal that is recommended by MHLW.

Next, we ask all subjects to attach Actimarker on his/her waist and play each exercise video game 1 hour. After played the game, we use a questionnaire, which includes 3 questions, to ask subjects' using experience of doing exercise by the exercise video game. These users' answers are shown in Table 5.

From the result of this questionnaire, we find the problems as follows: 1. If nobody alerts the user, he/she will probably forget to do exercise. 2. If a user exercise alone, he/she will easily lose interest and feel bother to exercise. 3. The user needs a goal and objective to do exercise. 4. Encourage is needed. 5. The user often wants to feel a sense of accomplishment.

Table 5. Answers from subjects after the experiments

Questions	Game 1 (Street Fighter II)	Game 2 (Skipping Rope Dance)
Do you want to play the game again?	Yes: 4 No: 2 (all Female) Reason: This content and motions of the game are intense and violent.	Yes: 5 No: 1 (Female) Reason: This game is just repeat same action as the real skipping rope, which made me feel very boring.
Did you enjoy the game?	Yes: 6 No: 0	Yes: 5 No: 1 (Female) Reason: Real skipping rope game is more interesting than this game.
Do you want to play the game every day?	Yes: 1 (Male) No: 5 Reason: 1. Game patterns are few that makes me lost interest gradually. 2. I want friends play together with me. 3. Because there are several actions have to be remembered. I guess I will forget how to play this game.	Yes: 0 No: 6 Reason: 1. Motion patterns are few that makes me lost interest gradually. 2. Motion patterns are very simple that makes me feel boring. 3. I need feel cheering up from games. 4. I want to know the level of achievement.

4   Exercise Support Robotic System

We designed an exercise support robotic system based on the result of pre-experiment as described in above. In this system, the authors adapt a robot as an assistance to help maintain the user's exercise motivation, moreover, to help user make exercise a habit.

4.1   System Outline

This system consists from Kinect and a cute robot, which named ApriPoco made by TOSHIBA. Fig. 3 shows the overall concept of this system. This system briefly divides into 3 stages: 1. Start of exercise; 2. Doing exercise; 3. End of exercise. In stage 1, the system observes a user's behaviors, and then, gets interactive with the user by the robot. In stage 2, the robot exercises with the user, and according to the user's actions to communicate with him/her. When the user exercises, the robot shows a facial expression as it's charging battery. In stage 3, the system synchronizes its database with user's exercise frequency and running days. If running days and exercise frequency decrease, the expressions and behaviors of the robot will look tired, on the contrary, if running days and exercise frequency increase, the expressions and behaviors of the robot will look healthiness. The authors refer previous work [8][9] to decide the facial expressions as shown in Fig. 4. In this research, we mainly encourage the continuation of the exercise using the attachment theory. In the next section, we describe the detailed implementation process and methods of the exercise support robotic system.

1. Start of exercise

Recognize user's activity. If a user did not exercise and stay sitting, the robot would say "Let's play"

2. Doing exercise

Robot recommends a suitable exercise pattern for the user, and do the exercise together.

3. End of exercise

Robot summarize a user's exercise and cheer up the user.

Let's play!

OK!

He has been sitting for 3 hours.

Calculate
• Coordinate of joint
• Time

OK!

Warming up!

Would you play skipping rope dance?

The robot recommends a game.
The user select a game.

Finnish!
You have been play the exercise games every day!

The robot grows up according user's exercise time

Fig. 3. Proposed system consists of Kinect and a robot. This system has 3 stages: 1. Start of exercise; 2. Doing exercise; 3. End of exercise.

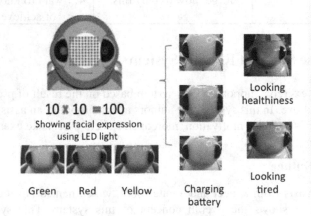

10 × 10 = 100
Showing facial expression using LED light

Green   Red   Yellow   Charging battery

Looking healthiness

Looking tired

Fig. 4. The robot shows different expressions using LED light installed on its face

4.2   Methods for Proposed Robotic System

The authors assume the system is installed in living room of a user's home. Kinect is set near TV to observe the user's behaviors. If he/she has been sitting over 3 hours, the robot will invite the user to do exercise by saying, "Let's play!" automatically, which is the first stage of this system. Fig. 5 is the scene that the robot invites the user to exercise. If the user doesn't want to exercise and leave the seat, when he/she comes back, the robot will invite the user to exercise again. If the user wants to exercise and stands up, the system will move to next stage.

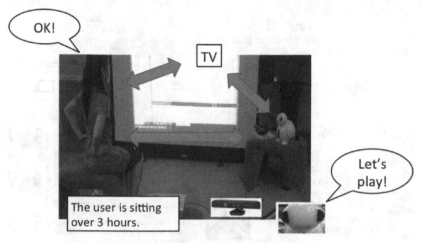

Fig. 5. The scene of stage 1. If the user has been sitting for a long time (over 3 hours), the robot will say: "Let's play!" to invite user to do exercise.

In this stage, which is stage 2, robot depends on the user's action to communicate with user and recommend a suitable exercise. Firstly, the system starts a reaction time test game. A circle appears randomly on the TV screen, and when the user raises his/her hand over his/her head to touch the circle, it disappears. If the user raises his/her hand and the height doesn't over the user's head, even the user touch the circle, it won't disappear. The test game executes 10 times. The system calculates and recodes the reaction time 10 times, and obtains the average time. If the user's motions are small or slowly, and the average reaction time is longer than his/her recodes before, the system judges the user might be tired. In this case, the robot will recommend light exercise. On the contrary, if user's motions are large or quick, and the average reaction time is same or shorter than his/her recodes, the robot will recommend intensity exercise. Moreover, if the user doesn't like the recommended game, he/she can also choose other games. If the user likes the recommended game, the exercise starts, the robot also starts exercise with user. The left side of Fig. 6 shows the image of choosing a game. Here we notice that although the system has 2 games in present, because we adapt the virtual keyboard and mouse in our system, numbers of games can increase easily.

When starting to exercise with the robot, the robot shows different facial expressions to represent emotions for encouraging the user. When the user exercises, the robot shows a facial expression as it's charging battery. If user plays in good performance over 15 times, the robot will say "Amazing!" to the user. Moreover, when user exercises, the system also analyzes the change of user's hand coordination to detect the weariness of the user. If the movement become smaller, the system judges the user is tired, and then, the robot will recommend he/she take a rest or to play light exercise video game. The right side of Fig. 6 shows the image when a user exercises using a game with the robot.

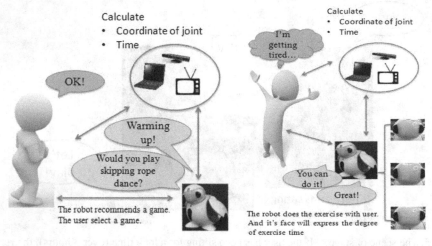

Fig. 6. The images in stage 2. The robot recommends an exercise game based on the result of reaction time test game (Left), and encourages the user during the exercise (Right).

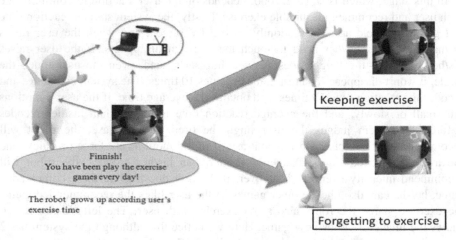

Fig. 7. The robot summarizes according to the user's exercise history

After the exercise, the system moves to stage 3. The robot summarizes the exercise and makes a comment about the exercise in this time, such as "You did it!". If the user keeps exercise, for example, every day in one week, the robot will encourage the user with comment and facial expression as shown in Fig. 7.

5   Experiments of the Robotic System

In this section, we experiment with the effectiveness of every stages of the exercise support robotic system, which are the experiments of the robot invites a user to exercise of stage 1, the relationship between reaction time and weariness of stage 2, and system effectiveness of stage 3.

5.1   Experiment of the Robot Invites a User to Exercise

There are 6 participants join this experiment. We let each subject sit in front a TV and watch movies. Kinect sensor observes a subject and the robot invites him/her to exercise. When robot invites, the subject decides whether to exercise or not freely. After the experiment, we ask subjects several questions about their thinking of being invited to exercise by the robot. The answers are shown in Table 6. As the results, every subject accepts the invitation what shows the effectiveness of using robot to invite. However, there are 2 subjects accept the invitation at second times because they need more intense words. This experiment shows that it maybe necessary for the robot to change intense words when user doesn't want to exercise.

Table 6. Answers from subjects after the experiments

Subject number	Do you accept the invitation?	How many times are you invited before accept it?	How do you think about the expressions of robot?
1	Yes	1	Good. The facial expressions attract me.
2	Yes	2 Reason: The invitation is received when movie enter the climax. I don't want to exercise at that time.	Normal. I need more intense words.
3	Yes	1	Good. The changes of facial expression is interesting
4	Yes	2 Reason: Same as Subject 2.	Normal. I need more intense words.
5	Yes	1	Good. The facial expression of doubt is interesting. If robot has more movements, it will be more attractive.
6	Yes	1	Good. The facial expressions are cute.

5.2   Experiment of the Relationship between Reaction Time and Weariness

The experiment has 7 participants to join to. The subjects are 4 males and 3 females, who are 20's college students. Every subject takes the experiment twice. First time is at 14:00, and second time is at 19:30. Before the experiment, the subject takes a questionnaire about their weariness firstly. And then, in the experiment, the subject is asked to do reaction time test game 20 times. Before the experiment, every subject's answers of weariness are shown in Table 7. From this table we knows that all subject's weariness increase with time. The more intense of the weariness change, the faster the reaction time decreased.   Moreover, as the results shown in Fig. 8, which is the average reaction time of subjects, the reaction time of night (19:30) is certainly longer than afternoon (14:00). Furthermore, we ask a subject to do the experiment 5 hours a time for a day, as shown in Fig. 9, the result proof the hypothesis that the reaction time increase along with weariness.

Through this experiment, we have the following conclusions: 1. The reaction time of night is longer than morning. 2. The reaction time increase along with weariness 3. When a subject feels tired, his/her average reaction time will over 700ms. 4. When a subject feels tired, the reaction time is longer than he/she doesn't feel tire. 5. Even if the subject doesn't feel tire, after he/she did an intense physical activity, the reaction time become longer.

Table 7. Answers from subjects before the experiments

Subject	What did you just do?	Are you tired?	Average reaction time at 14:00 (ms)	What did you just do?	Are you tired?	Average reaction time at 19:30 (ms)	The difference between average vale
Male 1	Study	A little	687.45	Housework	Yes	739.95	52.5
Male 2	Study	No	687.75	Study	Yes	754.26	66.51
Male 3	Study	A little	730.8	Study	Yes	744.85	14.05
Male 4	Study	No	587.2	Study	A little	647.2	60
Female 1	Study	A little	687.5	Study	Yes	720.6	33.1
Female 2	Study	No	690.4	Part time work	Very tired	784.6	94.2
Female 3	Study	No	686	Shopping	Very tired	799.1	113.1

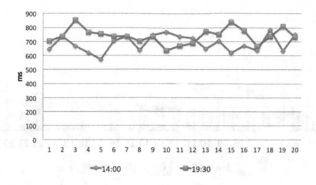

Fig. 8. The average reaction times of 7 subjects in 14:00 and 19:30

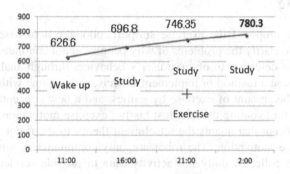

Fig. 9. The change of reaction times in a day

5.3   Experiment of System Effectiveness

For verify the system effectiveness, 2 subject join this experiment. We ask the subjects to do the experiment 28 days (about a month), and ask their comments after the experiment. One of the subject's exercise data is shown in Fig. 10. The subject keeps take 3Ex exercises at the first week, although the day that does exercise over 3Ex decreased, but the subject still keeps exercise because the effect of the robot. According to the subject's comments, the reasons that let them keep exercise are as followings: 1. It is interesting to exercise with the robot. 2. When the robot look tired, it let the user feel guilty and want to exercise for letting the robot look healthiness.

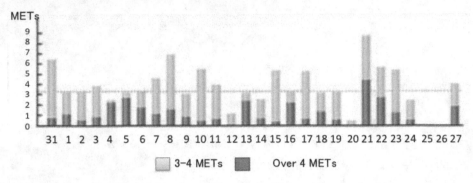

Fig. 10. A Subject's exercise data in 28 days (about a month)

6   Summary

In this paper, we proposed the exercise support robotic system based on detecting user's motion. We clarify the problems of why people cannot continue exercising, and implement motion detection to observe user's behaviors through analysis the human motions. We present a method of implement exercise video game with Kinect sensor, what calculates the amount of exercise by games, and a new concept of noticing the user's weariness for avoiding the user lost his/her exercise motivation. Furthermore, we propose a concept that adapts the attachment theory to the robot for maintaining the user's exercise motivation. The robot not only communicates with the user, but also analyzes the collected daily life activity data to provide service appropriately. The collected data and communication situation also be seen as a part of big data for constructing social network for users. The experiments show the feasibility and potential of this research. Because the system has a large space to extend, in the future, we will focus on the expression of the robot for increasing the effective of attachment theory. More exercise video games are also added for providing more choose to user. Moreover, because the simple motion patterns game is easily to let the user less interesting and feels boring, thus we have to discuss what kind of motion patterns are effective for exercise and can be recognized with high accuracy.

References

1. 2010 National Health and Nutrition Survey Japan, Ministry of Health, Labour and Welfare, http://www.mhlw.go.jp/stf/houdou/2r98520000020qbb.html (in Japanese)
2. Shibano, T., Murakami, K., Fujimoto, Y., Yamaguchi, T.: Support system for mental and physical functions by utterance and synchronous action using a robot. In: SICE System Integration Division Annual Conference (SI 2009), pp. 1193–1194 (2009)

3. Shibano, T., Ho, Y., Kono, Y., Fujimoto, Y., Yamaguchi, T.: Daily Support System for Care Prevention by Using Interaction Monitoring Robot. In: The 2010 IEEE/RSJ International Conference on Intelligent Robots and Systems (IROS 2010), Taipei, Taiwan, pp. 3477–3482 (October 2010)
4. Sato-Shimokawara, E., Murakami, K., Ho, Y., Ishiguro, S., Yamaguchi, T.: Clustering action data based on amount of exercise for use-model based health care support. In: The 2012 IEEE World Congress on Computational Intelligence (IEEE WCCI 2012), June 10-15, pp. 791–803. Brisbane Convention & Exhibition Centre, Brisbane (2012)
5. Ministry of Health Labour and Welfare Japan, "Exercise guidelines for health promotion 2006-Prevention lifestyle related disease (Exercise Guide 2006)" (in Japanese), http://www.mhlw.go.jp/bunya/kenkou/undou01/pdf/data.pdf
6. http://www.game-tm.com/action/street-fighter-ii-ce.html
7. http://www.wowgame.jp/game_html/45.html
8. Zhen, J., Aoki, H., Sato-Shimokawara, E., Yamaguchi, T.: Sino-Japanese Culture Differences in Interaction Robot System. In: The Fifth Symposium on System Integration (SII 2012), Kyushu University, Fukuoka, Japan, December 16-18, pp. 295–300 (2012)
9. Ho, Y., Sato-Shimokawara, E., Yamaguchi, T., Tagawa, N.: Interaction Robot System Considering Culture Differences. In: 2013 IEEE Workshop on Advanced Robotics and its Social Impatcs, ARSO 2013, FrC1.6 (November 2013)

Empowerment Intervention in a Ward: Nurses' Professional Commitment and Social Networks

Yi-Horng Lai[1], Hsieh-Hua Yang[1], and Li-Se Yang[2]

[1] Department of Health Care Administration, Oriental Institute of Technology
220 New Taipei City, Taiwan
FL006@mail.oit.edu.tw, yansnow@gmail.com
[2] Department of Nursing, En Chu Kong Hospital, New Taipei City, Taiwan
11705@km.eck.org.tw

Abstract. The aim of this study is to discuss the effect of intervention on job satisfaction, professional commitment and social networks. A quasi-experimental design was used in which one pre-intervention survey and one post-intervention survey was used to collect data from the 20 nurses in a ward. The score of job satisfaction in the domain of human relationship and professional commitment were significantly improved. The other 4 domains of job satisfaction had not been changed. The advice network centralization was decreased after intervention, but not friendship network. The symmetric dyad of advice network was no longer affected by position after intervention. It is concluded that empowerment intervention may be used to increase professional commitment, but not job satisfaction.

Keywords: Friendship, empowerment, intervention, professional commitment, social networks.

1 Introduction

In the present context of visible resource constraints, a hospital's management is crucially dependent upon the positive professional commitment of staff. To believe much of the related literature, mobilizing the necessary input is a relatively straightforward matter of internal team dynamics. Certainly the only way to deliver a responsive service is empowering staff to reinforce professional commitment and team coordination.

Since 1995, the implementation of national health insurance system has leaded to the rapid growth of the overall demand for medical services, and the health care organizations expansion in Taiwan. In order to curb rising medical costs, global budget, then case payment system had been implemented successively. Hospitals are feeling pressure from the internal dynamics of decentralization as well as from external forces of hospital accreditation. Hospitals have to adopt organizational change through empowerment. In order to mitigate the impact of the change, the empowerment plan is initiated in a volunteer ward. The aim of this study is to describe the intervention effect in a single ward.

L.S.-L. Wang et al. (Eds.): MISNC 2014, CCIS 473, pp. 268–281, 2014.
© Springer-Verlag Berlin Heidelberg 2014

2   Literature Review

2.1   Empowerment

Empowered employees were more effective in getting their work done and contributing to organizational productivity goals [1,2,3] and better performance in nursing practice [4]. The positive associations between empowerment, job satisfaction, and organizational commitment have been tested [5,6,7]. Chang et al. [8] developed an exploratory model and indicated the strong direct effects of organizational empowerment on job satisfaction. Also, DeCicco et al. [9] indicated that empowered nurses are highly satisfied with their jobs and committed to their organizations. A literative review [10] identified the huge potential exists for improvement through nursing empowerment to strengthen and support nurses' influential role in the quality and safety movement. Other researchers suggested that empowerment has mediating or moderating effect on the relationships between the work environment dimensions and work-related outcomes. Cai et al. [11] revealed that the empowerment mediated the impact of job characteristics on internal work motivation and general job satisfaction. Aryee and Chen [12] revealed that the empowerment fully medicated the relationship between leader-member exchange and the work outcomes.

Notions of empowerment are derived from theories of participative management and employee involvement [13]. The original impetus for research on empowerment was performance oriented. Herein, organizational scholars emphasize cascading power, information, rewards, and training to the lowest level possible in the organizational hierarchy to increase worker discretion [14], and advocate that managers share decision making power with employees to enhance performance and work satisfaction [15,16,17].

However, recent conceptualizations of empowerment are extended by joining psychological state and focusing on the individual experience. Empowerment is viewed as a state of power and control within the individual that is enabling and facilitative of a state of intrinsic task motivation for the individual [18]. The internal nature of empowerment is motivational and consistent with the expectations of the individual. Empowerment as a psychological conceptualization is an internal, subjective experience. Commitment may be view as a control structure that allows for the empowerment of individuals. Empowered people choose their own course and are committed to it because they play a part in determining it.

The concept of empowerment, which was developed in studies of western organizations, is often discussed from organizational and psychological perspectives. Cheung et al. [19] indicated that empowerment in Asian cultures relates much more to the individual and his/her merits, in contrast to organizationally-driven empowerment in Western countries. From a service perspective, empowerment gives employees the authority to make decisions about customer service. In industrial and organizational psychology and management, empowerment is the enhancement of the autonomy of employees in their work or increased involvement that results in increased decision making more generally within the wider agenda and interests of the organization [20].

Organizational empowerment is the perceived presence or absence of empowering conditions in the workplace [21], and psychological empowerment is the psychological perception or attitudes of employees about their work and their organizational roles [22]. Nurses who are psychologically empowered felt they had access to strategic information and perceived support from managers in the organization.

Frans [23] suggested that empowerment is composed of five components: a self-concept which describes self-validation and self-esteem; a critical awareness of one's place in larger systems such as family, agency, or society; possession of knowledge and abilities to influence one's self or others; a propensity to act which describes the ability to initiate effective action for one's self or others and ; a sense of collective identity which describes sharing goals, resources, and aspirations of an identified social systems. These components can help nurses to be autonomous, assertive and accountable, and have the ability to communicate effectively and accept leadership roles in practice areas. An empowered nurse will have commitment to broaden and fulfill her vision. And empowerment is viewed as a panacea that assists nurses cope with recurring healthcare crises.

2.2   Professional Commitment

Professional commitment is a person's involvement, pledge, promise or resolution towards his/her profession [24]. Fang's [24] study of Singaporean nurses demonstrated that job satisfaction was significantly and positively related to professional commitment. In another study [25], professional commitment is positively related to the job satisfaction of nurses. Harrison et al. [26] tested that overall job attitude is fundamentally important for understanding work behavior. However, Ahmadi et al. [27] discovered that professional role conception and job satisfaction were strongly negatively related at the time of hire and posited that the newly graduated nurse, faced with discrepancies between her school-taught values and practices and the values and practices of her workplace, may develop alienations and job dissatisfaction. The importance of professional commitment has been widely advocated [28,29,30,31]. A review of the literature shows that three aspects of commitment have been applied to many domains. They are willingness to exert effort on its behalf, a desire to remain in the profession, and identifying with the profession by acceptance of its goals and values [32].

There is broad consensus that future careers will be less organizationally focused and more directed toward occupational or professional commitment. And it is proposed that the empowerment intervention will improve professional commitment.

2.3   Social Networks

Employees who have friends at work are more likely to be engaged in their jobs. A review of research suggests that workplace friendship is positively related to employees' job satisfaction, performance, team cohesion, and organizational commitment; it is negatively associated with employees' turnover intentions and negative emotions [33,34,35].

The cohesiveness of the ward nursing team was mentioned as a strong predictor of job satisfaction [36]. The basic element of cohesive team is friendship network. Literature documents that individuals will have benefits through their social networks. Takase et al. [37] found the issues in interpersonal relationships were frequently cited causes that made nurses consider leaving their jobs. Bjorvell and Brodin [38] found social support might reduce personnel turnover in hospitals. Greater social support and pay can reduce turnover by their positive impact on job satisfaction [39]. In Taiwan, Chang et al. [40] showed that interpersonal relationship was the highest factor of job satisfaction. And, the research [41] in a Norwegian population of nurses found that interaction, followed by pay and autonomy were the most important job factors. In organizational settings, Hodson [42] convincingly argued, the social relations of the workplace may make a key contribution to employees' job satisfaction, productivity, and well-being. It was reported that change was positively influenced by health workers having a sense of belonging, as well as mutual respect between supervisors and health workers [43].

Social networks within organizations can be understood as formal and informal relationships. The formal relationships are formed through formal interaction, and the informal relationships are often perceived as friendship. People often have multiplex relationships and the advice and friendship relationships are not mutually exclusive. In an organization, the formal and friendship relationships are partially overlapped [44,45]. The friendship network may play an invisible role in an organization [46]. Yang et al. [47,48] have tested the importance of both advice and friendship networks are important factors of nurses' satisfaction in a hospital.

And it is proposed that the process of empowerment intervention will improve social network. The aim of this study was to describe the empowerment effects on professional commitment and social networks.

3   Method

3.1   Setting and Participants

The 13th ward is a surgical ward of En Chu Kong Hospital. The total bed number is 44 with 80% occupancy rate. All of the 20 nurses working in the 13th ward were voluntary to participate in this study.

3.2   Intervention

Theoretical based intervention program was applied from May to October in 2012. First, a needs assessment was carried out. Then, key persons were interviewed and group discussions were held. Subsequently, the results were translated into specific objectives and used in intervention development for the 13th ward's staff.

Strengthening professional commitment and harmonic interaction was defined as objectives. And 16-hour training sessions were designed. The training sessions included courses, conferences and workshops. According to Frans' 5 themes, the content of sessions included facilitation skills, standards of service, developing self-esteem and assertiveness, realize your potential, discovering the secrets of self-confidence, smart thinking and smarter working, effective communication, and essentials of personal development.

3.3   Study Design

A quasi-experimental design was used in which one pre-intervention survey and one post-intervention survey was applied to collect data from the participants.

The empowerment intervention was followed up through a combination of surveys of professional commitment and social networks. In the January of 2012, prior to the start of the intervention, a first survey was made through a questionnaire to measure the job satisfaction, professional commitment, and social network. The second survey was implemented after the intervention had been going on for 10months. The most important parameters of the overall data collection were the change in the job satisfaction, professional commitment, and social networks. The collected data was analyzed in both individual and organizational levels.

3.4   Questionnaires for Intervention Evaluation

Social network nomination was applied to derive the participants' advice and friendship networks. The participants were asked to nominate up to 18 colleagues using two questions: "When you encounter difficulty in your job, from whom you ask for help?", and "Whom you talk your private affairs with in daily chat?". The advice network will construct from the first question, and friendship network from the second question.

Social network variables were calculated for advice and friendship network. These variables include outdegree, indegree, effective size, efficiency, constraint, and hierarchy for individuals and centralization for the whole network. UCINET 6 for window [49] was used to calculate the social network variables. The centralization of whole network is proposed by Freeman [50]. Centralization (Cx) measures calculate the sum in differences in centrality between the most central node in a network and all other nodes, and divide this quantity by the theoretically largest such sum of differences in any network of the same degree. In the individual level, the nomination number is defined as outdegree and the number of nominated as indegree. Effective size is the number of alters minus the average degree of alters within the ego network, not counting ties to ego [51]. Efficiency is effective size divided by observed size [51]. Constraint is a measure of the extent to which ego is invested in people who are invested in other of ego's alters [51]. Hierarchy is the extent to which constraint on ego is concentrated in a single alters [51]. The equations are shown as below.

$$C_x = \frac{\sum_{i=1}^{n}[C_x(p^*) - C_x(p_i)]}{\max \sum_{i=1}^{n}[C_x(p^*) - C_x(p_i)]}$$

$$Effective\ size\ of\ i's\ network = \sum_j \left[1 - \sum_q p_{iq} m_{jq}\right], q \neq i, j$$

Efficiency = effective size/ observed size

$$Constra\,\text{int} = (p_{ij} + \sum_q p_{iq} p_{qi})^2, q \neq i, j$$

Based on Aryee et al. [32], professional commitment scale was developed. It is a 6-point Likert type scale (1 = strongly disagree, 6 = strongly agree) with 3 items. These 3 items are 'Being a professional nurse is great helpful to my self-image', 'I am proud of being a professional nurse', and 'I am enthusiastic about nursing care'. The Cronbach's alpha was 0.83.

The NJSS was adopted from Lin et al. [52]. It was applied to measure job satisfaction in 5 domains of job environment, human relationship, feedback, benefit and promotion, and work load. The questions were scored on a 6-point Likert scale ranging from 1 point (very dissatisfied) to 6 points (very satisfied). The measurement has strong internal consistency with all multiple-item constructs achieving Cronbach's alpha ranged between 0.89 and 0.98, exceeding the 0.7 threshold commonly suggested for exploratory research [53].

4   Results

4.1   Description of Participants

All of the participants are females. Within these 20 participants, one is charge nurse, 16 nurses and 3 nurse aides. Eight of them are married. The average age is 34.05 years old with 21-60 range. Most of them have graduated from college or university.

4.2   The Effect on Professional Commitment and Job Satisfaction

Comparing the professional commitment and 5 domains of job satisfaction between time 1 and time 2, t-test was implemented. The result is shown as table 1. The score of human relationship and professional commitment are significantly improved. The other 4 domains of job satisfaction had not been changed. The 4 domains are job environment, feedback, benefit and promotion, and work load. Obviously, the working conditions have not been changed after empowerment intervention.

Table 1. Effect on job satisfaction and professional commitment

Variables	Time 1		Time 2		t	p-value
	Mean	SD	Mean	SD		
Job satisfaction						
job environment	12.05	1.88	11.75	1.25	0.946	0.356
human relationship	16.75	2.17	17.25	2.07	-2.127	0.047
feedback	11.70	2.34	11.95	1.50	-0.960	0.349
benefit and promotion	23.05	3.85	22.45	3.53	1.552	0.137
work load	14.50	2.84	14.05	3.28	0.975	0.342
Professional commitment	12.70	2.05	13.05	1.99	-2.333	0.031

4.3   The Effect on Social Networks

The whole network centralization between time 1 and time 2 is shown as table 2. The advice network centralization was decreased after intervention for both outdegree and indegree. The friendship network centralization was increased for outdegree, and decreased for indegree. Figure 1 to figure 4 illustrate the advice and friendship networks at time 1 and 2. At time 1, one nurse has only one linkage in the advice network. And after intervention, this nurse has connection with two other nurses. For friendship network, there are 2 isolated nurses at time 1, and after intervention they have linkages with others. In the individual level analysis, the constraint of advice network was decreased and the indegree of friendship network was increased as table 3.

Table 2. Comparison of whole network centralization

Variables	Time 1	Time 2
Advice network		
Outdegree centralization	40.443%	19.114%
Indegree centralization	51.524%	24.654%
Friendship network		
Outdegree centralization	26.593%	44.598%
Indegree centralization	32.133%	27.978%

Further analysis was applied on dyads of networks at time 1 and time 2. Nomination forms a dyad. If two persons nominate each other, the dyad is symmetric. In advice network, there were 114 dyads at time 1 and 121 at time 2. The dyads were increased after intervention. Regression model was applied to identify the factors of symmetric dyads. Dyads were included as dependent variable, and education, marriage, tenure, and position were included as independent variables. If the education, marriage, tenure, and position were the same for two nurses of symmetric nominations, it was recoded as 1. Otherwise, it was 0. The results are shown as table 4. Obviously, the symmetric advice dyads were significantly related to marriage, tenure, and position at time 1. It means that two nurses of the same marriage, tenure, or position are more likely to form symmetric advice relation. However, after intervention the effect of marriage and position diminished. But tenure is still a significant factor of forming symmetric advice relations. For friendship network, the dyads were not correlated with any of these variables.

Table 3. Comparison of social network variables between time 1 and time 2

Variables	Time 1		Time 2		t	p
	Mean	SD	Mean	SD		
Advice network						
outdegree	5.70	4.04	6.10	3.92	-1.405	0.176
indegree	5.70	4.46	6.10	4.39	-1.710	0.104
effective size	5.50	2.55	6.03	2.23	-1.899	0.073
efficiency	0.60	0.11	0.60	0.11	-0.054	0.958
constraint	0.31	0.17	0.27	0.10	2.428	0.025
hierarchy	0.12	0.21	0.06	0.03	1.155	0.262
Friendship network						
outdegree	4.20	3.09	4.95	3.25	-1.831	0.083
indegree	4.20	3.02	4.95	3.05	-2.775	0.012
effective size	4.49	2.43	4.93	2.23	-1.506	0.148
efficiency	0.64	0.24	0.64	0.12	0.029	0.977
constraint	0.31	0.20	0.35	0.18	-1.193	0.248
hierarchy	0.10	0.23	0.11	0.23	-1.406	0.178

Table 4. Factors influencing symmetric advice relation

Variables	Time 1			Time 2		
	B	SE	p	B	SE	p
Age	0.002	0.027	0.929	0.008	0.032	0.793
Education (same/diff)	-0.843	0.477	0.077	-0.513	0.443	0.247
Marriage (same/diff)	-1.075	0.470	0.022	-0.196	0.434	0.651
Tenure	0.150	0.045	0.001	0.125	0.047	0.007
Position (same/diff)	1.976	0.971	0.042	1.889	0.987	0.056
Constant	-1.082	0.914	0.236	-1.672	0.978	0.087
-2Log likelihood	118.326			131.595		

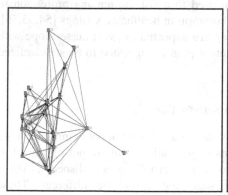

Fig. 1. Advice network at time 1

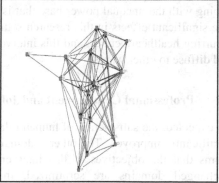

Fig. 2. Advice network at time 2

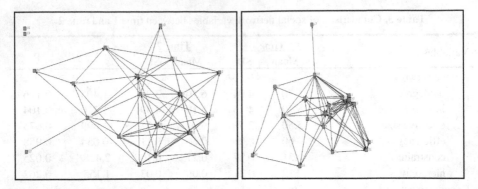

Fig. 3. Friendship network at time 1   **Fig. 4.** Friendship network at time 2

5   Discussion and Conclusion

Nurses' medical training emphasizes technical life-saving activities and skills. They need autonomy, assertiveness, accountability, and the ability to communicate and cooperate in a team. The aim of this empowerment intervention is improving these abilities. The effects on professional commitment and social network are discussed.

5.1   Effect of the Empowerment Intervention

The empowerment intervention has achieved its objectives. The professional commitment can be improved and the advice interaction can be modified, but not for all the domains of job satisfaction. Since professional commitment was tested to be related to job satisfaction [24], and the cohesiveness of the ward nursing team was mentioned as a strong predictor of job satisfaction [36], it is expected that the long-term effect will appear in the future. It needs to follow up.

The design of intervention is based on empowerment theory after needs assessment. Empowerment theory has been used to assist the nursing profession in coping with the unequal power base that is common in healthcare settings [54,55,56]. The significant effects in this research setting are expected to assist nurses cope with recurring healthcare crises, and this intervention plan is suggested to be standardized and diffuse to other wards.

5.2   Professional Commitment and Job Satisfaction

As expected, the satisfaction of human relationship and professional commitment are significantly improved. The other 4 domains of job satisfaction are not improved. It seems that the objectives of this intervention are partially accomplished. If the 4 unchanged domains are scrutinized, the statement would be different. These unchanged domains including job environment, feedback, benefit and promotion, and work load. These items are related to work conditions or regulations, and cannot be changed through empowerment. But the job satisfaction may be recognized as a psychological perception. And researchers [11] indicate the mediating effect of empowerment on the relationships between the job characteristics and job satisfaction. There is possibility to ignore the work conditions by increasing harmonic interpersonal relationships.

5.3   The Effect on Social Networks Is Significant

The whole network centrality is related to the compactness of graphs [50]. A compact network represents the distances between pairs of its members are short. The index, Cx, measures the degree to which Cx(p*) exceeds the centrality of all of the other points. Cx=0 if and only if all Cx(pk) are equal, and Cx =1 if and only if one point, p*, completely dominates the network with respect to centrality. The result showed the centrality is decreased after intervention. It means the cohesive of the ward team is increased.

Homophily is a major characteristic of friendship. Individuals tend to become and remain friends with others that are similar to them [57,58,59]. The homophilous effect is also found in this study. The symmetric relations exist between 2 persons of the same tenure and positions before intervention. It means that the different positions will hinder interaction. It is not suitable for cooperation. Fortunately, the homophilous effect of position disappeared after intervention. Baker et al. [60] found that empowerment involves reciprocity between the leader and follower. Since the homophilous effect of position vanished, the intervention purpose is accomplished.

Several researchers have emphasized the interdependence of formal and informal structures [61,62,63,64,65]. Especially for Chinese employees, Bozionelow & Wang [66] posited that that mentoring is an integral part of the Chinese culture and with the view that network ties or Guanxi in the Chinese society cannot exist in a purely instrumental form. The overlap of formal and informal relationship will be benefit for the network ties in a work setting.

5.4   Conclusion

In conclusions, the empowerment intervention has effects on professional commitment and social networks. Enhancing empowerment in a supportive environment would allow nurses to experience satisfaction with their jobs [7]. The effects of this intervention would be more significant if the hospital devote to create supportive conditions. Further, empowerment in Asian cultures relates much more to the individual and his/her merits, in contrast to organizationally-driven empowerment in Western countries [19]. The management should not ignore the interpersonal network for both formal and informal relationships. Further, work place interventions are complicated and complex and influenced by many factors. Ten months is a very short time for an intervention. The organization's complexity and the traditional hierarchy of hospitals do not conducive change, but the nurses' involvement is critical to the result of the staff's work. The importance of the experience of the staff of an improved work environment is however not to be diminished.

References

1. Laschinger, H.K.S., Wong, C.: Staff nurse empowerment and collective accountability: effect on perceived productivity and self-related work effectiveness. Nurs. Econ. 17(6), 308–317 (1999)
2. McNeeseSmith, D.: Increasing employee productivity, job satisfaction, and organizational commitment. Hosp. Health Serv. Adm. 41(2), 160–175 (1996)
3. Sigler, H.T., Pearson, C.M.: Creating an empowering culture: examining the relationship between organizational culture and perceptions of empowerment. J. Qual. Manage. 5, 27–52 (2000)
4. Manojlovich, M.: Linking the practice environment to nurses' job satisfaction through nurse-physician communication. J. Nurs. Scholarship 37(4), 367–373 (2005)
5. Manojlovich, M., Laschinger, H.K.S.: The relationship of empowerment and selected personality characteristics to nursing job satisfaction. J. Nurs. Admin. 32(11), 586–595 (2002)
6. Liu, C.H., Chang, L.C., Li, I.C., Liao, J.Y., Lin, C.I.: Organizational and psychological empowerment on organizational commitment and job satisfaction among primary health professionals. J. Evidence-Based Nurs. 2(1), 5–13 (2006)
7. Ning, S., Zhong, H., Libo, W., Qiujie, L.: The impact of nurse empowerment on job satisfaction. J. Adv. Nurs. 65(12), 2642–2648 (2009)
8. Chang, L.-C., Shih, C.-H., Lin, S.-M.: The medicating role of psychological empowerment on job satisfaction and organizational commitment for school health nurses: A cross-sectional questionnaire survey. Int. J. Nurs. Stud. 47, 427–433 (2010)
9. DeCicco, J., Laschinger, H., Kerr, M.: Perceptions of empowerment and respect: effect on nurses' organizational commitment in nursing homes. J. Gerontol. Nurs. 32(5), 49–56 (2006)
10. Richardson, A., Storr, J.: Patient safety: a literative review on the impact of nursing empowerment, leadership and collaboration. Int. Nurs. Rev. 57, 12–21 (2010)
11. Cai, C.F., Zhou, Z.K., Yeh, H., Hu, J.: Empowerment and its effects on clinical nurses in central China. Int. Nurs. Rev. 58, 138–144 (2011)
12. Aryee, S., Chen, Z.X.: Leader-member exchange in a Chinese context: Antecedents, the mediating role of psychological empowerment and outcomes. J. Bus. Res. 59, 793–801 (2006)
13. Spreitzer, G.M., Kizilos, M.A., Nason, S.W.: A dimensional analysis of the relationship between psychological empowerment and effectiveness, satisfaction, and strain. J. Manage. 23(5), 679–704 (1997)
14. Bowen, D.E., Lawler III, E.E.: The empowerment of service workers: what, why, how and when. MIT Sloan Manage. Rev. 33(3), 31–40 (1992)
15. Cotton, J.L., Vollrath, D.A., Froggatt, K.L., Lengnick-Hall, M.L., Jennings, K.R.: Employee participation: Diverse forms and different outcomes. Acad. Manage. Rev. 73, 103–112 (1988)
16. Miller, K.I., Monge, P.R.: Participation, satisfaction, and productivity: A meta-analytic review. Acad. Manage. J. 29, 727–753 (1986)
17. Wagner III, J.A.: Participation's effects on performance and satisfaction: A reconsideration of research evidence. Acad. Manage. Rev. 19, 312–330 (1994)
18. Thomas, K., Velthouse, B.: Cognitive elements of empowerment: An "interpretive" model of intrinsic task motivation. Acad. Manage. Rev. 15, 666–681 (1990)
19. Cheung, C., Baum, T., Wong, A.: Relocating empowerment as a management concept for Asia. J. Bus. Res. 65, 36–41 (2012)

20. Wall, T.D., Wood, S.J., Leach, D.J.: Empowerment and performance. Int. Rev. Ind. Organ. Psychol. 19, 1–46 (2004)
21. Kanter, R.M.: Men and Women of the Corporation. Basic Books, New York (1977)
22. Spreitzer, G.M.: Psychological empowerment in the workplace: Dimensions, measurement, and validity. Acad. Manage. J. 38(5), 1442–1462 (1995)
23. Frans, D.: A scale for measuring social work empowerment. Res. Social Work Prac. 3(3), 312–328 (1993)
24. Fang, Y.: Turnover propensity and its causes among Singapore nurses: an empirical study. Int. J. Hum. Resour. Man. 12(5), 859–871 (2001)
25. Lu, K.-Y., Lin, P.-L., Wu, C.-M., Hsieh, Y.-L., Chang, Y.-Y.: The relationship among turnover intentions, professional commitment, and job satisfaction of hospital nurses. J. Prof. Nurs. 18(4), 214–219 (2002)
26. Harrison, D.A., Newman, D.A., Roth, P.L.: How important are job attitudes? Meta-analytic comparisons of integrative behavioral outcomes and time sequences. Acad. Manage. J. 49(2), 305–325 (2006)
27. Ahmadi, K.S., Speedling, E.J., Kuhn-Weissman, G.: The newly hired hospital staff nurse's professionalism, satisfaction and alienation. Int. J. Nurs. Stud. 24(2), 107–121 (1987)
28. Aryee, S., Tan, K.: Antecedents and outcomes of career commitment. J. Vocat. Behav. 40, 288–305 (1992)
29. Bedeian, A.G., Kemery, E.R.: Career commitment and expected utility of present job as predictors of turnover intention and turnover behaviour. J. Vocat. Behav. 39, 331–343 (1991)
30. Blau, G., Lunz, M.: Testing the incremental effect of professional commitment on intent to leave one's profession beyond the effects of external, personal, and work-related variables. J. Vocat. Behav. 52, 260–269 (1998)
31. Aranya, N.A., Pollock, J., Amernic, J.: An examination of professional commitment in public accounting. Account. Org. Soc. 64, 271–280 (1981)
32. Aryee, S., Wyatt, T., Min, M.K.: Antecedents of organizational commitment and turnover intentions among professional accountants in different employment settings in Singapore. J. Soc. Psychol. 131, 545–556 (1991)
33. Berman, E.M., West, J.P., Richter, M.N.: Workplace relations: Friendship patterns and consequences (according to managers). Public Admin. Rev. 62(2), 217–230 (2002)
34. Morrison, R.: Information relationships in the workplace: Association with job satisfaction, organizational commitment and turnover intentions. New Zeal. J. Psychol. 33, 114–128 (2004)
35. Winstead, B.A., Derlega, V.J., Montgomery, M.J., Pilkington, C.: The quality of friendships at work and job satisfaction. J. Soc. Pers. Relat. 12, 199–215 (1995)
36. Lu, H., While, A.E., Barriball, K.L.: Job satisfaction among nurses: a literature review. Int. J. Nurs. Stud. 42, 211–227 (2005)
37. Takase, M., Oba, K., Yamashita, N.: Generational differences in factors influencing job turnover among Japanese nurses: An exploratory comparative design. Int. J. Nurs. Stud. 46, 957–967 (2009)
38. Bjorvell, H., Brodin, B.: Hospital staff members are satisfied with their jobs. Scand. J. Caring Sci. 6(1), 9–16 (1992)
39. Price, J.L.: Reflections on the determinants of voluntary turnover. Int. J. Manpower 22(7), 600–624 (2001)
40. Chang, F.M., Lin, C.J., Kuo, H.W., Teng, Y.K., Lee, J.N.: Appraisal and satisfaction among staff nurses in the health stations of Taiwan province China. J. Public Health 14, 78–87 (1995)

41. Bjork, I.T., Samdal, G.B., Hansen, B.S., Torstad, S., Hamilton, G.A.: Job satisfaction in a Norwegian population of nurses: A questionnaire survey. Int. J. Nurs. Stud. 44, 747–757 (2007)

42. Hodson, R.: Group relations at work: Solidarity, conflict and relations with management. Work Occupations 24, 426–452 (1997)

43. Sennun, P., Suwannapong, N., Howteerakul, N., Pacheun, O.: Participatory supervision model: building health promotion capacity among health officers and the community. Rural Remote Health 6(2), 440 (2006)

44. Yang, H.H., Lai, Y.H., Chao, W.C., Chen, S.F., Wang, M.H.: Social networks and job satisfaction of nurses. WSEAS Transactions on Communications 8(7), 698–707 (2009)

45. Yang, H.H., Lai, Y.H., Chao, W.C., Chen, S.F., Wang, M.H.: Propensity to connect with others, social networks and job satisfaction of nurses. In: Proceedings of the 13th WSEAS International Conf. on Communication, pp. 178–184 (2009)

46. Cross, R., Parker, A.: The Hidden Power of Social Networks. Harvard Business School Press, Boston (2004)

47. Yang, H.H., Yang, H.J., Yu, J.C., Hu, W.J.: Multinominal regression model for in-service training. International Journal of Mathematical Models and Methods in Applied Sciences 2(1), 74–80 (2008)

48. Yang, H.H., Huang, F.F., Lai, Y.H., Hsieh, C.J., Liao, Y.S., Chao, W.C., Chen, S.F.: Perceived organizational culture, professional commitment, advice network and job satisfaction of novice nurses. WSEAS Transactions on Communications 9(9), 595–604 (2010)

49. Borgatti, S.P., Everett, M.G., Freeman, L.C.: UCINET 6 for Windows: Software for Social Network Analysis. Analytic Technologies, Harvard (2002)

50. Freeman, L.C.: Centrality in social networks conceptual clarification. Soc. Networks 1, 215–239 (1978/1979)

51. Burt, R.S.: Structural Holes: The Structure of Competition. Harvard University Press, MA (1992)

52. Lin, C.J., Wang, H.C., Li, T.C., Huang, L.C.: Reliability and validity of nurses' job satisfaction scale and nurses' professional commitment. Mid-Taiwan Journal of Medicine 12, 65–75 (2007)

53. Nunnally, J.C.: Psychometric Theory. McGraw-Hill, New York (1978)

54. Manson, D.J., Backer, B.A., Georges, C.A.: Toward a feminist model for the political empowerment of nurses. Image: J. Nurs. Scholarship 23, 72–77 (1991)

55. Sabaston, J.A., Spence-Lashinger, H.K.: Staff-nurse work empowerment and perceived autonomy. J. Nurs. Admin. 25(9), 42–49 (1995)

56. Wilson, B., Laschinger, H.K.S.: Staff-nurse perception of job empowerment and organizational commitment: a test of kanter's theory of structural power in organizations. J. Nurs. Admin. 24(4), 39–47 (1994)

57. Cohen, J.: Sources of peer group homogeneity. Sociological Education 50, 227–241 (1997)

58. Kandel, D.B.: Homophily, selection, and socialization in an adolescent friendships. Am. J. Sociol. 84, 427–436 (1978)

59. McPherson, M., Smith-Lovin, L., Cook, J.M.: Birds of a feather: homophily in social networks. Annu. Rev. Sociol. 27, 415–444 (2001)

60. Baker, C.M., McDaniel, A.M., Fredrickson, K.C., Gallegos, E.C.: Empowerment among Latina nurses in Mexico, New York and Indiana. Int. Nurs. Rev. 54, 124–129 (2007)

61. Ranson, S., Hinings, B., Greenwood, R.: The structuring of organizational structures. Adm. Sci. Quart. 25, 1–17 (1980)

62. Boster, J.S., Johnson, J.C., Weller, S.C.: Social position and shared knowledge: Actors' perceptions of status, role, and social structure. Soc. Networks 9, 375–387 (1987)
63. Monge, P.R., Eisenberg, R.M.: Emergent communication networks. In: Jablin, F.M., Putnam, L.L., Roberts, K.H., Porter, L.W. (eds.) Handbook of Organizational Communication: An Interdisciplinary Perspective. Sage, Beverly Hills (1987)
64. Shrader, C.B., Lincoln, J.R., Hoffman, A.N.: The network structures of organizations: effects of task contingencies and distributional form. Hum. Relat. 42, 43–66 (1989)
65. Brajkovich, L.F.: Sources of social structure in a start-up organization: work networks, work activities, and job status. Soc. Networks 16, 191–212 (1994)
66. Bozionelos, N., Wang, L.: The relationship of mentoring and network resources with career success in the Chinese organizational environment. Int. J. Hum. Resour. Man. 17, 1530–1545 (2006)

Effects of Emotion and Trust on Online Social Network Adoption toward Individual Benefits: Moderating Impacts of Gender and Involvement

Yi-Jie Tsai, Chien-Hsing Wu, and Chian-Hsueng Chao

National University of Kaohsiung
700, Kaohsiung University Rd., Nanzih District, 811. Kaohsiung, Taiwan, R.O.C.
joytsaiyj@gmail.com, {chwu,cchao}@nuk.edu.tw

Abstract. For the past decade, the advanced development of Online Social Network (OSN) has demonstrated a considerable contribution to the industries and society. The current research proposes and examines the research models that incorporate types of emotion and trust as the antecedents of OSN adoption to describe individual benefits (IB) for positive emotion (PE) and negative emotion (NE) groups. A salient consideration is to examine the moderating effects of gender and involvement on the relationships between independent and dependent variables for both groups. Based on the analysis of 522 valid samples, research results show that (1) for PE group, attentive and active show significant effects on OSN adoption, respectively. (2) for NE group, only ashamed presents a significant influence on OSN adoption. (3) moderators of PE group report that alert, attentive, and active are significant, indicating that female reveals significant effects while male reveals insignificant effects for both alert and attentive, and low involvement reveals significant effects for inspired and active while high involvement reveals significant effects for attentive and active. (4) moderators of NE group report that upset, ashamed, and afraid are significant, indicating that female reveals a significant effect for upset while male reveals a significant effect for ashamed, and high involvement shows significant effects for both ashamed and afraid. Discussion and implications are also addressed.

Keywords: Online Social Network, Emotion, Trust, Gender, and Involvement.

1 Introduction

For the past decade, the development of Online Social Network (OSN) has extended the notion of social networks on Web 2.0 network applications. Therefore, to emphasize this articulated social network as a critical organizing feature of these sites, researchers label them Social Network Sites (SNSs) [1, 2]. For example, Facebook and Twitter are trends that are sweeping the world. Current studies have shown that OSN adoption remains the mainstream of interaction research for now and near future in the SNSs filed, and also focused on research areas which include SNSs' applications and developments, usage behaviors and satisfactions, key factors of

L.S.-L. Wang et al. (Eds.): MISNC 2014, CCIS 473, pp. 282–296, 2014.
© Springer-Verlag Berlin Heidelberg 2014

success, the effects on different traits or types of SNSs, friendship performances, networks management, and issues of online and offline connectivity and privacy, etc. [3, 4, 5, 6, 7, 8, 9]

Previous literature discussed factors and examined the use willingness of SNSs, IT use, and social capital [3, 10, 11, 12, 8]. First, technology acceptance model (TAM) argues that behavioral intention is influenced by psychological status that mainly includes various types of emotion, such as inspired, upset, active, hostile, etc. Second, trust has been studied in many disciplines including sociology, psychology, economics, and computer science [13, 14, 15, 16, 22]. And, there are literatures that illustrate the relationship between trust and OSN adoption in SNSs fields [17, 18, 19, 20, 21]. However, literature paid limited attention to the effect of emotion on OSN adoption toward individual benefit. The current study adopts the positive and negative emotion elements as the antecedents of OSN use, and at the same time argues that trust can influence the using behavior on OSN. From literature review of gender and involvement effects, this study also examines the moderation effect of these two variables on the relationship between emotions, trust, OSN adoption, and individual benefits. Therefore, the research objectives of this study are: (1) to develop and examine research models that describes the effects of emotion and trust on OSN adoption toward individual benefits. (2) to examine the moderating effects of gender and involvement on the relationship between independent variables and dependent variables.

2   Related Concepts

The concept of social networks was introduced by Barnes (1954) [1], who described them as connected graphs where nodes represented entities and edges. Nodes are individuals, groups, organizations, or government agencies; Edges are interactions, invitations, trades, values, etc. For the past decade, the emergence of online social network such as Myspace and Facebook has extended the notion of social networks in terms of their sizes [23]. The public accessibility of OSN using mobile phones makes such platforms ubiquitous [24]. Hanchard (2008) [25] showed that user retention rates of social networks were as good as online banking at high nineties. Previous literatures have shown that social network remains the mainstream of interaction research for now and near future [26].

2.1   Online Social Network Adoption and Individual Benefits

Literature indicated that there are reasons or benefits for people to adopt social network, in this study, there are three sub-factors of individual benefits including relationship maintenance and development, information and knowledge sharing, and use satisfaction.

First, in an online social network individuals are consciously able to construct an online representation of self, such as online dating profiles and online games (e.g. LoL- League of Legends, WoW- World of Warcraft, and StarCraft). To address the concept in more depth, the SNS has been developing an important research context for scholars to investigate processes of impression management, self-presentation, and relationship performance [27]. Second, prior studies [28, 29, 30, 31] have provided evidence demonstrating the importance of information and knowledge sharing in enhancing the performance, and provide mechanisms to support the information and knowledge sharing on OSN. Third, as the number and types of social networking sites continue to grow and compete for users' attention, the question of how to make users satisfied with their use of social network becomes a critical issue. Satisfaction is defined as a pleasing experience with the social networking experience in terms of technology and social interaction.

Previous literature has shown that factors have used to examine the use willingness of SNSs, IT use, and social capital which includes multidimensional construct: civic participation, political engagement, life satisfaction, and social trust, etc. [3, 10, 11, 12, 8]. According the Technology Acceptance Model (TAM) [32], two particular beliefs: perceived usefulness and perceived ease of use, were primary two elements for computer acceptance behaviors; TAM argued that computer usage was determined by behavioral intention to use, which was affected by the attitude toward ease of use and perceived usefulness toward use intention and actual use. This study adopts the concept of TAM and regards that SNS adoption will be influencing the individual benefits and therefore the first hypothesis is formed as follow.

Model1 H1: Online social network adoption is significantly related to individual benefits.
Model2 H6: Online social network adoption is significantly related to individual benefits.
(H1abc and H6abc: Individual benefits are relationship maintenance and development, information and knowledge sharing, and use satisfaction)

2.2   Emotion and Trust

Human's emotion is diverse. Thompson (2007) [33] developed The International Positive and Negative Affect Schedule Short Form (I-PANAS-SF), an internationally useable 10-item version that include PE: Alert, Inspired, Determined, Attentive, and Active; NE: Upset, Hostile, Ashamed, Nervous, and Afraid. This new measure schedule I-PANAS-SF, that (1) accounts for shortcomings highlighted above, (2) reflects items qualitatively assessed to be easy to understand and unambiguous in meaning across different populations of nonnative English speakers, (3) exhibits strong psychometric properties concerning reliability, cross-sample and temporal stability, and convergent and criterion-related validity, and (4) provides evidence of cross-national structural equivalence [33]. I-PANAS-SF schedule also can applied in many different research areas to measure the positive and negative affect, those emotions can related the thinking and behavior, combine other effects and variables to analysis some results for the theory assumption. Previous references are found that the positive and negative emotions are extensively used to explore their impacts on performance in various domains, such as Educational psychology, social behavior, sport science, medicine, etc. [34, 35, 36, 37, 38]. Importantly, positive and negative emotions are the main categories to describe emotions.

Trust has been studied in many disciplines including sociology, psychology, economics, and computer science [13, 14, 15, 16]. Each of these disciplines has defined and considered trust in using online social network from different perspectives, but their definitions may not be directly applicable to social networks. In general, trust is a measure of confidence that an entity or entities will behave in an expected manner, Sherchan et al. (2013) [26] pointed that literature concerning trust can be categorized three criteria: (1) trust about information collection: attitudes, behaviors, and experiences; (2) trust about value assessment: graph, interaction, and hybrid; (3) trust about value dissemination: based recommendation and visualization models.

This study review the definitions and measurements of trust from different disciplines that focus on online social networks. Accordingly, this study adopts the positive and negative emotional elements as the antecedents of OSN use, and also assumed that trust can influence the using behavior on OSN. Therefore, the study argues that the emotions and trust are the factors affecting use behavior of online social network. Hypotheses are then defined as follows.

Model1 H2: Positive emotions are significantly related to the online social network adoption.

H3: Trust are significantly related to the online social network adoption.

Model2 H7: Negative emotions are significantly related to the online social network adoption.

H8: Trust are significantly related to the online social network adoption.

(H2abcde: Positive emotions are alert, inspired, determined, attentive, and active)

(H7abcde: Negative emotions are upset, hostile, ashamed, nervous, and afraid)

2.3 Gender and Involvement

Venkatesh et al. (2003) [22] developed the Unified Theory of Acceptance and Use of Technology (UTAUT) model to consolidate previous TAM related studies. The UTAUT model attempts to explain how individual differences influence technology use. More specifically, the relationship between perceived usefulness, ease of use, and intention to use can be moderated by age, gender, and experience. The research design of moderating effects are from UTAUT model, to measure the effects between human's emotions, trust, the behavior of OSN adoption, and individual benefits, the moderators of this study are gender and involvement. There are two parts of this sections including literatures reviews of gender and involvement as following.

First, literature indicates that male is more likely than female to adopt a new technology earlier [39, 40, 41]. More males used the Internet in its nascent years than females [42]. A recent report by Crunchies (2008) [43] indicated that more women used the Internet than their male counterparts in the U.S. as of 2008 and the trend will continue in the near future. The current study proposes that there is a gender difference in using social networking site. Most users of SNSs are young individuals. SNSs are considered to play an active role in younger generation's daily life [44]. The relationship between the usage behavior and other factors such as gender and frequency has been studied in many researches that focused on young people's online activities [45, 46, 47].

Second, the original concept of involvement from social judgment theory was proposed by Sherif and Cantril (1947) [48]. They argued that ego-involvement was determined by internal factors such as motivations or emotional states by personal attitudes, and by human perception or interpretation of the external situations. There are many studies dealing with the involvement definition. For example, Zaichkowsky (1994) [49] defined that the involvement is a motivational construct which partly relies on the antecedent factor of the person's values and need. Involvement is also applied on the consumer behavior study; it shows how consumers process information and learn, and also form attitudes to make purchase decision. Moreover, the consequence of involvement may be relevant or interesting, and therefore may play a strong role in determining what is relevant or interesting to consumers [50].

From those references review of gender and involvement effects, this study also use these two variables to moderate the relationship between emotions, trust, OSN adoption, and individual benefits, to examine the different results in various reaction. Therefore, hypothesis H4, H5, H9 and H10 are defined as follows:

Model1 H4: Gender has a significant moderating effect on H1, H2, and H3.

H5: Involvement has a significant moderating effect on H1, H2, and H3.

Model2 H9: Gender has a significant moderating effect on H6, H7, and H8.

H10: Involvement has a significant moderating effect on H6, H7, and H8.

3   Method

3.1   Research Model

According to hypothesis 1 to 10 defined above, the research models are illustrated in Figure 1 and Figure 2. Model 1 contains the five major components: (1) Positive Emotions (PE), (2) Trust, (3) Online Social Network (OSN) Adoption, (4) Individual Benefits (IB), (5) moderators are Gender and Involvement. There are also five major components in Model 2 which are (1) Negative Emotions (NE), (2) Trust, (3) Online Social Network (OSN) Adoption, (4) Individual Benefits (IB), and (5) moderators include Gender and Involvement. In this study, research models include three independent variables: (1) the sub-factors of PE are Alert, Inspired, Determined, Attentive, and Active. (2) NE has five sub-factors including Upset, Hostile, Ashamed, Nervous, and Afraid. (3) Trust. The dependent variables are OSN adoption, and IB: Relationship Maintenance and Development, Information and Knowledge Sharing, and Use Satisfaction. Figure 1 and 2 are research models, and there are ten major hypotheses defined accordingly. They are presented as follows.

3.2   Sampling Plan and Measurement

The population in this study focuses on the people who are experienced in using online social network. The questionnaire was available in online community that included Facebook, Twitter, Google+, Line, WeChat, and PPT. The maximum variance obtained from the pre-test was 0.85 from the variable of negative emotion. It was assumed that the sample error was 0.1 and significance level was set to be 95%. It was also assumed that the valid returned rate was 0.7. This implied that there were at least about 400 questionnaires returned to meet these requirements. To increase the response rate, it was announced that 20 of the respondents who returned valid questionnaires win a 100-TWD voucher each based on a lucky draw activity.

The questionnaire was designed with five major parts: (1) positive emotions (PE), (2) negative emotions (NE), (3) online social network adoption, (4) individual benefits (IB), and (5) moderators. Each variable was measured based on the Likert rating scale (from 1 to 5, and only trust variable from 1 to 6) using bipolar descriptors for each question. The basic information included age and education level. There are 51 questions in total for variables that were included in the research model. The data analysis was divided into three parts: (1) descriptive statistics, (2) reliability and validity analysis, and (3) structure equation model using partial least-square technique for hypothesis testing.

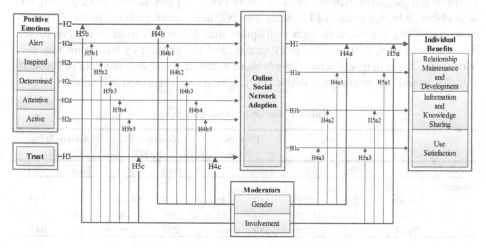

Fig. 1. Research Model 1 (PE group)

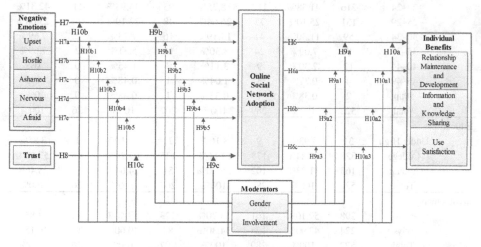

Fig. 2. Research Model 2 (NE group)

4   Data Analysis Results and Discussion

4.1   Descriptive Statistics

The demographic data of the samples included gender, ages, educational level, and years of using online social network (involvement). Questionnaire was available online from January 1 to 31. Invitations were posted in the online communities (e.g., Facebook, and twitter). The number of samples collected was 522 in total, which was more than the expected number of samples. This study examines the behavior internet adoption between positive emotions (PE) group and negative emotions (NE) group. The valid samples were divided into PE and NE group. Based on the International Positive and Negative Affect Schedule Short Form (I-PANAS-SF) [33], participants who obtained higher points of PE items than NE items were grouped in the PE group. Conversely, the sample with higher NE pints than PE points went to the NE group. Finally, there were 289 samples in PE group, and the number of NE group was 207. Moreover, 26 participants whose scores were equal, it mean that the data of those 26 samples would not be analyzed in this study. Details are presented in Table 1.

Table 1. The descriptive statistics

Items	Scales	All N=522 Freq.	(%)	PE = NE N=26 Freq.	(%)	Group1: PE N=289 Freq.	(%)	Group 2: NE N=207 Freq.	(%)
Gender									
	Female	255	48.85%	130	45%	113	54.60%	12	43.15%
	Male	267	51.15%	159	55%	94	45.40%	14	53.85%
	Total	522	100%	289	100%	207	100%	26	100%
Ages									
	0~19	61	11.69%	22	7.61%	34	16.43%	5	19.23%
	20~24	216	41.38%	112	38.75%	93	44.93%	11	42.31%
	25~29	131	25.10%	77	26.64%	48	23.19%	6	23.08%
	30~34	59	11.30%	41	14.19%	16	7.73%	2	7.69%
	35~39	37	7.09%	24	8.30%	11	5.31%	2	7.69%
	40~44	12	2.30%	9	3.11%	3	1.45%	0	0%
	45~49	4	0.77%	3	1.04%	1	0.48%	0	0%
	50up	2	0.38%	1	0.35%	1	0.48%	0	0%
	Total	522	100%	289	100%	207	100%	26	100%
Education									
	Under High	31	5.93%	9	3.10%	21	10.10%	1	3.85%
	College	328	62.84%	178	61.60%	135	65.20%	15	57.69%
	Graduate	163	31.23%	102	35.30%	51	24.60%	10	38.46%
	Total	522	100%	289	100%	207	100%	26	100%
Involvement									
	<=5ys	298	57.10%	161	55.70%	123	59.40%	14	53.85%
	>5ys	224	42.90%	128	44.30%	84	40.60%	12	46.15%
	Total	522	100%	289	100%	207	100%	26	100%

4.2   Reliability and Validity

The exploratory factor analysis (EFA) results are presented in Table 2. The factor loading of each variable exceeds 0.5 which is an acceptable value [51]. The reliability

Table 2. The Results of All Constructs Reliability and Validity

Comp.	Factors	Items	Factor Loading	Facet Alpha	Facet AVE	Facet CR	Comp. Alpha	Comp. AVE	Comp. CR
Positive Emotions	Alert	PE101	0.895	0.880	0.805	0.925	0.688	0.417	0.771
		PE102	0.896						
		PE103	0.901						
	Inspired	PE201	0.872	0.856	0.775	0.912			
		PE202	0.862						
		PE203	0.907						
	Determined	PE301	0.956	0.880	0.791	0.919			
		PE302	0.867						
		PE303	0.841						
	Attentive	PE401	0.859	0.916	0.833	0.937			
		PE402	0.967						
		PE403	0.909						
	Active	PE501	0.923	0.909	0.846	0.943			
		PE502	0.881						
		PE503	0.953						
Negative Emotions	Upset	NE101	0.950	0.861	0.767	0.908	0.841	0.595	0.878
		NE102	0.838						
		NE103	0.834						
	Hostile	NE201	0.575	0.903	0.632	0.833			
		NE202	0.853						
		NE203	0.915						
	Ashamed	NE301	0.935	0.935	0.881	0.957			
		NE302	0.913						
		NE303	0.967						
	Nervous	NE401	0.945	0.895	0.819	0.932			
		NE402	0.866						
		NE403	0.903						
	Afraid	NE501	0.906	0.921	0.863	0.950			
		NE502	0.914						
		NE503	0.967						
Trust		TRU1	0.745	0.668	0.602	0.819	0.668	0.602	0.819
		TRU2	0.830						
		TRU3	0.751						
Online Social Network Adoption		OSNA1	0.765	0.725	0.646	0.845	0.725	0.646	0.845
		OSNA2	0.772						
		OSNA3	0.870						
Individual Benefits	Relationship Maintenance And Development	IB101	0.895	0.821	0.739	0.894	0.799	0.714	0.882
		IB102	0.778						
		IB103	0.900						
	Information And Knowledge Sharing	IB201	0.875	0.860	0.782	0.915			
		IB202	0.858						
		IB203	0.919						
	Use Satisfaction	IB301	0.904	0.861	0.783	0.915			
		IB302	0.832						
		IB303	0.916						

Table 3. Correlation of All Constructs with AVE Validity

	PE01	PE02	PE03	PE04	PE05	NE01	NE02	NE03	NE04	NE05	TRU	OSN	IB01	IB02	IB03
PE01	*0.897*														
PE02	0.234	*0.881*													
PE03	0.308	0.306	*0.889*												
PE04	0.284	0.181	0.482	*0.912*											
PE05	0.134	0.339	0.472	0.322	*0.920*										
NE01	0.209	-0.008	-0.152	0.039	-0.135	*0.876*									
NE02	0.275	-0.041	-0.119	-0.103	-0.209	0.448	*0.795*								
NE03	0.203	0.074	-0.105	0.053	-0.085	0.465	0.300	*0.939*							
NE04	0.244	-0.015	-0.105	0.006	-0.171	0.627	0.481	0.485	*0.905*						
NE05	0.193	-0.021	-0.151	0.073	-0.169	0.585	0.488	0.492	0.765	*0.929*					
TRU	0.126	0.117	0.215	0.251	0.223	0.012	0.002	0.035	0.028	0.000	*0.776*				
OSN	0.073	0.137	0.098	0.040	0.271	0.052	0.011	0.064	0.032	0.042	0.322	*0.804*			
IB01	0.017	0.182	0.209	0.206	0.340	0.011	-0.025	0.035	-0.001	0.002	0.360	0.558	*0.860*		
IB02	0.108	0.165	0.093	0.069	0.141	0.040	0.080	0.081	0.062	0.039	0.359	0.507	0.486	*0.884*	
IB03	0.084	0.143	0.138	0.131	0.182	0.059	-0.004	0.055	0.001	0.008	0.364	0.518	0.510	0.713	*0.885*

NOTE: The squared root of the average variance extracted (AVE) value is in bold italic type on the diagonal; PE01: Alert, PE02: Inspired, PE03: Determined, PE04: Attentive, PE05: Active; NE01: Upset, NE02: Hostile, NE03: Ashamed, NE04: Nervous, NE0s: Afraid; TRU: Trust; OSN: Online Social Network Adoption; IB01: Relationship Maintenance and Development, IB02: Information and Knowledge Sharing, IB03: Use Satisfaction.

Table 4. The Results of Path Analysis (PE group)

Hyp.	Path	R^2	Beta	T-value	P-value	Sig.	Decision
Layer One							
H1	OSN → IB	0.434	0.658	16.218	0.000	***	Support
H2	PE → OSN	0.145	0.238	4.159	0.000	***	Support
H3	TRU → OSN		0.253	4.351	0.000	***	Support
Layer Two							
H1a	OSN → IB01	0.344	0.586	13.309	0.000	***	Support
H1b	OSN → IB02	0.311	0.558	12.819	0.000	***	Support
H1c	OSN → IB03	0.337	0.581	11.910	0.000	***	Support
H2a	PE01 → OSN	0.184	0.096	1.465	0.143		Not Support
H2b	PE02 → OSN		0.088	1.538	0.124		Not Support
H2c	PE03 → OSN		0.071	0.601	0.548		Not Support
H2d	PE04 → OSN		-0.174	2.320	0.021	**	Support
H2e	PE05 → OSN		0.235	3.700	0.000	***	Support
H3	TRU → OSN		0.280	4.543	0.000	***	Support

NOTES: PE: Positive Emotions; TRU: Trust; OSN: Online Social Network Adoption; IB: Individual Benefits; PE01: Alert, PE02: Inspired, PE03: Determined, PE04: Attentive, PE05: Active; TRU: Trust; OSN: Online Social Network Adoption; IB01: Relationship Maintenance and Development, IB02: Information and Knowledge Sharing, IB03: Use Satisfaction. *p<0.1, **p<0.05, ***p<0.01

(Cronbach's alpha) of each construct is more than 0.6 which exceeds the acceptable level [52]. Composite reliability (CR) values show that the reliability of each construct is higher than the suggested value of 0.7, indicating that the proposed model has good construct reliability [53]. Average variation extracted (AVE) values for all facets are greater than 0.5, which also indicates a model with convergent validity [51]. In Table 3, the method of measuring discriminant validity, in which a correlation between two constructs should be lower than the squared root of the AVE value. According these suggestions, all constructs present discriminant validity.

4.3   Path Analysis

First, in Table 4, it presents the path results of layer one in research model 1, all hypotheses were supported. The data analysis results included four components: positive emotions, trust, OSN adoption, individual benefits. In layer two, the path analysis results were also showed in Table 4. There were nine hypotheses for the sub-factor analysis; of which six results were significant. Ten sub-factors were alert, inspired, determined, attentive, active, trust, OSN adoption, relationship maintenance and development, information and knowledge sharing, and use satisfaction. Then, table 5 presents the path analysis results of moderators which included gender and involvement. For the first layer, there were six hypothesis, of which the relation between PE and OSN adoption shows significant only for male. This implies that gender moderates the relation between PE and OSN adoption. For the moderation of involvement, it shows not supported. For the second layer, there were 18 hypothesis, and four of them were significant, results showed in Table 5.

Table 5. The Path Results of Moderators (PE group)

Hyp.	Path	R^2	Beta	T	P	Sig.	R^2	Beta	T	P	Sig.	Dec.
				Female					**Male**			
H4a	OSN → IB	0.302	0.549	7.488	0.000	***	0.519	0.721	16.934	0.000	***	N
H4b	PE → OSN	0.175	0.315	0.916	0.360		0.193	0.305	4.095	0.000	***	S
H4c	TRU → OSN		0.284	3.544	0.000	***		0.251	3.388	0.001	***	N
H4a1	OSN → IB01	0.159	0.399	4.203	0.000	***	0.478	0.691	16.231	0.000	***	N
H4a2	OSN → IB02	0.240	0.490	7.921	0.000	***	0.360	0.600	10.596	0.000	***	N
H4a3	OSN → IB03	0.232	0.482	5.050	0.000	***	0.414	0.643	12.643	0.000	***	N
H4b1	PE01 → OSN	0.212	0.182	1.837	0.066	*	0.226	0.026	0.261	0.794		S
H4b2	PE02 → OSN		0.094	0.926	0.355			0.111	1.289	0.198		N
H4b3	PE03 → OSN		-0.197	1.069	0.285			0.107	1.366	0.172		N
H4b4	PE04 → OSN		-0.210	2.503	0.012	**		-0.097	1.181	0.238		S
H4b5	PE05 → OSN		0.192	1.858	0.064	*		0.256	2.698	0.007	***	N
H4c	TRU → OSN		0.282	3.505	0.000	***		0.280	3.426	0.001	***	N
				<=5ys					**>5ys**			
H5a	OSN → IB	0.505	0.710	14.817	0.000	***	0.373	0.611	10.577	0.000	***	N
H5b	PE → OSN	0.158	0.253	2.365	0.018	**	0.153	0.242	2.267	0.024	**	N
H5c	TRU → OSN		0.277	3.218	0.001	***		0.247	3.307	0.001	***	N
H5a1	OSN → IB01	0.356	0.597	9.102	0.000	***	0.335	0.578	9.464	0.000	***	N
H5a2	OSN → IB02	0.423	0.650	13.692	0.000	***	0.212	0.460	6.338	0.000	***	N
H5a3	OSN → IB03	0.381	0.618	9.502	0.000	***	0.299	0.546	7.645	0.000	***	N
H5b1	PE01 → OSN	0.186	0.107	0.815	0.415		0.232	0.117	1.305	0.192		N
H5b2	PE02 → OSN		0.131	1.862	0.063	*		0.028	0.309	0.758		S
H5b3	PE03 → OSN		0.022	0.176	0.860			0.152	0.868	0.386		N
H5b4	PE04 → OSN		-0.157	1.602	0.109			-0.251	1.975	0.049	**	S
H5b5	PE05 → OSN		0.201	2.475	0.013	**		0.307	3.109	0.002	***	N
H5c	TRU → OSN		0.293	3.182	0.002	***		0.301	3.463	0.001	***	N

NOTES: PE: Positive Emotions; TRU: Trust; OSN: Online Social Network Adoption; IB: Individual Benefits; PE01: Alert, PE02: Inspired, PE03: Determined, PE04: Attentive, PE05: Active; TRU: Trust; OSN: Online Social Network Adoption; IB01: Relationship Maintenance and Development, IB02: Information and Knowledge Sharing, IB03: Use Satisfaction. *p<0.1, **p<0.05, ***p<0.01; S/N: Support/not support.

Second, in Table 6, it presents the path results of layer one in model 2. There were three hypotheses in this step, and two results were significant. The data analysis results included four components: negative emotions, trust, OSN adoption, individual benefits. In the layer two, the path analysis also showed in Table 6, five out of nine hypotheses were supported. There are ten sub-factors presented: upset, hostile, ashamed, nervous, afraid, trust, OSN adoption, relationship maintenance and development, information and knowledge sharing, and use satisfaction. Then, table 7 presents the path results of moderators which included gender and involvement. In layer one of research model two, there were six hypothesis, and one of them was significant. The results included four components: negative emotions, trust, OSN adoption, individual benefits. In layer two of research model one, there were eighteen hypothesis, and four of them were significant.

Table 6. The Results of Path Analysis (NE group)

Hyp.	Path	R^2	Beta	T-value	P-value	Sig.	Decision
Layer One							
H6	OSN → IB	0.353	0.594	11.663	0.000	***	Support
H7	NE → OSN	0.171	0.158	1.458	0.145		Not Support
H8	TRU → OSN		0.379	6.270	0.000	***	Support
Layer Two							
H6a	OSN → IB01	0.245	0.495	7.188	0.000	***	Support
H6b	OSN → IB02	0.214	0.463	6.984	0.000	***	Support
H6c	OSN → IB03	0.239	0.489	7.054	0.000	***	Support
H7a	NE01 → OSN	0.181	-0.008	0.078	0.938		Not Support
H7b	NE02 → OSN		0.040	0.446	0.656		Not Support
H7c	NE03 → OSN		0.155	2.278	0.023	**	Support
H7d	NE04 → OSN		0.004	0.036	0.971		Not Support
H7e	NE05 → OSN		0.050	0.503	0.615		Not Support
H8	TRU → OSN		0.379	6.172	0.000	***	Support

NOTES: NE: Negative Emotions; TRU: Trust; OSN: Online Social Network Adoption; IB: Individual Benefits; NE01: Upset, NE02: Hostile, NE03: Ashamed, NE04: Nervous, NE05: Afraid; TRU: Trust; OSN: Online Social Network Adoption; IB01: Relationship Maintenance and Development, IB02: Information and Knowledge Sharing, IB03: Use Satisfaction. *p<0.1, **p<0.05, ***p<0.01

Table 7. The Path Results of Moderators (NE group)

Hyp.	Path	R^2	Beta	T	P	Sig.	R^2	Beta	T	P	Sig.	Dec.
				Female					Male			
H9a	OSN → IB	0.328	0.573	8.422	0.000	***	0.407	0.638	9.774	0.000	***	N
H9b	NE → OSN	0.265	0.168	0.811	0.418		0.152	0.304	2.093	0.037	**	S
H9c	TRU → OSN		0.469	6.532	0.000	***		0.294	2.728	0.006	***	N
H9a1	OSN → IB01	0.274	0.523	5.729	0.000	***	0.229	0.478	4.996	0.000	***	N
H9a2	OSN → IB02	0.166	0.408	4.252	0.000	***	0.288	0.537	6.703	0.000	***	N
H9a3	OSN → IB03	0.160	0.400	3.617	0.000	***	0.359	0.599	7.946	0.000	***	N
H9b1	NE01 → OSN	0.321	-0.281	2.097	0.036	**	0.186	0.206	1.634	0.103		S
H9b2	NE02 → OSN		0.045	0.448	0.654			0.073	0.495	0.621		N
H9b3	NE03 → OSN		0.102	1.062	0.289			0.238	2.318	0.021	**	S
H9b4	NE04 → OSN		0.090	0.689	0.491			0.066	0.394	0.693		N
H9b5	NE05 → OSN		0.131	0.923	0.356			-0.105	0.776	0.438		N
H9c	TRU → OSN		0.497	6.706	0.000	***		0.285	2.519	0.012	**	N
				<=5ys					>5ys			
H10a	OSN → IB	0.379	0.615	9.900	0.000	***	0.320	0.566	7.741	0.000	***	N
H10b	NE → OSN	0.166	0.196	1.526	0.127		0.239	0.252	0.849	0.396		N
H10c	TRU → OSN		0.362	4.491	0.000	***		0.336	3.430	0.001	***	N
H10a1	OSN → IB01	0.241	0.491	5.320	0.000	***	0.254	0.504	5.345	0.000	***	N
H10a2	OSN → IB02	0.270	0.519	7.075	0.000	***	0.149	0.386	3.447	0.001	***	N
H10a3	OSN → IB03	0.242	0.492	6.465	0.000	***	0.227	0.477	3.498	0.000	***	N
H10b1	NE01 → OSN	0.192	-0.021	0.153	0.879		0.267	0.125	0.688	0.492		N
H10b2	NE02 → OSN		-0.111	0.863	0.388			0.131	1.050	0.294		N
H10b3	NE03 → OSN		0.088	0.717	0.474			0.217	1.805	0.071	*	S
H10b4	NE04 → OSN		0.013	0.110	0.913			-0.034	0.218	0.828		N
H10b5	NE05 → OSN		0.180	1.559	0.119			-0.194	1.773	0.076	*	S
H10c	TRU → OSN		0.375	4.418	0.000	***		0.342	3.551	0.000	***	N

NOTES: NE: Negative Emotions; TRU: Trust; OSN: Online Social Network Adoption; IB: Individual Benefits; NE01: Upset, NE02: Hostile, NE03: Ashamed, NE04: Nervous, NE05: Afraid; TRU: Trust; OSN: Online Social Network Adoption; IB01: Relationship Maintenance and Development, IB02: Information and Knowledge Sharing, IB03: Use Satisfaction. *p<0.1, **p<0.05, ***p<0.01; S/N: Support/not support.

5   Concluding Remarks

There are positive and significant relationships between positive emotion, trust, OSN adoption, and individual benefits which include relationship maintenance and development, information and knowledge sharing, and use satisfaction. Considering the moderating effects, negative emotion shows a positively significant with OSN adoption. Moreover, most sub-factors of emotions are significantly influences on the path testing, which include attentive, active, and ashamed. With the effects on moderators: gender and involvement, there are other four significant emotions including alert, inspired, upset, and afraid. This study provides concepts on the importance of emotions to understand user reactions to the implementation on SNSs. But this research only indicates some insights about the relationships between positive and negative emotions, OSN adoption, and individual benefits. There are also uncovers several variables, factors, and questions that need to be addressed. Therefore, future research may focus on those ideas and results to explore and stimulate researches about different emotions and OSN adoption in the SNSs field.

References

[1] Barnes, J.A.: Class and committees in a Norwegian island parish (1954)
[2] Donelan, H.M., Kear, K., Ramage, M. (eds.): Online communication and collaboration: A reader (2010)
[3] Ellison, N.B., Steinfield, C., Lampe, C.: The benefits of Facebook "friends:" Social capital and college students' use of online social network sites. Journal of Computer-Mediated Communication 12(4), 1143–1168 (2007)
[4] Waters, R.D., Burnett, E., Lamm, A., Lucas, J.: Engaging stakeholders through social networking: How nonprofit organizations are using Facebook. Public Relations Review 35(2), 102–106 (2009)
[5] Ganley, D., Lampe, C.: The ties that bind: Social network principles in online communities. Decision Support Systems 47(3), 266–274 (2009)
[6] Pfeil, U., Arjan, R., Zaphiris, P.: Age differences in online social networking–A study of user profiles and the social capital divide among teenagers and older users in MySpace. Computers in Human Behavior 25(3), 643–654 (2009)
[7] Roblyer, M.D., McDaniel, M., Webb, M., Herman, J., Witty, J.V.: Findings on Facebook in higher education: A comparison of college faculty and student uses and perceptions of social networking sites. The Internet and Higher Education 13(3), 134–140 (2010)
[8] Kim, Y., Sohn, D., Choi, S.M.: Cultural difference in motivations for using social network sites: A comparative study of American and Korean college students. Computers in Human Behavior 27(1), 365–372 (2011)
[9] Lin, K.Y., Lu, H.P.: Why people use social networking sites: An empirical study integrating network externalities and motivation theory. Computers in Human Behavior 27(3), 1152–1161 (2011)
[10] Valenzuela, S., Park, N., Kee, K.F.: Is There Social Capital in a Social Network Site?: Facebook Use and College Students' Life Satisfaction, Trust, and Participation1. Journal of Computer-Mediated Communication 14(4), 875–901 (2009)
[11] Shi, N., Lee, M.K., Cheung, C., Chen, H.: The continuance of online social networks: how to keep people using Facebook? In: 2010 43rd Hawaii International Conference on System Sciences (HICSS), pp. 1–10. IEEE (January 2010)
[12] Beaudry, A., Pinsonneault, A.: The other side of acceptance: studying the direct and indirect effects of emotions on information technology use. MIS Quarterly 34(4), 689–710 (2010)
[13] Cook, K.S., Yamagishi, T., Cheshire, C., Cooper, R., Matsuda, M., Mashima, R.: Trust building via risk taking: A cross-societal experiment. Social Psychology Quarterly 68(2), 121–142 (2005)
[14] Hughes, D., Coulson, G., Walkerdine, J.: Free riding on Gnutella revisited: the bell tolls? IEEE Distributed Systems Online 6(6) (2005)
[15] Huang, F.: Building social trust: A human-capital approach. Journal of Institutional and Theoretical Economics (JITE)/Zeitschrift für die gesamte Staatswissenschaft, 552–573 (2007)

[16] Maheswaran, M., Tang, H.C., Ghunaim, A.: Towards a gravity-based trust model for social networking systems. In: 27th International Conference on Distributed Computing Systems Workshops, ICDCSW 2007, p. 24. IEEE (June 2007)

[17] Acquisti, A., Gross, R.: Imagined communities: Awareness, information sharing, and privacy on the facebook. In: Danezis, G., Golle, P. (eds.) PET 2006. LNCS, vol. 4258, pp. 36–58. Springer, Heidelberg (2006)

[18] George, A.: Living online: The end of privacy? New Scientist, 2569 (September 18, 2006), http://www.newscientist.com/channel/tech/mg19125691.700-living-online-the-end-of-privacy.html (retrieved August 29, 2007)

[19] Kornblum, J., Marklein, M.B.: What you say online could haunt you. USA Today (March 8, 2006), http://www.usatoday.com/tech/news/internetprivacy/2006-03-08-facebook-myspace_x.htm (retrieved August 29, 2007)

[20] Dwyer, C., Hiltz, S., Passerini, K.: Trust and privacy concern within social networking sites: A comparison of Facebook and MySpace. In: AMCIS 2007 Proceedings, vol. 339 (2007)

[21] Jagatic, T.N., Johnson, N.A., Jakobsson, M., Menczer, F.: Social phishing. Communications of the ACM 50(10), 94–100 (2007)

[22] Venkatesh, V., Morris, M.G., Davis, G.B., Davis, F.D.: User acceptance of information technology: Toward a unified view. MIS Quarterly, 425–478 (2003)

[23] Golbeck, J.: The dynamics of web-based social networks: Membership, relationships, and change. First Monday 12(11) (2007)

[24] Humphreys, E.: Information security management standards: Compliance, governance and risk management. Information Security Technical Report 13(4), 247–255 (2008)

[25] Hanchard, S.: Measuring trust and social networks - Where do we put our trust online? (2008), http://weblogs.hitwise.com/sandra-hanchard/2008/09/measuringtrustandsocialnet.html (retrieved)

[26] Sherchan, W., Nepal, S., Paris, C.: A survey of trust in social networks. ACM Computing Surveys (CSUR) 45(4), 47 (2013)

[27] Boyd, D., Ellison, N.B.: Social network sites: Definition, history, and scholarship. Journal of Computer-Mediated Communication 13(1), 210–230 (2008)

[28] Wang, S., Noe, R.A.: Knowledge sharing: A review and directions for future research. Human Resource Management Review 20(2), 115–131 (2010)

[29] Yang, S.J., Chen, I.Y.: A social network-based system for supporting interactive collaboration in knowledge sharing over peer-to-peer network. International Journal of Human-Computer Studies 66(1), 36–50 (2008)

[30] Zhang, Y.,Tanniru, M.: An agent-based approach to study virtual learning communities. In: Proceedings of the 38th Annual Hawaii International Conference on System Sciences, HICSS 2005, p. 11c. IEEE (January 2005)

[31] Wasko, M.M., Faraj, S.: Why should I share? Examining social capital and knowledge contribution in electronic networks of practice. MIS Quarterly, 35–57 (2005)

[32] Davis, F.D.: A technology acceptance model for empirically testing new end-user information systems: Theory and results (1985)

[33] Thompson, E.R.: Development and validation of an internationally reliable short-form of the positive and negative affect schedule (PANAS). Journal of Cross-cultural Psychology 38(2), 227–242 (2007)

[34] Oliver, E.J., Markland, D., Hardy, J.: Interpretation of self‐talk and post-lecture affective states of higher education students: A self-determination theory perspective. British Journal of Educational Psychology 80(2), 307–323 (2010)

[35] Wong, Y.J., Ho, R.M., Li, P., Shin, M., Tsai, P.C.: Chinese Singaporeans' lay beliefs, adherence to Asian values, and subjective well-being. Personality and Individual Differences 50(6), 822–827 (2011)

[36] Felton, L., Jowett, S.: What do coaches do" and "how do they relate.: Their effects on athletes' psychological needs and functioning. Scandinavian Journal of Medicine & Science in Sports 23(2), e130–e139 (2013)

[37] Brogan, A., Hevey, D.: Eating styles in the morbidly obese: restraint eating, but not emotional and external eating, predicts dietary behavior. Psychology & Health 28(6), 714–725 (2013)

[38] Sanchez, X., Moss, S.L., Twist, C., Karageorghis, C.I.: On the role of lyrics in the music–exercise performance relationship. Psychology of Sport and Exercise 15(1), 132–138 (2014)

[39] Dutton, W.H., Rogers, E.M., Jun, S.H.: Diffusion and social impacts of personal computers. Communication Research 14(2), 219–250 (1987)

[40] LaRose, R., Atkin, D.: Satisfaction, demographic, and media environment predictors of cable subscription. Journal of Broadcasting & Electronic Media 32(4), 403–413 (1988)

[41] Jeffres, L., Atkin, D.: Predicting use of technologies for communication and consumer needs. Journal of Broadcasting & Electronic Media, 40(3), 318–330 (1996)

[42] Young, E.: The second annual ernst & young internet shopping study: the digital channel continues to gather steam, New York, 128 (1999)

[43] Crunchies, T.: Internet statistics and numbers (2009)

[44] Lenhart, A.: Adults and social network websites (2009)

[45] Lenhart, A., Madden, M.: Social networking websites and teens: An overview (2007)

[46] Zywica, J., Danowski, J.: The Faces of Facebookers: Investigating Social Enhancement and Social Compensation Hypotheses; Predicting Facebook^{TM} and Offline Popularity from Sociability and Self-Esteem, and Mapping the Meanings of Popularity with Semantic Networks. Journal of Computer-Mediated Communication 14(1), 1–34 (2008)

[47] Pempek, T.A., Yermolayeva, Y.A., Calvert, S.L.: College students' social networking experiences on Facebook. Journal of Applied Developmental Psychology 30(3), 227–238 (2009)

[48] Sherif, M., Cantril, H.: The psychology of ego-involvements: Social attitudes and identifications (1947)

[49] Zaichkowsky, J.L.: The personal involvement inventory: Reduction, revision, and application to advertising. Journal of Advertising 23(4), 59–70 (1994)

[50] Hawkins, D., Mothersbaugh, D.: Consumer behavior building marketing strategy (2009)

[51] Hair, J.F., Anderson, R., Tatham, R., Black, W.: Multivariate data analysis, 5th edn. Prentice-Hall, Upper Saddle River (1998)

[52] Murphy, K.R., Davidshofer, C.O.: Psychological testing: Principles and applications. Prentice-Hall, Englewood Cliffs (1988)

[53] Nunnally, I.C., Bernstin, I.H.: Psychometric theory, 3rd edn. McGraw-Hill, NY (1994)

[54] Fornell, C., Larcker, D.: Evaluating structural equation models with unobservable variables and measurement error. Journal of Marketing Research, 18(1) (1981)

Emerging Issues in Cloud Storage Security: Encryption, Key Management, Data Redundancy, Trust Mechanism

Daniel W.K. Tse, Danqing Chen, Qingshu Liu, Fan Wang, and Zhaoyi Wei

City University of Hong Kong
iswktse@cityu.edu.hk

Abstract. Cloud computing is one of the most cutting-edge advanced technologies around the world. According to the recent research, many CIOs who work for famous corporation mentioned that security issues have been the most critical obstacles in the adoption of cloud technology. As one of the prominent application of cloud computing, cloud storage has attracted more concerns; however, many security problems existing in cloud storage need to be resolved. This applied research paper briefly analyzes the development of cloud storage security, comprehensive discussion of several general solutions to those problems introducing several non-technical issues, such as third-party issues, trust mechanism in cloud computing. Finally, some possible improvement to those solutions is provided.

Keywords: Cloud storage, security, encryption, data redundancy, key management, trust mechanism.

1 Introduction

Cloud computing has provided enormous flexibility of its computation power to numerous users, including individuals, corporations and governments. The benefits brought by cloud services are obvious in terms of time and cost effectiveness. It is widely accepted that the cloud service will be a utility service after electrical utility services, water and gas services and telecommunication infrastructures.

Cloud storage is the cornerstone of cloud computing. Data security in cloud includes storage security, transmission security and processing security while storage security is a critical component. Gartner's survey indicates that the security problems have been the greatest obstacles on the way to cloud. From our intensive literature review exercise, most past research work done for cloud storage security is either too superficial or too theoretical. Because of this, we found the need to explore the nature of the problems by dissecting all critical scenarios and solutions comprehensively. This paper describes the loss of control in confidentiality, integrity and availability and then provides the solution to earn trust. Hopefully, some useful and high-impact improvements can be derived.

2 Research Methodology

Since the purpose of this research is to dissect traditional approaches to address the security issues in cloud and analyze some emerging techniques such as encryption,

L.S.-L. Wang et al. (Eds.): MISNC 2014, CCIS 473, pp. 297–310, 2014.
© Springer-Verlag Berlin Heidelberg 2014

key management, trust mechanism and data distribution, these are very practical and complicated by nature, the traditional research methodologies are not appropriate to cope with such fast-technology world. Our research starts from some empirical studies and then drill down into the inner-working or mechanism of threats and their solutions. Thus, an applied research methodology, which is more practical and agile but weak in traditional research formality, is adopted.

3   Root Problems Analysis

3.1   Trust

Trust is a very important research area in cloud computing, especially in the field of cloud storage security. In traditional architectures, trust was enforced by an efficient security policy, which addressed constraints on internal control, constraints on access by external systems and adversaries including programs and access to data by people. In cloud architecture, this perception is totally obscured. Using public clouds, control is delegated to the organization owning the infrastructure to enforce a sufficient security policy which guarantees that appropriate security activities are being performed to ensure that risks have been reduced. All the policies and processes are behind the wire, users may think they are losing the control of their data.

3.2   Confidentiality

Confidentiality refers to data only being accessed by authorized people. Firstly, using the cloud to store data means delegating the authorization issue to cloud. Protecting user account from not being stolen and provide strong authentication can just lead to an external security. Avoiding internal threat is important too. The data in the cloud is facing the threats.

3.3   Integrity

Integrity is a key part of electronic information which means data can just be modified by authorized people. Within the local infrastructure, it is easy to limit and monitor the access of data so that integrity can be guaranteed. However, cloud is owned by other party, all the data uploaded to cloud is recorded and controlled by cloud service provider, users need a method to get control of the data's integrity.

3.4   Availability

Users want to get what they need when they want. Even though the requirement is not being satisfied, they have to know what is happening, why they cannot access, how and when it will be fixed. Using cloud, it seems that availability is never being a problem because cloud service provider has more server than you, so that they can perform backup and can hire professional person to manage the data center. As the same situation stated before, the backup mechanism is processed in the cloud and user can never know whether there is a backup or not. It also makes user feel like loss of control.

4   Problems Analysis and Existing Solutions Discussion

4.1   The Basic Technology of Data Security -- Encryption

Encryption is a traditional and fundamental method to protect data. In this cloud era, it is still the first choice to safeguard the sensitive information.

In a multi-tenant environment [1], one of the basic security tools to protect data is encryption. When cloud service users do not have full control over the power in the cloud computing environment, cloud computing encryption allows users to protect the data effectively. This is extremely valuable for both private cloud and public cloud computing, especially when users share with other users a server which contains different levels of sensitive data repository.

When using cloud computing for data storage, usually virtual private storage architecture [2] should be used. Before sending the data to the cloud computing, encrypting the data, and decrypting them when they are sent back. For example, when cloud computing is a backup service, before storing the data in cloud, the backup software will use a local secret key to encrypt data in local. Because it is the users themselves who are in charge of encrypting operation and the secret key, keeping the key in the form of safety backup copy is one of the tools for data protection.

Volume label can be a useful medium to finish the encryption [2]. For encrypting operation to the data which are stored in IaaS application, the users can use equipment's volume label (volume label actually is not a file, but only a directory entry, it does not use extra disk space) to finish the encryption, and save the data in the second encrypted volume label. Due to the secret key and the encryption engine are all stored in local equipment, this is not the safest method but this method can protect the data from unauthorized access effectively. For example, assume that the encrypted volume label has been created correctly, the cloud computing providers will not be authorized to operate systems or applications (this is a typical default setting). Also, they cannot get the secret key and gain access to the encrypted volume label.

As more and more concerns are focusing on the encryption for cloud computing, there discovered several senior ways to improve the encryption method, one of them is using the third party broker [3]. We apply a third party broker rather than the users or cloud computing services providers to finish the encryption and decryption. The broker encrypts the clients' information, partitions the information into multiple segments and transfers them to corresponding virtual machines (VMs) of the cloud storage providers. The client information's integrity is ensured by a trusted third party called a cloud broker and it provides a more time-saving algorithm to encrypt, decrypt and detect the information.

Though encryption is significant for the security of implementing cloud computing, we should not treat it as the amulet because several problems [4] still need to be solved by the cloud computing services provider to ensure the clients' information security. Firstly, protection during the encrypting: Services providers always encrypt sensitive data running in the cloud but if the data is not encrypted and being used or stored at the same time, reducing or preventing data from being destroyed will be their duty. Secondly, management of secret key: More and more internet users take part in the cloud services, along with the growing of the user group. The number of secret keys keeps increasing; it is a big challenge for services providers to protect data and a large number of secret keys. Thirdly, access control: Sometimes safety means complication. When layers and layers are added to the users data, it becomes not so convenient to get the data as the original intention of cloud storage. To ensure the user convenience so that user can use the data anytime anywhere they need in the safe way, it is necessary to build up an access control system.

For protection during the encryption process, there is another traditional way to realize the protection to the data being used - firewall. Deploying firewall which is dedicated for that kind of data, block the illegal access to both server side and client side. For secret key management, service providers should formulate severe management flow paths and rules and provide the secret key management system based on the hardware, like using independent encryption chips just for secret keys. For access control, the clients should require the encryption provider to give them enough power to access and management control and stronger authentication, such as two-factor authentication, access management and separate duties of safety management, such as security, networking, and operation maintenance.

4.2   Third Party Key Management Software

Cloud storage service is provided by big private company and it is self-supervised, so there is a high risk to have a policy fault or inner fraud in the company which can cause the data leakage. If the pernicious data is sent to the user's competitor or someone else who is hostile to this user, it will create a catastrophic damage. Obviously, the solution is to encrypt the data, enhance the difficulty for unauthorized people to get the useful information. There comes another question: How to transfer the key without the untrusted cloud? Using another cloud is a good idea. We need to make it more automatic, process the encryption and key storage in the background. It is a bridge between cloud and cloud user, called Key Management Software (KMS). KMS should be developed by professional companies which know about encryption, control the integrity checking, key generation, key distribution, periodic key changing and key destroy. Besides, KMS has to be verified by cloud service provider. The perfect mutual restriction is: Cloud takes charge of storage and the KMS takes charge of key management.

With such a restriction, cloud just needs to concern about the storage and lost recovery, the encryption efficiency and encryption deployment are transferred to KMS. It gives the chance for security companies to provide their service in a competition relationship which makes the cloud data decryption more difficult. For normal user, who just wants to share data and does not care about the confidentiality of data, he can use cloud storage service straightly. For small business companies who have the security concern, they can select which KMS they want according to their budget and security requirement. For big company, they can even develop the KMS themselves.

Furthermore, KMS can do more than key management. As the storage and key management are separated, which means KMS and cloud are not with one-to-one relationship. A user can use different KMS to do different encryption based on the files security level. More importantly, the KMS can manage multi-cloud to provide an ultimate storage scheme: (1) Optimizing the download speed - Based on the network condition and server location, one cloud service provider cannot supply the best service anytime anywhere. But if user is using multi-cloud, he can test immediately and then find out the best performance he can get; (2) Never lose data - Cloud look like having infinite space for storage but users can never know whether their data has been backed up or not. At the worst situation, what if the company bankrupt? Not everyone can bear the risk of losing data. The best way to make user trust with is to store data in different clouds. This can be done by KMS and also KMS can check the data periodically; (3) Data splitting - This means cutting data into several pieces and storing them in different place, making bad guy hard to collect the entire data.

4.3 Distributive Storage in Cloud

4.3.1 Availability Threats

Availability means that the data stored in the remote cloud is available whenever the users request for it. There are many factors that may compromise data availability: Lack of replication and failure of servers, etc.

There are several kinds of risks related to cloud storage, like flood attacks (including direct DOS and indirect DOS) [5], fraudulent resource consumption attack, Single Point of Failure (SPOF) and the single point compromise, etc. Flood attacks, fraudulent resource consumption attacks and SPOF would harm the availability of data. Single point compromise would harm the confidentiality of data. Different data distribution scheme has different impact on both two aspects.

4.3.2 Defense Strategy: Redundancy Distribution Scheme

Redundancy
According to [6], in the traditional storage scheme, only one copy of the data is stored in the cloud. Once the only copy of data is corrupted or the server fails, the data will be temporarily or permanently unavailable to users. No data redundancy can meet the needs of data availability nowadays. The redundancy distribution scheme uses different techniques to guarantee that the data is available all the times.

For example, three clients store their data in the CSP1, CSP2 and CSP3 respectively. When CSP1 fails to function due to power-off or other reasons, client 1 cannot retrieve the data stored in CSP1. What's more disastrous is that if CSP1 cannot be repaired, the data stored in it will be gone forever. So no data redundancy can do great harm to both users and CSP. Many people turn to an alternative for help: Data is broken into several blocks and every block is stored in different servers. This approach is vulnerable to collusions among cloud service providers, which means the providers holding different data blocks of the user can collude to reconstruct the original data. Consequently, encryption is still of great importance in distributed storage.

4.3.3 Different Distribution Schemes

Algorithm Comparison

A formula can be used to describe how different schemes influence the availability of data. The data distribution threshold algorithm [7] provides a comparison among those schemes. In the algorithm, three characters n, m and p are used as parameters. Character n means that in this scheme, data would be cut into n shares. Character m tells us under this scheme, m shares of original data are needed to reconstruct the data. Character p is the value with less than which the confidentiality of data would not been disclosed.

Table 1. Distribution Schemes

Algorithm	Parameters (n-m-p)
Non-Distribution	1-1-1
Replication	n-1-1
Data Striping	n-n-1
Splitting	n-n-n
Information Dispersal	n-m-1
Secret Sharing	n-m-m

a) Non-Distribution

It refers to traditional no-redundancy storage method. Data is solely stored in one server. Availability of data heavily relies on the availability of the server. Also, it is obvious that when the server is compromised, all information would be disclosed.

b) Replication

As the name implies, replication refers to the complete copy of the original data. It, to some extent, solves single point of failure (SPOF). Several replications of the original files guarantee data redundancy and effectively increase the availability of the original files. There are some inevitable and unpredictable reasons that may cause network inaccessibility or machine outage, which will result in data loss or render the data stored in cloud unavailable. Generally speaking, more replicas of data equals to higher availability of data. Besides, making replications for users' data can not only enhance the availability of data but also boost system performance by reasonably allocating replicas to storage nodes and realizing the nearest visit via certain configuration of router. Additionally, parallel visits can be realized when clients request for some data which is replicated and stored in several nodes. In this way, users can access data with less time-lag and the load of every node is relieved as well.

However, it is economically unfeasible to replicate all files as much as we want. Numerous replicas mean enormous consumption of system storage resources and complicated file management. So a good balance must be struck among availability, system performance and cost. Another concern is that when the data is manipulated, all copies need to be modified. The last problem may be the increased risk of information disclosure. Because each copy contains the complete information of original data, single server compromising is fatal to confidentiality of entire data set. In cloud environment, replication is even useless, because customers would like the CPS to learn as least as their information and hold as least copies as possible.

c) Decimation (Striping)

RACS [8] identifies two kinds of failures in data storage: Outages and economic failures (vendor lock-in). In order to address the two problems, a redundant stripping mechanism is recommended. The technique adopted in redundant striping is called erasure code and the basic idea is that the data to be stored can be broken into N shares, each of which is 1/N size of the original data. By utilizing some coding techniques, the N shares are transformed to K blocks of data and then they are stored in K nodes. The total overhead factor of such storage is K/N. One only needs N available blocks to retrieve the original data.

The advantages of this approach are: It allows K-N blocks to be out of work; Provide higher fault tolerance rate and lower storage complexity. Disadvantage is that it requires complicated coding process and more computation. Compared to replication, the erasure coding is more likely to meet the needs of data storage which requires higher reliability and less storage cost.

d) Splitting

Data splitting is an approach to prevent critical information or confidential data against unauthorized users. The data is split into N shares and each share is encrypted and stored in different servers. To retrieve the data blocks from the servers, the user must know the locations of the nodes storing the parts. After all the split data is retrieved, the parts are combined together and decrypted. All these steps are completed on the basis that the user is an authorized one. All the n shares of data will be needed to reconstruct the original data. If one part is missing, the original data is gone forever.

e) Information Dispersal

Mo [9] proposed a well-known information distribution scheme - the Information Dispersal Algorithm (IDA). The IDA has been introduced in several cloud infrastructures. Bowers [10] proposed the High-Availability and Integrity Layer (HAIL), which can prove a user IDA distributed data's integrity in cloud. It uses the IDA to distribute files among several CPSs. A so-called "Dispersal Code" is used for checking data integrity.

Under this algorithm, information is cut into n pieces and distributed among n servers. Unlike data striping or splitting algorithm, IDA allows reconstruction of original information by a smaller m pieces than all n pieces. It means that even if several servers down, as long as at least m servers are alive, the original data can also be obtained. The benefit of IDA compared to data striping and splitting is obvious - It increases the availability of distributed data.

f) Shamir's Secret Sharing

Secret Sharing is a method created to solve secret information distribution among a group of parties. The Shamir's Secret Sharing scheme is widely used in information distribution area. In [11], a Multi-clouds Database Model based on the Shamir's Secret Sharing algorithm was proposed. Data would be cut into several pieces and be stored in several databases, which are under control by a unified DBMS layer.

Using this method, one can distribute data among n servers. Similar to IDA, Shamir's Secret Sharing allows reconstruction of original data by a subset, like m, of all n pieces. The difference is that, any single piece of Secret Sharing reveals no information about the original data. One piece of IDA algorithm contains a part of original data's information which is not complete but makes sense to the user. The advantage of Shamir's Secret Sharing algorithm is that a single point compromise reveals nothing about the data which provides more confidentiality than IDA does.

4.3.4 Measurement

Availability Assessment

Suppose that, the probability of a single node's uptime is r. We assume the availabilities of single nodes are independent. Then we can use the following formulas to calculate the probability of failure for each distribution algorithm [7].

Table 2. Availability Assessment

Algorithm	Availability
Non-Distribution	r
Replication	$1 - (1 - r)^n$
Data Striping	r^n
Splitting	r^n
Information Dispersal	$\sum_{i=m}^{n} \binom{n}{i} r^i (1-r)^{n-i}$
Secret Sharing	$\sum_{i=m}^{n} \binom{n}{i} r^i (1-r)^{n-i}$

Risk of Disclosure

Suppose the probability of single node compromise is p. We assume the disclosure risks of single nodes are independent. The risk of information disclosure for each algorithm can be calculated by the following formulas:

Table 3. Risk of Disclosure

Algorithm	Risk of disclosure
Non-Distribution	p
Replication	$p \times n$
Data Striping	$p \times n$
Splitting	p^n
Information Dispersal	$p \times n$
Secret Sharing	p^m

We cannot tell which scheme has the largest availability or the smallest disclosure risk unless specific parameters, like n, m and p, are given.

4.3.5 Integrity Realization

Sanitization

Zhang & et al [12] put forward a new concept---cloud shredder, in their paper published in 2011. They emphasized the danger of losing critical information caused by the lost or theft of laptops, PCs, tablets or even hard drives. The idea specified is that every file in our devices (PCs, tablets, laptops, etc.) is split into two parts, one of which is stored in the local device while the other one is uploaded and stored on a remote server in the cloud. In this way, even though the devices are lost or stolen, the data in them will still remain safe because all the thieves get from the devices are pieces of useless data. If the other half of data in the cloud is hacked by bad people, they need the devices to view the complete version of data. As a result, the cloud shredder concept makes the data safer than storing all our secrets in the local devices or in the cloud.

However, several drawbacks are obvious with such a system. Backup is hard to realize; if either part of the data is lost (due to cloud failure or machine failure), how can we retrieve the complete data? If the internet is out of reach, how can we check the data? We cannot take ourselves with our laptops everywhere and anytime, so what if we want to access our data then? Is the cloud shredder better than a thumb drive?

The value of cloud shredder is that it provides a way to delete the data. If the laptop has been stolen, all we need to do is to delete the data in the cloud to make sure the thief can never obtain our secrets via our laptop.

Active bundles [13] can protect sensitive or private data from being disclosed to unauthorized or un-trusted parties and from being maliciously disseminated. The basic idea of active bundles is that an active bundle comprises of three parts: Sensitive data (the data we want to protect), metadata, and a virtual machine. The metadata is used to specify how the entire or parts of sensitive data can be accessed and how the active bundle can be disseminated. The virtual machine (or VM) uses information provided by metadata to manage the use of the active bundle. The three main operations of the active bundles are evaporation, apoptosis and self-integrity check. Evaporation means that after the active bundle arrives at a host and gets the trust level of it, the bundle will decide whether the sensitive data can be presented to the host and which part of this sensitive data can be released. The rest of the sensitive data which the host is not authorized to access is evaporated. Apoptosis means that if a bundle checks the potential danger of data compromise, it will perform self-destruction and leave no trace for malicious attackers. Self-integrity check means that the active bundle can use its own algorithm to check whether the integrity within its data is being compromised or not. Actually, the active bundle may be an effective solution to data protection in cloud due to its structure, mechanism, algorithm, encryption technology and so on. Also, it achieves the goal of not using un-encrypted data to do authentication and protecting identity information from un-trusted hosts as well. However, the whole concept of the active bundle solution is based on the assumption that the virtual machine is secure against attacks. In fact, the virtual machine can be a weak link or a loophole in the whole solution. So how to realize this assumption becomes the problem.

Single Node Independence

When we are applying distribution scheme in cloud, there is a question that should we distribute data either among one CPS's server or among several clouds. In [3], data is distributed among CPSs. In [11], data is distributed among servers.

Amazon S3 uses the concepts of Availability Zone (AZ) and Availability Region (AR) to illustrate its redundancy storage strategy [14]. ARs are geographically separated, like data centers in multiple locations. AZs are physical separated, like isolated servers in a data center. CPSs are more likely to be "business independent", which means they are isolated in ownership.

We use this concept to illustrate the differences between storage locations. Different method provides different single point independence. As what is discussed above, a comprehensive comparison including other business concerns, like cost-benefit analysis, would be needed.

4.4 Trust Mechanism in Cloud Storage

4.4.1 The Development Status and Trust Issues of Cloud Computing

In today's competitive environment, cloud computing offers a range of services which are so convenient that a large number of enterprises have been attracted by its highly scalable technology. Although enterprises gain many opportunities provided by cloud computing, new challenges cannot be ignored. Trust issue is one of the most obvious challenges which are a paramount concern for most enterprises. Actually, the dearth of customer confidence is not only comes from technology itself, also comes from a lack of transparency, a loss of control over data assets and unclear security assurances. The issues can only be resolved from two aspects, technology and psychology.

The most significant issue is the sufficiency of data information which can be presented with services. Other issues which are also very important, including control, ownership, prevention and security need to be considered. According to the research, we know that people not having confidence on cloud computing is not because they do not trust the service provider, but they feel their data is out of control. The ownership of data assets also has an impact on trust. When the enterprise consigns their data to the service provider, both the enterprise and its client should trust the cloud provider. If the client suspects the provider, once the data is damaged, the enterprise must take full responsibility. As for prevention, it is more important than compensation. In cloud storage, data security must be guaranteed. Although the corporation can compensate for the economic loss of the client, a security breach of data is irreparable. Obviously, no amount of money can remedy the lost data or the enterprise's reputation. Thus, more attention should be paid to prevent failure than to post-failure compensation.

Security, which plays a crucial role in preventing service failures and cultivating trust in cloud computing is urgent to be improved. Much effort has been done to satisfy customers' demand but the security status is lack of transparency. It means that the client is not clear on how service provider secures and controls their data, so they still hold a skeptical attitude towards data security.

There are two main challenges we confront today: Diminishing control; and Lack of transparency. Some data provided by clients need to be processed or stored on various disks in multiple locations and possibly managed by third-party providers. In this situation, because of the loss of control over the data and processes, data confidentiality, integrity and availability cannot be guaranteed. Additionally, in some cloud service model, the service provider usually has complete control of data, while enterprises retain only partial control of their data, which they often find quite alarming. Therefore, enterprises should be permitted to trace their data processing procedure. Actually, they do not need to really control or process their data. They only want to know where the data is and how it is dealt with which can improve their trust level of cloud computing.

Generally, because of the lack of transparency, the consumers perceive that an in-house system is more secure compared with cloud computing services. If we improve the level of transparency, we can decrease consumes' perception. There are two issues existing in transparency: One is the physical location of storage and processing sites; the other is the security profiles of these sites. Therefore, in order to address the issue, the service provider must supply sufficient information of their data through which the client can trace the location and processing procedure of the data.

4.4.2   A Model Applied to Build a Trust Environment in Cloud Storage

Trust is a subjective measurable scale that can thrust decisions based on the beliefs of decisions [1] and it is widely used in social science to build human being's relationship. However, trust should not be restricted within the domain of philosophy, sociology and psychology. It needs to be addressed by all attempting good governance [2]. Nowadays, trust is becoming an essential part to form a secure distributed cloud computing environment, especially in cloud storage, because trust has many security properties, such as reliability, dependability, confidence, honesty etc. In cloud environment, trust issues increase because customers feel their data is not under control. As a customer's infrastructure is located at an off-site location and managed by a second or third party entity, the customer lacks the confidence of transparency, data security and unclear security assurance. Therefore, we can regard trust in cloud as the customers' level of confidence in using the cloud. In order to enhance customers' confidence level, we need to increase their competence trust and benevolence trust of cloud computing service by mitigating technical barriers and improving security management.

The above provides us a concept that how we can construct a trust mechanism in cloud computing to encourage people to use cloud computing service. "Trust decision usually consist of two parts, reasoning and feeling, both cognitive trust and emotional trust will influence the decision making process" [7]. In the emotional part of trust, it contains more about individual's feelings and emotion. [15] pointed out that "when people feel that the trustee is well-meaning and has the intention to work for trustor's benefit, he/she will think the trustee is benevolent" [15]. Such beliefs of benevolence is necessary for an emotional trustworthy relationship which make the trustor feel that he/she will not be cheated and have less risk, in turn enhance trustor's feelings of comfortable and secure. Thus, we propose that the benevolence trust has a positive influence on emotional trust.

[16] also indicated that "customer trust in trustee's competence means that the customer believe that the trustee has the capability to transmit professional and helpful information, which also mean that the customer will be more relying on the relationship between them for decision make and feel comfortable and secure" [16]. Thus, customer's cognitive trust in trustee's capability will influence their emotional trust in trustee, according to that; we propose that the competence trust has a positive influence on emotional trust. The level of customers' emotion trust will influence their consuming behavior.

In order to improve people's trust level of cloud computing service, we need to improve both benevolence trust and competence trust. Benevolence trust reflects a person's psychology statement and the lack of it comes from the uncertainty between two parties. If no uncertainty exists between two parties, it indicates that no risk or threat is found in future interaction between two parties [3]. However, we live in a real world in which we cannot absolutely eliminate the uncertainty and what we can do is to reduce uncertainty and to increase predictability on what other party will act in the future. The way to decrease uncertainty is the communication and the share of information between parties [4]. According to Berger [17], uncertainty about the other party is the "(in)ability to predict and explain actions". It means that uncertainty can be reduced by sharing information and obtaining the condition of the information [17]. The reason why customers do not trust the service provider is that they know nothing about their data processed by the provider. When the valuable data is put in total dependence of someone else, the customer feels unsafe and the level of uncertainty increases. Therefore, the provider of cloud storage needs to improve the level of transparency, which means they should let customers know what will be done on their data and keep customers tracking the situation of the data.

As for competence trust, it is a cognitive feeling to measure someone's ability in specific area. In cloud storage, if the service provider wants to improve customers' competence trust for them, they must enhance their secure technology to protect users' data from three aspects which are confidentiality, integrity and availability.

5   Conclusion

Cloud computing is considered to be one utility service after water, gas and electricity power. As an important component of cloud computing, cloud storage is facing various security challenges. To protect the valuable data stored in cloud, we proposed several possible solutions to cloud storage security problems in terms of confidentiality, availability and integrity.

Encryption as a traditional method for data confidentiality protection has been proved very effective in cloud environment. Although, problems remains, such as manipulating encrypted data in cloud cause computational overhead. Several intelligent solutions to those problems have been proposed. Encrypting data before it is sent to the storage pool is the most suggested way used to solve untrusted CPS problem.

Other than confidentiality, the availability of data in cloud also attracts attention. As a utility service, cloud should be able to provide reliable service at any given time. To meet with their own needs, users can use those straightforward comparisons combined with non-technical concerns to make a better decision.

One important concern on data integrity in cloud is how users can know their data is integrated in cloud. In the past, people did not pay more attention to build a perfect trust mechanism and more concerns are concentrated on developing technology. Recently, we found that various techniques have been created to protect the data security, but no matter how advanced the technology is, the customers still suspect the security of the data. Therefore, we must use the knowledge of psychology and information technology to build the trust mechanism. Integrity of data can be delivered by both non-technical and technical methods. A well-designed trust mechanism provides users the basic integrity protection in cloud. In cloud environment, key management is related to different parties. To enhance the trust between those parties, a third party scheme of key management is proper.

References

[1] Brohi, S.N., Bamiah, M.A., Brohi, M.N., Kamran, R.: Identifying and analyzing security threats to Virtualized Cloud Computing Infrastructures. In: 2012 International Conference on Cloud Computing Technologies, Applications and Management (ICCCTAM), December 8-10, pp. 151–155 (2012) ISBN: 978-1-4673-4415-9

[2] CIOTIMES. 云计算安全架构中的加密. TechTarget China (October 19, 2012), http://www.ciotimes.com/cloud/cjs/72702.html

[3] Han, S., Xing, J.: Ensuring data storage security through a novel third party auditor scheme in cloud computing. In: 2011 IEEE International Conference on Cloud Computing and Intelligence Systems (CCIS), September 15-17, pp. 264–268 (2011) ISBN: 978-1-61284-203-5

[4] Brodkin J. : Gartner: Seven cloud-computing security risks. NetworkWorld (July 2, 2008), http://www.networkworld.com/article/2281535/data-center/gartner-seven-cloud-computing-security-risks.html

[5] Whitman, M.M., Mattord, H.J.: Principles of information security, 4th edn. Cengage Learning (January 1, 2011) ISBN: 978-1111138219

[6] Singh, Y., Kandah, F., Zhang, W.: A secured cost-effective multi-cloud storage in cloud computing. In: 2011 IEEE Conference on Computer Communications Workshops (INFOCOM WKSHPS), April 10-15, pp. 619–624 (2011) ISBN: 978-1-4577-0249-5

[7] Wylie, J.J., Bakkaloglu, M., Pandurangan, V., Bigrigg, M.W., Oguz, S., Tew, K., Williams, C., Ganger, G.R., Khosla, P.K.: Selecting the Right Data Distribution Scheme for a Survivable Storage System. Carnegie Mellon University (May 2001), http://www.pdl.cmu.edu/ftp/Storage/CMU-CS-01-120.pdf

[8] Abu-Libdeh, H., Princehouse, L., Weatherspoon, H.: RACS: a case for cloud storage diversity. In: SoCC 2010 Proceedings of the 1st ACM Symposium on Cloud Computing, pp. 229–240 (2010) ISBN: 978-1-4503-0036-0

[9] Rabin, M.O.: Efficient dispersal of information for security, load balancing, and fault tolerance. Journal of the ACM (JACM) 36(2), 335–348 (1989), doi:10.1145/62044.62050

[10] Bowers, K.D., Juels, A., Oprea, A.: HAIL: a high-availability and integrity layer for cloud storage. In: Proceedings of the 16th ACM Conference on Computer and Communications Security, pp. 187–198(2009) ISBN: 978-1-60558-894-0

[11] AlZain, M.A., Soh, B., Pardede, E.: MCDB: Using Multi-clouds to Ensure Security in Cloud Computing. In: 2011 IEEE Ninth International Conference on Dependable, Autonomic and Secure Computing (DASC), pp. 784–791 (2011) ISBN: 978-1-4673-0006-3

[12] Zhang, N., Jing, J., Liu, P.: CLOUD SHREDDER: Removing the laptop on-road data disclosure threat in the cloud computing era. In: 2011 IEEE 10th International Conference on Trust, Security and Privacy in Computing and Communications (TrustCom), November 16-18, pp. 1592–1599 (2011) ISBN: 978-1-4577-2135-9

[13] Ranchal, R., Bhargava, B., Othmane, L.B., Lilien, L., Kim, A., Kang, M., Linderman, M.: Protection of identity information in cloud computing without trusted third party. In: 2010 29th IEEE Symposium on Reliable Distributed Systems, October 31- November 3, pp. 368–372 (2010) ISBN: 978-0-7695-4250-8

[14] Amazon EC2. Availability Region and Availability Zone concept, http://docs.aws.amazon.com/AWSEC2/latest/UserGuide/using-regions-availability-zones.html

[15] McKnight, D.H., Choudhury, V., Kacmar, C.: The impact of initial consumer trust on intentions to transact with a web site: a trust building model. Journal of Strategic Information Systems 11, 297–323 (2002)

[16] Komiak, S.Y.X., Benbasat, I.: The effects of personalization and familiarity on trust and adoption of recommendation agents. MIS Quarterly 30(4), 941–960 (2006)

[17] Dean, J., Ghemawat, S.: MapReduce: Simplified Data Processing on Large Clusters. Communications of the ACM 51(1), 107–113 (2008), doi:10.1145/1327452.1327492

An Investigation of How Businesses Are Highly Influenced by Social Media Security

Daniel W.K. Tse, Derek HL To, Xin Chen, Zhongyi Huang, Zhenlin Qin,
and Shaneli Bharwaney

City University of Hong Kong
iswktse@cityu.edu.hk

Abstract. Social media plays an immensely important role in business nowadays. With the social media platform and interactions continuously growing, an increasing number of organizations are engaging into it. The idea of collaboration via social networks is also a growing trend. Although social media benefits the companies, it also poses certain threats. Companies have the likelihood facing privacy invasion as well as encountering leakage of confidential and sensitive data. As a result, firms tend to bear a reputational risk within this social media era which is undoubtedly caused by the lack of information security.

This paper discusses the information security risks caused towards social media. It examines three important aspects: privacy, data leakage and human factors. The study uses interviews and surveys as research instruments in order to prove that these issues are critical and overlooked. In the discussion part, it highlights the reasons for which the aforementioned concerns are generated by the companies and the existing problems. Furthermore, in order to ensure information security and to eliminate the risks, a management strategy containing policy, SETA (security education, training and awareness) and technical control are proposed.

Keywords: Social Media, information security, data leakage, privacy, security training, security policy.

1 Introduction

Nowadays, social media plays an integral part in large enterprises and small business. According to the Statistics Sweden [1], there were sixty-nine percent of large enterprises that allow their employees use social media. On the other hand, [2] Zupan's statistics also showed that there were thirty-seven percent of the enterprises that are using social media to keep in touch with their customers. This shows the rising popularity of using the social media today.

However, some companies find it hard to control social media security within their organizations. According to the ISACA survey [3], social media has a high risk to be invaded. For instance, 45,000 Facebook accounts have been hijacked and 760 companies were attacked by the hackers. In addition, 8 out of 10 companies tend to

L.S.-L. Wang et al. (Eds.): MISNC 2014, CCIS 473, pp. 311–324, 2014.

use Twitter as their main channel of communication. Also, [1] reported that there were only thirteen percent of small businesses having policy on using the social media, security threats are quite prevalent in enterprises and companies have unawareness.

On the other hand, Liau's study [4] found out that the top five social media threats include the malware, spam, targeted attack, phishing as well as the data leakage. Besides, Sophos [5] showed that seventy-four percent agreed that using the personal devices would increase the threat on company's data loss.

Therefore, the aim of this study investigates how social media affects the operation of large enterprises and small business. It will examine the different kinds of social media's threats to the companies, analyze both positive and negative impacts of social media and suggest a framework including solutions to reduce and avoid the security matters on social media. In addition, the study would use an exploratory study which is suggested by Lewis [6].

2 Literature Review

Nowadays, social media is affecting the security level of company. Three major risks of information security are identified below. In order to help companies gain benefits from social media and minimize any potential risks, these three threats have been analyzed and the corresponding solutions have been established.

2.1 Privacy and Security threats on Social Media in Business

According to the ISACA report [7], social media privacy and security risk can be categorized into three aspects. These include "communication with customers and other business partners", "employees' access of social media websites in their working environment" and their "use of business smartphones in order to access social media". All these security risks tend to exist in many companies.

Balzarotti [8] investigated several hacking social networks experiments to prove that they are unsafe. It examines the way in which hackers use a fake account to pretend to be someone else, and utilize this identity for business purposes. This study has stated a clear point to the common hacking principle with social engineering method. It aroused that many social network users do not scan and filter before they accept the friend's request. It increases the chance for hacker to steal their information easily. The cloning attack is one of the social media privacy and security threat to the company.

Although social media is widely used to network with family, friends and expand one's business links, there are several trust issues linked to privacy. Revelation of too much information on the basis of its location also includes privacy issues. Rose [9] indicated the way in which one's confidentiality is more exposed as our society changes. McCullagh [10] exhibited 40% of social media users disable their privacy settings. Therefore, the impact on their profile security is on a lower rating than a profile consisting of high discretion. It also exhibits the power of influence we reveal to other users, decreasing the privacy level [9].

According to Barbier [11], it states that the root cause of exploiting privacy via the social media is the vulnerable friends. This study has conducted with the quantitative research methodology with investigation on 2 million users to prove assumptions. Users tend to face a high degree of privacy breach which is mainly due to their improper privacy settings. In a nutshell, the vulnerability of a user transparently indicates the level of intention they have to protect their social media profiles. What characterizes the user's exposure on social media networking sites is categorized under two attributes: Individual and Community. Therefore, taking full control on the revelation of information on social media, and how much one would like to disclose to their friend or unknown friends, solely depends on the user.

Lastly, in relation to the aforementioned, a research by H. J Kim [12] highlights the use of privatizing one's account, or limiting the views to a certain number of people, can help eliminate any strange encounters with unknown people who may get hold of one's personal information. However, as the Internet is commonly used and trusted by people of all ages, one may not pay attention to such precautions.

2.2 Human Factor

The human factor in social media security refers to the effect of employees' behavior on social media platform. [13] Human factor of social media security refers to the impact of emotions, perceptions and fear on peoples' behavior, and these are crucial for an effective alignment. However, the human aspect of security is always overshadowed by the emphasis of technological means [14].

Employees' behavior on social media can generate positive or negative impact for an organization. The easiness of using social media to communicate with customers can increase customer retention and loyalty. Business relationships are strengthened by the open dialogue among employees and business partners. [13] However, the interactions over the Internet can linger and be seen by others. Once improper information is published through social media, people can record it. This will cause negative influence such as reputational problems or data leakage, and the enterprise will be at an amplified subject to be attacked. Whether the companies can gain benefit from social media depends on the human factors on the online platforms. If an organization can control the human aspects, they will not only eliminate the security problems but also benefit from social media. The existing problems include:
1. Human factor ignorance and no policies implementation
The key threat to information security is generated by careless employees but a number of companies do not realize the risk when they apply social media. Oehri [13] stated that companies usually ignore the human aspect and there is no policy for information security. Moreover, it is because social media has radically changed our world and a myriad of people are quite involved in the use of it. Even though the organization has not adopted the use of social media, their employees have, and this means that employees may have revealed certain data into the social media world. In this case, the companies should also develop certain policies to educate their employees and secure their data.

2. Employees' non-compliance with the policy

Siponena [15] suggested that the factors affecting employees' intention to comply with information security policies are environmental effect, threat appraisal and self-efficacy. Environmental effect stands for behavior of others, the visibility of information security actions of the organization and the security measure outside the enterprise. Threat appraisal refers to employees' assessment of whether the security threats would harm their company. If people do not think that they will be confronted with the negative consequences, they will not adhere to the policy. Self-efficacy represents employees' beliefs on their ability to comply. The research implicated that these factors have remarkable influence on individual's intention of compliance. If the enterprises cannot take advantage of these factors, their employees will simply ignore the policies.

3. Lacking of social and technology combination

Using social power to conduct the policies is not enough to manage information security, as Inglesant [16] said. The enforcement of information security requires interplay of policies and technology forces, which is a new challenge to conventional security thinking. Monitoring and assessments are necessary to undertake information security policies, and many companies may consider using human actor for supervision. However, this kind of social power is difficult to control whereas technical power can offer effective tools. Technical force may be mandatory but can efficiently form the social media culture in the working environment.

2.3   Data Leakage

Data loss plays an increasingly important role in the information security. It is an essential component in the issue of data loss. According to the study by Kuhn [17], data leakage refers to the data that is no longer under the control of the organization.

The phenomenon of data leakage has become a worldwide issue. With the development of the information technology, the geographical limitation has been broken down by the mobility technology. According to Hasimoto [18], the various ranges of the definition of data leakage has been analyzed. The issue of data leakage varies from confidential information of one customer being exposed to some of the strategy related files having been sent to the competitors in a deliberate yet unknown way. Hasimoto [18] also considered that data loss can be classified into three categories: Data in motion; Data at rest and Data at endpoint. Data in motion refers to the data moving through the Internet to outside. Data at rest refers to the data stored by some storage methods, including the database and some files systems. Data at endpoint denotes the loss of data on devices such as USB, external drives and laptops.

Focusing on the part of data in motion, the company adopting the growing use social media within their enterprise can encounter serious data leakage issues. The study by Benson [19] showed that some enterprise may block social media sites within their company network's Internet. Some sensitive information is included on the sites; however the data is not protected in a legal way – copyrights, patents and the incorporation of intellectual property strategic planning. Confidential information is always related to the core business process in any company whereas the loss of intellectual property and proprietary is a serious issue that revolves around. Companies are at a disadvantage as they are bound to lose their competitive advantage as well as being challenged with a certain level of financial turmoil affecting the whole enterprise. Moreover, it can damage existing and/or potential investors' confidence. A variety of social network site (SNS) services available are related to this issue which is dramatically growing recently. The adoption of social media also enacts as a threat on some personal information. However, certain information is unavoidable to disclose when one is in the process of communicating and/or discussing matters with another party, resulting in identification theft.

A study [7] indicated that several risks and mistakes by the employees are the reasons of data leakage which is the main trigger of a risky behavior. It shows that risky behaviors are classified into five items including unauthorized application use, misuse of corporate computer, unapproved physical and network access and remote worker security, misuse of password and unregulated log in/log out procedure.

The report [20] also revealed why several employees bring some risks to the enterprise. 40 percent of employees of IT staffs tend to view unauthorized sites. The study [7] also showed that the reason behind risky behavior is just the lack of education, awareness and the ineffectiveness of security policies. Furthermore, communication failure between the end user and IT decision makers causes different understandings and perceptions on security policies. The study [7] demonstrated that the way of broadcasting news and regulations also have a significant influence on the effectiveness of the security policies.

3   Research Methodology

In this research, an exploratory study is applied. This research methodology is conducted in collaboration with Lewis study [6]. It focuses on three main ways: a review literature, an interview with an experienced person in the related field and subject, and a survey.

3.1   Interview

With the intention of familiarizing with existing trends and delivering the effective information on social media impact to the company, an interview was conducted with five persons with IT management level background from medium to large of financial enterprises as well as the educational organizations.

The following are the characteristics of the interviewers.

Interviewee	Institutions	Job Titles
1	Medium financial enterprises	Senior IT manager
2	Educational organizations	Computer Officer
3	Large financial enterprises	Senior Technical Officer
4	Medium financial enterprises	Security Manager
5	Educational organizations	Chief Technical Officer

The interview has conducted with series of questions. The objective is to learn more about the understanding of social media they may relate to.

Interview questions:
1. Is social media commonly used in your company?
2. What do you think of social media security, related to your business?
3. What kind of social media threats tend to occur in your company?
4. Define the risk level of privacy, human and data leakage combined in your company (0- 10).
5. What kind of measures does your company take to control social media security?
6. How effective are these measures?

Survey
The 2013 information security technical report which is published by *Price Waterhouse Coopers* with *Info security Europe association* [21] was examined. The report was utilized as an example of our analysis as it entails the number of famous companies and employees in different. The following is a set of survey that is related to social networking or media in 2013:

1. How confidential is the data that interviewees store on the public website?
2. How significant is the use of social media sites or networking to the institutions?
3. How do interviewees avoid employees' abuse of the web and social media or networking sites?
4. What procedures have interviewees taken to alleviate the risks related with employees using smartphones or tablets?
5. How many interviewees have an officially standardized information security rules or guidelines?
6. What type of security incidents do organizations plan for, and how effective are those contingency plans?

4   Research Results Analysis

4.1   Analysis of Interview Results

1) Privacy
Both financial enterprises and educational institutions presented a high level concern about private information. Clients' personal data and students' records were both regarded as top confidentiality. Besides, all organizations admitted that the usage of social media raised potential risks of normal operation. Several security measures have been carried out by these organizations. Financial institution took comprehensive governance in both policies (e.g. censorship, penalty rules) and IT controls (e.g. monitor access restriction and anti-malware software). However, in comparison, educational organizations are more vulnerable.

2) Human
Human concern was normally regarded as a subordinate to privacy and leakage in terms of its level. However, during the interviews the five organizations mentioned

that most employees lack the awareness of adequate security, even though they are periodically trained. They admitted that inadvertent use of social media would pose potential threats on information security. Besides, policy implementation is one of the most disturbing predicaments in human respect. Three interviewees indicated that majority of errors and losses were made by employees who were always violated to comply with the policies.

3) Data Leakage
Data leakage is an increasingly common concern since the growing use of social media. All of the interviewees agreed that social media usage raised the risks of data leakage and the existing security control still needed to be strengthened. Moreover, the educational organization regarded normal use of social media as acceptable while financial institution thought general use may cause data loss any time. It was observed that there was a tighter control for financial organization.

4) Challenge
All five organizations failed to identify the real risk level of human aspect. Employees, as the most direct controller of social media, were effectively the highly critical factor to social media security. The other challenge was an uneven cognitive in security awareness and controlled levels of different organizations, which indicated that effective solutions for systematic and comprehensive security control was needed.

4.2   Analysis of Survey Findings [21]

1) Ubiquitous privacy threats among organization
Large amount of confidential data stored in the public website. Over 55% of large enterprise and small business respectively have stored medium confidential data on public website. In addition, there were nearly 20 percent of companies that stored highly confidential information. In total, 82 percent of large enterprises have stored data on public websites while 75 percent of small business did so. This data showed that large enterprises were more risky in terms of privacy security, to an extent.

2) Social media usage and the risk
Over half of the respondents consider social media as *unimportant* to their organization and approximately 36 percent of them thought that networking was *quite* important. At the same time, only 15 respondents regarded it as *very* important. The same status was showed between large enterprise and small business regarding this question.

3) Security control in access restriction and monitor respects
Both access restriction and monitor logs of websites were popularity applied in social media security control. It showed that the data of block access to inappropriate websites doubled the data of block access to social media, which meant the former method was more commonly used. Besides, in large enterprises, the data of monitor logs of which websites were much higher than the data of monitor the contents that staff posted on social networking sites, which indicated the former method is more acceptable.

4) More measures for controlling social media risks (caused by smartphones or tablet)

The situation of using social media without taking any measures was more common in small business than large enterprise. Nevertheless, large enterprises took more full-scale protective measures in both Internet control (access restriction of remotely connection, email protection, encryption) and policy governance (security strategy defined, policy issued). 110 large enterprises have issued a policy on mobile computing. However, only 38 of them trained staff on the threats associated with mobile devices. In comparison, 52 small businesses complied with the policy and 48 have trained their employees correspondingly. It could be indicated that small business focused more on staff training and management implementation of mobile device.

5) Documented policy and guidelines of social media security

In terms of officially standardized policy and guidelines setting, large enterprises performed much better than small business. 150 respondents in larger companies were more confident when documenting information security policies implementation while only 82 were identified in small business.

6) The object and effectiveness of organizations' secure plan

Three key security incidents that organizations planned for were identified as *confidentiality breach*, *staff misuse of information systems* and *virus infection or disruptive software*. This means that data leakage and human dimension have become a high threat that subordinates virus infections only. Meanwhile, for all of the threats except the systems failure of data corruption, ineffective contingency plan only represent tiny part in range of 0 to 11%. It meant implement security plan was imperative and effective to most of threat control.

5   Discussion

5.1   Policies and Guidelines

From the data showed in research analysis, 67 percent of organizations conduct a plan to deal with the issue about staff misuse of information. However, the percentage of staff-related incidents of small business increases by 12 percent from 2012 to 2013. Moreover, the number goes up to 84 for large organizations. It shows that the plan is ineffective to eliminate the risks caused by employees [22]. As a result, it is necessary for organizations to plan and implement more detailed and effective policies.

Policy is defined as a course of actions used by organizations to convey instructions from management to those who perform duties. It is also the law of an organization. Security policy is the *cheapest* control for enterprises to implement information security but it is also the most difficult one. [22] Therefore, companies should give value to it. In addition, guidelines should be included in the policy management and be used to enforce compliance with policy. Guidelines always affect the culture of a company, which creates the safe climate within the working environment [13].

Step 1: Risk Assessment

Before planning the security policy, the company should have a clear cognition of what need to be protect in social media platform and the corresponding method. This is the reason why the companies should conduct risk assessment. With this step, the policy will be more comprehensive. Risk assessment can be divided into following actions:

1. Identify the asset which may be influenced by social media, for example, people's ID, documentation, data.
2. Recognize the potential threat to each asset. For instance, the conversation between staff and client on social media platform may contain confidential data.
3. Identify the vulnerabilities of the company's existing security policies.
4. Propose the methods, guidelines, tools and techniques to deal with the threats, such as firewall.

Because the social media and company remain changing, the risk assessment should be updated periodically.

Step 2: Plan

Security policies should define what assets are considered valuable for a company and what techniques and methods are taken to safeguard the assets. In addition, it is necessary to consist of guidelines for different assets and rules for staff to follow. The policy for social media usage can be drawn with subtitles such as *Content Publication Policies*, *Password Policies*, and *Data Transmission Policies*.

The following factors need to be considered when organizations plan their security policies for social media:

1. Give a clear structure of the whole policy. Because every employee may have a chance to use social media in business, the content should be simple and detailed so that every reader is able to understand [18].
2. Security policy should be integrated into the company culture. The enterprises need to think about their common way of using social media and select the best practice for their own philosophy.
3. Define the responsibility of employees. Specifically, whether the behavior has violated the security policy or not. The security policies should explain the potential risks of certain behavior and describe the penalty which the staff will receive. [7]
4. Appoint policy administrators. Monitoring and assessment are necessary to undertake information security policies. The policy administrators are the people who have a certain degree of authority, for example, senior managers and supervisors. They should monitor their subordinates and evaluate the implementation performance. The performance of the security policy needs to be included in employees' evaluation, which will influence the year-end bonus.
5. Using policy administrators to conduct with the policies is not enough for managing information security. The enforcement of information security requires interplay of policies and technology forces [22]. Technical force may be mandatory but can efficiently form the security culture in the working environment.

6. Develop a schedule to review the policies and receive recommendation from employees.
7. Ensure the visibility of security policies and sufficient communication between IT decision makers and other department.[7]

Step 3: Implementation
The company should make a partnership with the employee to implement the security policy, safeguard the employee to follow the policy initiatively and integrate the security policy into the daily practice. [7]

To minimize people's resistance to security policy implementation, the Lewin Change Model [22] is introduced. When implementing the new security policy, the enterprises can lower the requirement of current general policies and rules to relax the employees, and then put forward the new security policy with low standard. This will increase the confidence for the employees to believe that they are good for the new policy. After that, the company can gradually improve the demanding to normal level. As a result, the Lewin Change Model can increase self-efficacy and reduce the resistance of security policy. Sustaining commitment and support always play an important role in the company security management. [18] After putting the new policy into implementation, policies must be managed and the performance must be monitored by policy administrator and technical tools.

Step 4: Evaluation
Valuable information about the performance and efficiency of security policies should be measured. The evaluation can help the companies to review their policies and keep up-to-date improvement.

5.2   Safety Precaution to Avoid and Minimize the Social Media in Security Risk

According to the Almeida study [23], it lists out several technical methods to tackle and minimize social media security risks and threats. All these technologies can be applied to the aforementioned parts of privacy, human aspect as well as the data leakage of social media. In this part, it would subtract and apply some of the solutions to discuss.

Firstly, the application traffic shaping is limited by the employees' computers bandwidth on accessing the social network site. The methodology is limiting the bandwidth on accessing the social media site to less than 100 kilobytes per second and applies this rule in the firewalls. From the human aspect point of view, it is impossible to restrict and block all social media network sites in enterprises. Some employees require communicating with their client through the social media.   Therefore, restricting the bandwidth on accessing the social media site does not only allow the employees using it, but also avoid large amount of data to leak from the companies. It is a win-win strategy to both employee and company [23].

In addition, from the privacy view, the browser setting is the alternative way to keep the transmitted data through the network in private and confidential way. The administrators should set the browser and security settings in the highest level and restrict the employees to make any changes on those advanced settings. For example, Mozilla Firefox and Opera browsers could allow surfing the website anonymously. Also, the HTTPS should be always in the first priority on security features. It can secure the communication channels and avoid the man in middle attack [24].

Besides the browser setting, social media privacy sites' setting is also required. [11] Barbier et al. suggested three methods to avoid the vulnerable friend. The first is to set up the privacy settings. It can protect the information to limit the friends' access the user's information. The second is filtering the new friends before accepting the friend request. Several [9], [11], [7] studies mentioned that one of the serious faults on social media is accepting the friends without filtering. Some users do not make any privacy settings on limiting their personal information and post. Once the vulnerable friend's has been accepted, most personal information is exposed. Finally, removing the vulnerable friends is the final issue to minimize the privacy and security risk on social media. Hence, the privacy setting is one of the biggest issues against the vulnerable friends.

Moreover, one of the most effective ways to tackle the data leakage problem is *data loss protection* (DLP). The methodology of DLP is based on central policies, such as applicable laws and regulations [25], to recognize, monitor and protect the data under three states. They include data in rest, in motion as well as in use. The advantage of DLP is that it can protect organizational information in integrity and confidentiality. The organization should classify the data in different privileges as well as confidential levels. Especially the busy traffic in web 2.0, this policy could allow the companies to understand their data in different categories and prevent from the confidential data loss.

Besides the above precautions, the company is necessary to have further content protection. For example, they should use application control to filter the network traffic, such as IPsec and SSL-encrypted method. In addition, the company should provide the content protection and encryption to the document, such as enhancing the authentication (token and biometric). Finally, it should install anti-virus and anti-malware program to avoid the virus infection.

5.3   Security Education, Training and Awareness Program (SETA)

Employee error and risky behavior are one of the most serious threats of information security in social media. The security education, training and awareness (SETA) program aims to reduce the incidence of security policy violation by employees. The purpose of SETA is to raise the awareness of protecting some confidential information and cultivate more computer skills to perform their job in a more secured way.

Security education program
Security education refers to gaining a formal information security –related coursework in some institution [26]. After acquiring security education, the employee will have a deeper insight and understanding in information security.

During the process of course design, different courses are designed to deliver based on different learning objects. The expected results of the different learning objects give a direction of the knowledge structure to be taught in the course. The knowledge map has been clearly defined that people in different position have various required level of information security knowledge. Moreover, the depth of knowledge can be classified into five levels, including understanding, accomplishment, proficiency and mastery [27]. The last step of course design is to identify the prerequisite knowledge of each class so that we can have a scientific course design.

Security training program

Security training provides some detailed information and some practical instruction to employees to perform more securely in their daily activity [22]. In the business practice, the organization needs to clarify their expectation on data protection, which is documented in the security policy and needed to emphasize in the training program. In the training program, the organizations also need to give some practical skills on the information security. For instance, they should train their employee considering about security issue when connecting to social media and browsing some website. The format of training can be organized in formal class, computer- based training, distance training/ web seminars, user support group and on-the- job training.

Security awareness program

Security awareness gives a general picture of information security and enables them to distinguish the dos and don'ts in the working activity [22]. In reality, security program is one of the least frequent but most cost-effective programs implemented by the organization. The security policy can be implemented successfully when the employees have understood and follow the policy in the working procedure [7]. The purpose of the awareness program is to remind the importance of the information security in the employee's mind. The program is not complicated, nor costly. In the view of business practices, organizations can arrange some activities related to security awareness in the new-hire orientation events [20]. Moreover, some simple slogans can be printed on objects that are frequently used, as a reminder and/or alert. The awareness program is the *least* expensive but *influential* programs, reminding people to adhere to the importance of information security.

6   Conclusion

In summary, this paper comprehensively evaluates and stresses the importance of the growing use of social media in enterprises today and how it is highly influenced by various security issues.

Privacy, data leakage and human factors are the contiguous triggers on social media in businesses today. With the popularity and reliance on such communication tools, studies by several researchers have highlighted the various risk posed by the use of social media sites and its element of networking. Individuals and organizations need to be well aware of the idea of cloning, via fake accounts and identity thefts over the Internet, as a larger number of unknown external parties may be targeting a certain person or their company. The over exposure of information and the lack of privatization on one's account also provides hints to potential attackers. In relation to this, the impact of human factors is another main elicit in the risk of using social media. The information revealed by organizational staff also plays a role in the level of security one has on their firm. In addition, data leakage should be carefully monitored as well as complying with company policies which will be a stepping stone to decrease the threats caused by social media.

An interview with employees from 5 large enterprises and a survey by *Price Waterhouse Coopers* gave an insight to the use of social media within their companies, and how it affects them on a day-to-day manner in terms of any security threats and issues faced. A series of recommendations were also obtained in order to improve the security aspect of social media. These were further evaluated by 4 aspects: privacy, human, data leakage and challenges.

Last but not least, policies and guidelines should be implemented by organizations in order to avoid any misuse of information by employees. A risk assessment, plan, implementation, and evaluation of the entire policies and guidelines will strategically take charge of the control of information provided by the firms. In addition, safety precautions such as limiting bandwidths on the Internet, protecting browser views as well as applying laws and regulations should help to decrease potential social media challenges. Furthermore, introducing the SETA program can create awareness within the firm, and have an impact over staff to work in a safer and secure Internet environment.

References

[1] Statistics Sweden. ICT usage in enterprises 2013: Enterprises use social media to develop their image (October 29, 2013), http://www.scb.se/en_/Finding-statistics/Statistics-by-subject-area/Business-activities/Structure-of-the-business-sector/ICT-usage-in-enterprises/Aktuell-Pong/15318/Behallare-for-Press/ICT-usage-in-enterprises-2013/

[2] Zupan G.: Usage of social media in enterprises, Slovenia, 2013 – final data. Statistical Office of the Republic of Slovenia (October 7, 2013). http://www.stat.si/eng/novica_prikazi.aspx?ID=5799

[3] ISACA, Social Media: Business Benefits and Security, Governance and Assurance Perspectives. An ISACA Emerging Technology White Paper (2010), http://www.isaca.org/Groups/Professional-English/security-trend/GroupDocuments/Social-Media-Wh-Paper-26-May10-Research.pdf

[4] Liau, Y.Q.: Top 5 social networking business threats. ZDNet (February 1, 2010), http://www.zdnet.com/top-5-social-networking-business-threats-2062060912/

[5] Sophos. Security Threat Report 2012. Sophos Ltd. (2012), http://www.sophos.com/medialibrary/PDFs/other/SophosSecurityThreatReport2012.pdf

[6] Saunders, M., Lewis, P., Thornhill, A.: Research Methods for Business Students. Financial Times/Prentice Hall (2007) ISBN: 9780273701484

[7] Cisco. Data Leakage Worldwide: Common Risks and Mistakes Employees Make. Cisco Systems, Inc. (2008), http://www.cisco.com/c/en/us/solutions/collateral/enterprise-networks/data-loss-prevention/white_paper_c11-499060.pdf

[8] Bilge, L., Strufe, T., Balzarotti, D., Kirda, E.: All Your Contacts Are Belong to Us: Automated Identity Theft Attacks on Social Networks. In: The 18th International World Wide Web Conference (WWW 2009), April 20-24, pp. 551–560 (2009) ISBN: 978-1-60558-487-4

[9] Rose, C.: The Security Implications of Ubiquitous Social Media. International Journal of Management and Information Systems 15(1), 35–40 (2011)

[10] McCullagh, D.: Why no one cares about privacy anymore. CNet News (March 12, 2010), http://www.cnet.com/news/why-no-one-cares-about-privacy-anymore/

[11] Gundecha, P., Barbier, G., Liu, H.: Exploiting Vulnerability to Secure User Privacy on a Social Networking Site. In: 17th ACM SIGKDD International Conference on Knowledge Discovery and Data Mining (KDD 2011), pp. 511–519 (2011) ISBN: 978-1-4503-0813-7

[12] Kim, H.J.: Online Social Media Networking and Assessing Its Security Risks. International Journal of Security and Its Applications 6(3) (July 2012)

[13] Oehri C., Teufel S.: Social Media Security Culture. In: Information Security for South Africa (ISSA), August 15-17, pp 1–5 (2012) ISBN: 978-1-4673-2160-0

[14] Briggs, R. Edward, C.: The Business of Resilience: Corporate security for the 21st century. Demos (2006), http://www.demos.co.uk/files/thebusinessofresilience.pdf

[15] Siponena, M., Pahnila, S., Mahmoodb, A.: Factors Influencing Protection Motivation and IS Security Policy Compliance. Innovations in Information Technology, p. 105 (November 2006) ISBN: 1-4244-0674-9

[16] Inglesant, P., Sasse, M.: Information Security as Organizational Power: A framework for re-thinking security policies. In: 2011 1st Workshop on Socio-Technical Aspects in Security and Trust (STAST), pp. 9–16 (September 8, 2011) ISBN: 978-1-4577-1182-4

[17] Liu, S., Kuhm, R.: Data loss prevention. IT Professional 12(2), 10–13 (2010) ISBN: 1520-9202

[18] Hashimoto, G.T., Rosa, P.F., Filho, E.L., Machado, J.T.: A Security Framework to Protect against Social Network Services Threats. In: Fifth International Conference on Systems and Networks Communications (ICSNC), August 22-27, pp. 189–194 (2010) ISBN: 978-1-4244-7789-0

[19] TechNet. Best Practices for Enterprise Security. Microsoft, http://technet.microsoft.com/en-us/library/cc750076.aspx

[20] Cisco. Data Leakage Worldwide: The Effectiveness of Security Policy. Cisco, http://www.cisco.com/c/dam/en/us/solutions/collateral/enterprise-networks/data-loss-prevention/Cisco_STL_Data_Leakage_2008_.pdf

[21] Information Security Breaches Survey 2013. Department for Business, Innovation and Skills, https://www.pwc.co.uk/assets/pdf/cyber-security-2013-technical-report.pdf

[22] Whitman, M.E., Mattord, H.J.: Principles of Information Security, 4th edn. Course Technology, Cengage Learning (2012) ISBN: 978-1-111-13821-9

[23] Almeida, F.: Web 2.0 Technologies and Social Networking Security Fears in Enterprises. International Journal of Advanced Computer Science and Applications (IJACSA) 3(2), 152–156 (2012)

[24] Xia, H., Brustoloni, J.C.: Hardening Web Browsers against Man-in-the-Middle and Eavesdropping Attacks. In: Proceedings of the 14th International Conference on World Wide Web (WWW 2005), pp. 489–498 (2005) ISBN:1-59593-046-9

[25] Thomson, G.: BYOD: enabling the chaos. Network Security 2012(2), 5–8 (2012), doi:10.1016/S1353-4858(12)70013-2

[26] Putchala, S.K., Bhat, K., Anitha, R.: Information security challenges in social media interactions: Strategies to normalize practices across physical and virtual worlds. In: 2013 DSC Best Practices Meet (BPM), pp. 1–4 (July 12, 2013) ISBN: 978-1-4799-0637-6

[27] Jones, K.J., Bejtlich, R., Rose, C.W.: Real Digital Forensics: Computer Security and Incident Response. Addison-Wesley Professional (2005) ISBN: 0321240693

Data Quality Assessment on Taiwan's Open Data Sites

Cathy S. Lin and Hsin-Chang Yang

National University of Kaohsiung, Kaohsiung, Taiwan
{cathy,yanghc}@nuk.edu.tw

Abstract. Open data has emerged to be a hot topic recently and attracted lots of attention from both researchers and practitioners. Many governments around the world are publishing government open data for public well-being, information disclosure, service innovation, and so on. Enormous amount of data and applications have been released and built in recent years. In accordance to such trend, authorities of Taiwan government are also striving to establish platforms for users to access and employ open data. However, the status and usage of these platforms are still unclear. This study conducts an exploratory investigation on the open data platforms in Taiwan to reveal their feasibility. Besides, we also try to assess the data quality of these platforms by examining medical facilities datasets published in these platforms. We hope that this research can provide a step stone for building better open datasets as well as platforms in Taiwan.

Keywords: Open Data, Open Data Platform, Data Assessment, Data Quality.

1 Introduction

Research on open data has emerged to be a hot topic for governments around the world start to release their data to the public. The term 'open data' primarily refers to the data released by governments. In this aspect, we often adopt 'open data' as the short of 'open government data' rather than other types of open data such as scientific open data. Governments release the data to the public intending to enhance the data interchange facility among departments, promote the efficiency of government, and fulfill the need of the public. Huijboom and Van den Broek [1] indicated that there are three strategies of reusing open data in different countries which can then achieve the following goals:

1. **Democracy enhancement and political participation** Open data can help the public to exercise her civil rights. For example, British government stated that citizens should have the right to supervise the government for its operations and spending. The US government publicized several datasets to disclose the spending and distribution of Hurricane Sandy funds [1].

---

[1] http://www.recovery.gov

L.S.-L. Wang et al. (Eds.): MISNC 2014, CCIS 473, pp. 325–333, 2014.
© Springer-Verlag Berlin Heidelberg 2014

2. **Service provision and product innovation** Open data can help creating innovations. Information and communication industries can use open data to develop novel applications or services and create both commercial and social benefits.
3. **Law enforcement** Both British and US open data policies recognize that open data can help improving security and law enforcement. There already have applications using open data related to safety aspects nowadays which allow the public to monitor and participate in security issues.

The importance of open data has also been recognized nowadays. A Gartner report suggests open data can be more valuable to businesses than collecting big data [2]. The value of open data relies not only on the data itself, but also the potential benefit brought from using it. Therefore, many governments have devoted to release their data and encouraged people to create new applications on these data. However, the quality of released open data is still far from perfect. There are many errors existed in lots of open datasets. For example, British government released the locations of some 300,000 bus stops which contain about 18,000 (6%) erroneous items [3].

The deployment and development of open data started at 2009 in Taiwan by non-government organizations. The Taiwan Government started promoting open data during 2009 and 2011. In 2011, Taipei City's open data platform went online and became the first open data platform in Taiwan. The official open data platform of Taiwan central government was established later on 2013. Several local governments have also established their open data platforms later in various progresses and details.

The quality of open data should affect the correctness, credibility, and applicability on its applications. Poor quality data will definitely result in erroneous outcome. Therefore, data quality assessment for open datasets have been noticed recently in various aspects and methodologies. In this work, we will conduct surveys on several open data sites in Taiwan to explore the following issues:

- Taiwan's central and local governments have established several open data sites in these years. In this research, we will try to apply a standard on these sites to reveal their quality in various aspects.
- The datasets in Taiwan's open data sites are still far from perfect. One of the major deficiencies is possible difficulty in using the data. In this research, we will try to assign grades to medical facility datasets across various sites according to their quality on data usage.

The remaining text is divided into following sections. We first review some articles that are related to this research in Sec. 2. After the literature review, we will discuss the research method in Sec. 3. The result of the method applied to Taiwan's open data sites is presented in Sec. 4. Finally, we give conclusions and suggestions in the last section.

2   Related Work

2.1   Development of Open Data

Open data has become one of the hottest issues around the world in past few years. However, making data public is not new. The concept of e-government arisen from 1980s allows the public to get information of government. Meanwhile, several governments also legislated acts such as the Freedom of Government Information Law in Taiwan [4] to force or encourage the governments to open their data. Open Knowledge Foundation announced the Open Data Index in 2013 to measure the usability and openness of government open data all over the world [5]. The result ranking is depicted in Table 1. It is noted that Taiwan ranks 33 in this survey.

Table 1. The ranking of open data index surveyed by Open Knowledge Foundation

Rank	Country	Score
1	United Kingdom	940
2	United States	855
3	Denmark	835
4	Norway	755
5	Netherlands	740
6	Finland	700
7	Sweden	670
8	New Zealand	660
9	Australia	660
10	Canada	590
⋮	⋮	⋮
33	Taiwan	420

2.2   Data Quality

The problem of measuring the quality of open data is still open. Although there are various schemes proposed to conquer this problem, there is still no gold standard. We will mention two schemes here for evaluating open data sites and sets here.

Open Data Sites Evaluation. Opquast [2] collects and announces a number of checklists for evaluating Web quality, including open data sites. The checklist for open data sites includes 72 items and is called 72 best practices OpenData [3]. The full test consists of 72 best practices which are divided into 13 themes and 3 levels. Higher level practices stand for higher applicability while lower level ones stand for basic requirements. Table 2 summarizes these practices.

---

[2] http://opquast.com/en/

[3] A complete list is in http://checklists.opquast.com/en/opendata

Table 2. The summary of open data dataset quality test

Theme	no. of practices for		
	Level 1	Level 2	Level 3
Animation	3	2	4
API	0	2	7
Applications	0	0	3
Metadata	5	3	3
Format	3	3	2
Versions	0	1	1
Identification	2	2	0
License	2	2	1
Linkeddata	0	1	3
Naming	0	2	0
Transparency	3	3	1
Usability	2	1	3
Privacy	1	1	0
Total	21	23	28

The checklist provides a comprehensive examination on the usability and completeness of open data. Moreover, it allows the users to understand the concepts behind open data through these evaluations.

Open Datasets Evaluation. Tim Berners-Lee, one of the founders of WWW, devised the 5 star rating system to rate (linked) open data [6]. The system resembles to the Maslow's Hierarchy of Needs which the lowest level (1 star) reflects the most fundamental need while the highest level (5 stars) reflects the greatest achievement. This scheme can help the authorities to evaluate the quality of data and improve their usability. Table 3 depicts the criteria of the rating.

3 Research Method

3.1 Selection of Research Method

This work conducted a preliminary investigation on the open data sites in Taiwan to understand the current development status of them. Moreover, we also conducted a survey on the data quality of these sites using medical facility datasets as example.

3.2 Research Subject

Our objective is to provide a broad view on the current state of open data in Taiwan. Therefore, the subjects of this work include all established open data sites by Taiwan's authorities in June 2014. However, only few sites have been established by Taiwan's central and local governments on the time of survey (5 out of 21 local governments). Table 4 lists the open data sites surveyed in this work.

Table 3. The 5 star rating scheme

Rating	Requirement	Example format
1 star:★	Available on the web (whatever format) but with an open licence, to be Open Data	PDF, JPG
2 stars:★★	Available as machine-readable structured data (e.g. Excel instead of image scan of a table)	XLS
3 stars:★★★	As 2 stars plus non-proprietary format (e.g. CSV instead of excel)	CSV, XML
4 stars:★★★★	All the above, plus: Use open standards from W3C (RDF and SPARQL) to identify things, so that people can point at your stuff	SPARQL
5 stars:★★★★★	All the above, plus: Link your data to other peoples data to provide context	JSON-LD

Table 4. The surveyed open data sites in Taiwan

Level	Abbr.	Site Name	Site URL
Central government	TW	DATA.GOV.TW	http://data.gov.tw/
	TP	Taipei City Open Data Platform	http://data.taipei.gov.tw/
	NT	New Taipei City Open Data Platform	http://data.ntpc.gov.tw/NTPC/
Local government	TC	Taichung City Open Data Platform	http://data.taichung.gov.tw/
	KH	Kaohsiung City Open Data Platform	http://data.kaohsiung.gov.tw/Opendata/
	YL	Yilan County Open Data Platform	http://opendata.e-land.gov.tw/

3.3 Survey on Taiwan's Open Data Sites

To reveal the quality of open data sites in Taiwan, we adopted the Open Data site quality test announced by Opquast described in Sec. 2.2 to evaluate various aspects of quality issue in the subject sites. In this work, we only adopted the level 1 and 2 practices since all subject sites were established in past 3 years so they are not likely to pass any level 3 practices. However, the 'application' theme does not include any level 1 or 2 practices, as shown in Table 2. Therefore, we added all level 3 practices under this theme in the evaluation process and resulted in a total of 47 practices. Three denotations are used to annotate the practices for different levels of completeness, namely $\bigcirc$, $\triangle$, and $\times$ for fully, partially, and not meeting the practices, respectively.

3.4 Survey on Data Quality of Open Data Sets

To evaluate the quality of open datasets, we adopted the five-star criteria devised by Tim Berners-Lee [6] to measure the quality of datasets. Several datasets of medical facility from the sites listed in Table 4 were selected as examples. Only three sites provide such datasets with various scopes and details and are listed in Table 5.

Table 5. The medical facility datasets used in the survey

Platform	Title of dataset	URL
DATA.GOV.TW	Hospital General Information Dataset	http://data.gov.tw/node/6143
Taichung City Open Data Platform	List of Hospitals' Addresses and Phone Numbers	http://data.taichung.gov.tw/GipOpenWeb/wSite/ct?xItem=3033&ctNode=230&mp=1
Kaohsiung City Open Data Platform	Kaohsiung's Medical Facility Data	http://data.kaohsiung.gov.tw/Opendata/DetailList.aspx?CaseNo1=AG&CaseNo2=2&Lang=C

4   Research Result

4.1   Result of Survey on Taiwan's Open Data Sites

The survey result of Taiwan's open data sites listed in Table 4 using criteria addressed in Sec. 3.3 is shown in Table 6. We give score to each practice as follow. If the site meets the requirement of the practice, it will obtain a score of 2. However, if the site partially fulfils the requirement, it will get a score of 1. Otherwise, it will obtain a score of 0 for not meeting that requirement. In this regard, a 'perfect' site will have a maximum score of 94. We should normalized the scores to have maximum value of 100 to meet the common practice. It is clear that all sites did not perform well in most of the themes. However, Taipei City's open data platform performs better than other sites for a normalized score of 62.77. This may due to the fact that it is the first open data sited in Taiwan. The average score of each theme is shown in the last column and is obtained by averaging the total scores across all platforms and practices. The score will approach 2 if all platforms perform well in this theme.

After further inspection, we can observe that all platforms did not provide any API for application development. In other words, these sites still play a passive role as data providers, rather than a much active role as data coordinators. The application developers should find no regulation or standard to follow and thus spend more time and cost for developing applications. In the mean time, the 'Identification' themes which provide data-related information such as providers and identification numbers are also poorly satisfied in all sites. The other poorly performed themes are 'Versions', 'Usability', and 'Privacy'. Overall speaking, none of the subject sites is good enough for providing quality data and services to the public and need to be improved by both central and local governments.

4.2   Result on Data Quality Survey of Medical Facility Datasets

All three medicalfacility datasets listed in Table 5 did not annotate their datasets with quality indicators. Therefore, we adopted the well-known 5 star rating system proposed by Tim Berners-Lee [6] to evaluate these datasets. The result is shown in Table 7.

Table 6. The result of survey on Taiwan's open data sites

Practice #	Theme	Level	TW	TP	NT	TC	KH	YL	Average Score
1		1	×	×	×	×	×	×	
2		1	○	○	○	○	○	○	
3	Animation	1	○	×	×	×	○	×	1.200
4		2	○	○	×	○	○	○	
5		2	○	○	○	×	○	○	
6	API	2	×	×	×	×	△	×	0.167
7		2	×	×	×	×	△	×	
8		3	×	×	×	×	×	×	
9	Applications	3	○	○	×	△	○	○	0.889
10		3	○	○	×	△	○	×	
11		1	×	○	×	×	×	×	
12		1	○	○	○	○	○	○	
13		1	○	○	○	○	○	○	
14	Metadata	1	○	○	○	○	×	×	
15		1	×	○	○	○	△	○	1.438
16		2	○	○	○	○	○	○	
17		2	○	○	○	○	○	○	
18		2	○	×	×	×	×	×	
19		1	×	×	×	×	×	×	
20		1	○	○	○	○	○	○	
21	Format	1	○	○	○	○	○	○	
22		2	○	○	○	○	○	○	1.444
23		2	○	○	○	○	○	○	
24		2	○	○	×	×	×	×	
25	Versions	2	×	×	×	×	×	×	0.000
26		1	×	○	×	○	△	×	
27	Identification	1	×	○	×	△	×	×	
28		2	×	×	×	×	×	×	0.667
29		2	○	×	×	×	○	○	
30		1	○	○	×	×	○	×	
31		1	○	○	×	○	×	○	
32	License	2	○	○	○	○	○	○	1.167
33		2	×	○	×	×	×	×	
34	Linked data	2	○	○	○	○	○	○	2.000
35		2	○	○	○	○	○	○	
36	Naming	2	×	×	×	×	×	×	1.000
37		1	△	△	△	△	△	△	
38		1	△	△	△	△	△	△	
39	Transparency	1	○	○	×	○	×	○	
40		2	×	×	×	×	×	×	1.167
41		2	○	○	○	○	○	○	
42		2	○	×	○	○	○	○	
43		1	×	×	×	×	×	×	
44	Usability	1	○	×	○	×	×	○	0.333
45		2	×	×	×	×	×	×	
46	Privacy	1	△	△	△	△	△	△	0.5
47		2	×	×	×	×	×	×	
		○	27	28	17	17	18	23	
Total number checked		×	17	16	26	21	19	21	
		△	3	3	1	6	7	3	
Score			57	59	35	40	43	49	47.17
Score normalized to 0-100			60.64	62.77	37.23	42.55	45.74	52.13	50.18

Table 7. The result of data quality of medical facility datasets

Platform	Title of dataset	Format	Grade
DATA.GOV.TW	Hospital General Information Dataset	CSV	★★★
Taichung City Open Data Platform	List of Hospitals' Addresses and Phone Numbers	XLS	★★
Kaohsiung City Open Data Platform	Kaohsiung's Medical Facility Data	Web service	★

Taiwan Central Government's dataset was graded as 3 stars since it uses non-proprietary CSV format. On the other hand, Taichung City's dataset uses proprietary format XLS so it get 2 stars. Finally, Kaohsiung City's dataset can only be access as web pages without any proper format. Therefore, it can only obtain 1 star.

5   Conclusions and Suggestions

Issues on open data have attracted lots of attention in the past few years. Governments around the world are encouraged to disclose their data to the public. Enormous amount of open datasets as well as derived applications are available for benefit of people and society. It should be emphasized that the quality of open data, both in provider and data aspects, should be justified to ensure better usability and outcomes. However, there are still few efforts striving on these issues. This deficiency is much serious in Taiwan since open data is still on its infancy. In this work, we tried to evaluate the quality of open data sites as well as datasets in Taiwan. The result shows that open data sites in Taiwan are still far from perfect. All subject sites are given scores less then 65 out of 100. For the quality of datasets, we selected medical facility datasets as example and obtained various grades, from 1 to 3 stars in 5 stars rating system, for datasets in various platforms. Both results show that the open data platforms in Taiwan did not recognize the importance of conforming to contemporary standards and thus need to be improved.

The quality specification schemes such as the 5 star rating scheme are still not introduced to Taiwan's open datasets. Such scheme will not only help the users using the data, but also allow the platforms meeting the international standards. We suggest the data providers, i.e. Taiwan's central and local governments, to conform these standards in publishing the data since it will not cause too much extra effort.

The other suggestion is that these platforms should provide more interactive interfaces to allow users to access the data easily. Besides, there are many deficiencies in different themes in surveying these platforms. The data providers should consult the result of this work to improve the data usability, presentation, applicability, privacy, as well as other related issues.

References

1. Huijboom, N., Van den Broek, T.: Open data: an international comparison of strategies. European Journal of ePractice (12), 1–13 (2011)
2. Eddy, N.: Big data is important, but open data is more valuable: Gartner (August 2012), http://www.eweek.com/c/a/IT-Management/Big-Data-Important-but-Open-Data-More-Valuable-Gartner-217111/ (accessed July 10, 2014)
3. Gray, J.: How can open data lead to better data quality? (2013), http://blog.okfn.org/2013/09/03/how-can-open-data-lead-to-better-data-quality/ (accessed July 20, 2014)
4. Ministry of Justice: The freedom of government information law (2005), http://law.moj.gov.tw/Eng/LawClass/LawAll.aspx?PCode=I0020026 (accessed July 20, 2014)
5. Open Knowledge Foundation: Open data index (2013), https://index.okfn.org/ (accessed July 20, 2014)
6. Berners-Lee, T.: Linked data (2010), http://www.w3.org/DesignIssues/LinkedData.html (accessed July 10, 2014)

Improvement of Achievement Level
Using Student's Relational Network

Junko Shibata[1], Koji Okuhara[2], and Shogo Shiode[3]

[1] Faculty of Economics, Kobe Gakuin University,
518 Arise, Ikawadani, Nishi-ku, Kobe, Hyogo, 651-2180, Japan
shibata@eb.kobegakuin.ac.jp
[2] Science and Technology, Osaka University, 2-1 Yamadaoka, Suita, Osaka,
565-0871 Japan.
[3] Faculty of Business Administration, Kobe Gakuin University,
518 Arise, Ikawadani, Nishi-ku, Kobe, Hyogo, 651-2180, Japan

Abstract. The teacher is requested to do a high-quality lecture to the solution of the issue of decline in academic ability. It is difficult to understand for the teacher the student's achievement level. Therefore, there is a difference of understanding between the teacher and the student. Then, to understand student's achievement level, we execute the questionnaire. And, we discuss the method of making the best use of the result from the analysis for the lecture improvement. As a result, the teacher can control the degree of progress of the class. Thereby, it is thought that the student can deepen understanding of the learning more. Furthermore, it is thought that we can expect the improvement of the achievement level by sending a student having a high achievement level of the group if I can constitute the social network of the friendly relations of the student.

Keywords: Lecture Improvement, Questionnaires, Multivariate Analysis, Student's Relational Network.

1   Introduction

In Japan, the issue of decline in academic ability was caused from about 2000 by the reduction of the entrance examination subject of the university, more relaxed education and decreased motivation for learning. It is thought that the decrease in younger population and the rising university advancement rate hasten the deterioration of students' scholastic abilities.

In such situation, the improvement of the education method to the achievement level of a changing student is necessary for the teacher every year. The result of the entrance examination is effective to know the scholastic ability of the student. The result of the entrance examination is effective in order to know the scholastic ability of the student. But, it does not know whether the student who passed it maintains the level until entrance to school because time before entering is long. After it begins to lecture, a lot of teachers understand student's achievement level by the attitude of

L.S.-L. Wang et al. (Eds.): MISNC 2014, CCIS 473, pp. 334–344, 2014.
© Springer-Verlag Berlin Heidelberg 2014

attending a lecture and the question. And they adjust the progress of the class. However, it is difficult for the teacher to grasp the student's achievement level because of a decrease of the student who voices his opinion.

To take the student's opinion directly to the lecture, various universities execute the questionnaire for the teaching evaluation. However, after lectures of 15 all end, most questionnaires are executed. Therefore, the student who is taking a course in now cannot feedback the opinion. Moreover, the analysis of most questionnaires is insufficient. It is difficult for the teacher to use the result in the lecture improvement.

In the future, the teacher will be requested to understand the **level** of the changing student, and to reflect the opinions of students in his class. Therefore, we think that the execution method and the content of the questionnaire for the current teaching evaluation are insufficient. In this paper, we aim to execute the questionnaire by a usual lecture, and to make the best use of the result for the lecture improvement.

We study with the lecture on economic mathematics for which one of the authors is responsible in Kobe Gakuin University. We analyze the result of the questionnaire in principal component analysis, and examine scattering of student's achievement level. After making the result of the report visible, we discuss the relation between the student's internal evaluation and external evaluation.

2   Multivariate Analysis of Questionnaire

2.1   Dataset of Questionnaire

The content of the lecture on economic mathematics has divided into many units "Number and expression" and "Various functions" according to the text. In each unit, after the teacher explains the concept, he explains the exercise. After that, the student solves the practice question. To examine the achievement level of two units of the fraction function and the irrational function, we executed the questionnaire to 56 students of attending a lecture. The question items of the questionnaire used in this study are shown in Table 1.

Table 1. Questions of questionnaire

Item	Question number	Content
1	Q1.3	Explanation of fraction functions
2	Q1.4	Practice question 2.1(1)
3	Q1.5	Practice question 2.1(2)
4	Q1.6	Practice question 2.1(3)
5	Q2.1	Explanation of graph of fraction function
6	Q2.2	Similar kind of problem (1)
7	Q2.3	Similar kind of problem (2)
8	Q2.4	Explanation of irrational function
9	Q2.5	Practice question 2.2(1)
10	Q2.6	Practice question 2.2(2)
11	Q2.7	Practice question 2.2(3)

The student evaluates the achievement level in each item by five stages (Table 2).

Table 2. Five stages of answer

Stage	Content
1	I don't understand.
2	I understand little
3	It's OK.
4	I understand a little
5	I completely understand

In addition, we use the result of the report in these units.

In this study, the dataset that we used is shown in Table 3. R_1 shows the number of the correct answers of four fractional function problems, R_2 shows the number of the correct answers of two irrational function problems.

Table 3. Example of the dataset

Student No.	Q1.3	Q1.4	Q1.5	Q1.6	Q2.1	Q2.2	Q2.3	Q2.4	Q2.5	Q2.6	Q2.7	R_1	R_2
1	2	4	3	2	3	4	5	4	3	3	3	4	2
2	5	5	5	5	5	5	5	5	5	5	5	3	1
3	4	4	4	3	4	4	3	4	4	4	1	4	2
4	2	3	1	1	2	2	2	2	1	3	2	2	1
5	2	5	1	1	2	2	1	1	1	1	1	4	2
...	...	...	...	...	...	...	...	...	...	...	...	...	...
56	4	5	5	5	5	5	5	5	5	5	5	4	2

2.2 Principal Component Analysis

The principal component analysis is a technique for summarizing information about many kinds of characteristic values with the correlation mutually, to a small number of non-correlated integrated characteristic values. The input data of the principal component analysis are comprised of the p items shown in Table 4. In this study, we assume that the number of questionable items is 11 except the report items, that is $p = 11$ and $n = 56$.

Table 4. Input data

		Item			
		x_1	x_2	...	x_p
Student	1	x_{11}	x_{12}	...	x_{1p}
	2	x_{21}	x_{22}	...	x_{2p}
	$\vdots$	$\vdots$	$\vdots$	$\ddots$	$\vdots$
	n	x_{n1}	x_{n2}	...	x_{np}

When the data of Table 4 is obtained, the covariance matrix of the matrix $x = [x_{ij}]$ and the eigenvalues are obtained. The eigenvector a_k^T arranged in descending order is a non-correlated element.

$$a_k^T = \begin{bmatrix} a_{k1} & a_{k2} & \cdots & a_{kp} \end{bmatrix} \quad 1 \le k \le p. \tag{1}$$

Therefore, the kth principal component of z_k is given by

$$z_k = a_{k1}x_1 + a_{k2}x_2 + \cdots + a_{kp}x_p. \tag{2}$$

The principal components can be requested up to the number of original characteristic values. However, because the variance becomes small, the last principal component can be rounded down. As a result, it is possible to show by characteristic values of a number that is less than an original characteristic value.

2.3   Cluster Analysis

The cluster analysis is a technique for dividing the data group into some clusters based on the degree of similarity. Under the given condition, the distance between each data in the cluster is minimized and the distance between each cluster is maximized. Here, the distance between clusters adopts the Ward's method used most. As a result, the variance in each cluster is minimized.

When data are observed for $x_1, x_2, \cdots, x_p$, the data n_i included in cluster C_i is shown as

$$x_r^{(i)} = \left(x_{r1}^{(i)}, x_{r2}^{(i)}, \cdots, x_{rp}^{(i)} \right), \quad r = 1, 2, \cdots, n_i. \tag{3}$$

The mean vector in C_i is

$$\bar{x}_l^{(i)} = \frac{1}{n_i} \sum_{r=1}^{n_i} x_{rl}^{(i)}, \quad l = 1, 2, \cdots, p. \tag{4}$$

The sum of squared deviation in around average in C_i is

$$E_i = \sum_{r=1}^{n_i} \sum_{l=1}^{p} \left(x_{rl}^{(i)} - \bar{x}_l^{(i)} \right)^2. \tag{5}$$

The increment of the sum of squared deviation when it creates a fusion of cluster C_i and cluster C_i , becomes

$$\Delta E_{ij} = \frac{n_i n_j}{n_i + n_j} \sum_{l=1}^{p} \left(x_{rl}^{(i)} - \bar{x}_l^{(i)} \right)^2. \tag{6}$$

This is regarded as the non-degree of similarity of cluster C_i and cluster C_j. The increment when $C_i \cup C_j$ fuses with other clusters C_k becomes

$$\Delta E_{(ij)k} = \frac{(n_i+n_j)n_k}{n_i+n_j+n_k} \sum_{l=1}^{p} \left(\bar{x}_l^{(i \cup j)} - \bar{x}_l^{(k)} \right)^2. \tag{7}$$

In Ward's method, the cluster that minimizes this is fused.

When C_i fuses with C_j, the non-degree of similarity of other clusters C_k and clusters $C_i \cup C_j$ is calculated by

$$d\left(C_i \cup C_j, C_k\right) = \frac{n_i + n_k}{n_i + n_j + n_k} d(C_i, C_k) + \frac{n_j + n_k}{n_i + n_j + n_k} d\left(C_j, C_k\right) \tag{8}$$
$$+ \frac{-n_k}{n_i + n_j + n_k} d\left(C_i, C_j\right)$$

where, in the initial state, each cluster consists of one individual respectively $C_i \equiv \{i\}, C_j \equiv \{j\}$, so $d\left(C_i, C_j\right) = d_{ij}$.

3   Numerical Example

At first, we show the fraction of the student's achievement level for the explanation's questions. Next, we discuss the results of the cluster analysis and the principal component analysis.

Item 1 and item 5 and item 8 are questions to examine the understanding of the content that the teacher explained. Figure 1 shows these results.

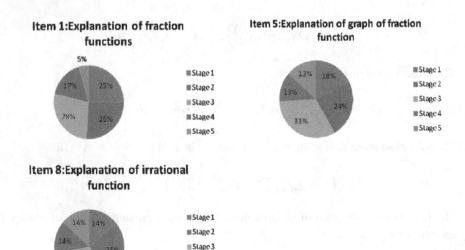

Fig. 1. The student's achievement level for the teacher's explanation

From Figure 1, the student who answered as stage 4 and stage 5 is below the half in item 5 and item 8. In the item 1, since there was a student who cannot understand fraction functions in more than half, the teacher gave again. As a result, it is thought that the students who have understood the item 5 increased in number. In this questionnaire, the item about a teacher's lecture technique is these three. From now on, a teacher will need a device which the student who has understood explanation increases.

Next, the result of cluster analysis is shown by Figure 2. Since the degree of similarity is calculated by the Euclid distance, the student with the same as degree of achievement level is considered to belong to the same cluster

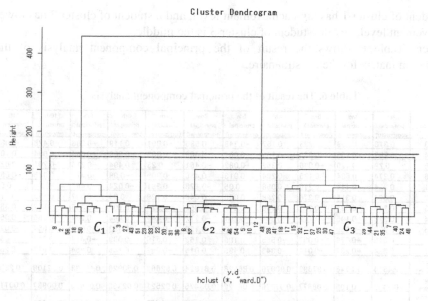

Fig. 2. The cluster dendrogram

In Figure 2, the student is divided into three groups by a red line. We describe the left cluster as C_1, the middle cluster as cluster C_2 and the right cluster as C_3. Then, cluster C_1 is composed of 14 students, cluster C_2 is composed of 21 students and cluster C_3 is composed of 21 students. And, the sample of student data belonging to each cluster is shown in Table 5.

Table 5. Degree of similarity

Cluster	Student No.	Q1.3	Q1.4	Q1.5	Q1.6	Q2.1	Q2.2	Q2.3	Q2.4	Q2.5	Q2.6	Q2.7
C_1	9	3	5	4	4	5	5	5	4	4	5	5
	2	5	5	5	5	5	5	5	5	5	5	5
	56	4	5	5	5	5	5	5	5	5	5	5
C_2	23	1	1	1	1	3	3	1	1	1	1	1
	33	1	2	2	1	2	2	2	1	1	1	1
	22	2	1	1	1	1	1	1	3	1	1	1
C_3	16	3	4	2	1	5	5	4	2	2	2	2
	17	2	1	1	1	5	5	5	1	1	1	1
	15	4	5	1	1	5	4	1	5	1	1	1

A student of cluster 1 has high achievement level, and a student of cluster 2 has lower achievement level, and the student of cluster 3 is the middle.

Then, Table 6 shows the result of the principal component analysis of the correlation matrix for the questionnaire.

Table 6. The result of the principal component analysis

Question number	1st principal component	2nd principal component	3rd principal component	4th principal component	5th principal component	6th principal component	7th principal component	8th principal component	9th principal component	10th principal component	11th principal component
Q1.3	0.313	0.259	−0.223	0.167	−0.348	0.585	−0.091	0.119	−0.286	−0.428	−0.089
Q1.4	0.269	0.158	−0.625	0.165	−0.067	−0.369	0.009	−0.566	−0.064	0.094	0.105
Q1.5	0.29	−0.429	−0.066	0.352	−0.082	−0.461	0.23	0.494	−0.215	−0.201	−0.029
Q1.6	0.274	−0.534	−0.135	0.284	0.013	0.456	−0.005	−0.08	0.417	0.385	−0.058
Q2.1	0.324	0.415	0.216	0.098	0.05	−0.123	0.391	−0.024	0.467	−0.09	−0.52
Q2.2	0.324	0.315	0.376	0.288	0.066	0.046	0.129	0.04	−0.007	0.193	0.714
Q2.3	0.325	0.027	0.267	0.165	0.442	−0.052	−0.524	−0.103	−0.396	0.17	−0.354
Q2.4	0.316	0.195	−0.22	−0.394	−0.201	−0.172	−0.447	0.484	0.284	0.265	0.059
Q2.5	0.225	−0.259	0.476	−0.197	−0.651	−0.156	−0.131	−0.383	−0.027	−0.057	−0.028
Q2.6	0.315	−0.076	−0.057	−0.555	0.109	0.163	0.519	0.009	−0.417	0.314	−0.059
Q2.7	0.324	−0.237	−0.011	−0.343	0.435	0.015	−0.092	−0.146	0.254	−0.611	0.253
Standard deviation	2.4514	1.0746	0.97699	0.90076	0.86626	0.61824	0.52663	0.50926	0.46823	0.32108	0.2802
Proportion of Variance	0.5463	0.105	0.08677	0.07376	0.06822	0.03475	0.02521	0.02358	0.01993	0.00937	0.00714
Cumulative Proportion	0.5463	0.6513	0.73804	0.8118	0.88002	0.91477	0.93998	0.96356	0.98349	0.99286	1

From Table 6, the 1st and 2nd principal components have collected about 65% of the information due to the fact that the value of each proportion of variance and the eigenvalue of the other principal components are less than one. Since all the values of the coefficient of the 1st principal component are positive, it is thought that the composite understanding in two units is shown. The 2nd principal component is considered that the understanding of only basic ability is shown.

Next, the graph of the category score in the 1st principal component and the 2nd principal component is shown in Figure 3.

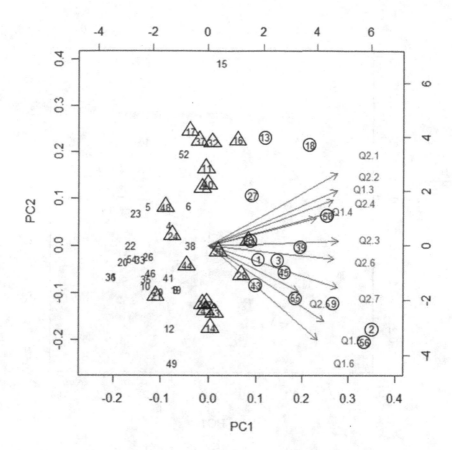

Fig. 3. The graph of the category score in the principal component analysis

Where, the student with ○ belongs to cluster 1 and the student with △ belongs to cluster 3, other students belong to cluster 2. This figure shows that the student of the cluster 1 is a higher understanding and the student of cluster 2 is a lower understanding.

Finally, we show the Figure 4 which superimposed the result of the graded report in Fig. 1. Here, the report consists of 6 questions (4: fractional functions, 2: irrational function). The gray number shows the entirely satisfactory student.

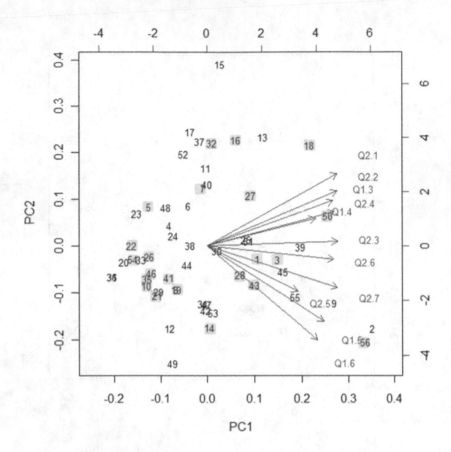

Fig. 4. The graph of the category score and the result of the graded report

From Figure 4, the students whose reports were perfect are scattered all around the place. One of the reasons is the some hints which the teacher gave. Since there were some students who think together with a friend since it is not an examination, the student of the low understanding also became perfect. To study together with a friend is required at the point of deepening a student's understanding. Here, we pay attention to the student 2 and the student 9. Although a total understanding is high, those students have not taken perfect in this report. Here, the student 2 has 4 points and the student 9 has 5 points. For the student 9, the error in calculation was the cause. However, the student 2 has the low understanding obtained from the result of the report compared with the understanding which he evaluated. That is, although the student can understand, he cannot understand so much in fact.

The result of the numerical examples is clarified the following three points.

1. We can show the degree of the achievement level for the explanation of the teacher. When there is little students' understanding, the teacher can add more detailed explanation, and a student can understand more about it.

2. We can clarify the achievement level of the student and the dispersion of it. The teacher can teach at the appropriate progress of the class.

3. Because the student made out a report with its friends, the reports of some students who have lower achievement level were good. This result suggests the possibility of the following. If I can constitute the social network (Figure 5) from the friendly relations of the student, we can arrange the student which has a high and low achievement level. It is thought the improvement of the achievement level of the whole student.

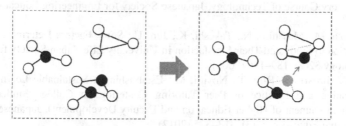

Fig. 5. The image of The Social Network

From this result, it is thought that effective group learning into consideration of human relations and the achievement level is enabled.

4   Conclusion

In this paper, to understand the student's achievement level in the lecture on economic mathematics, we execute the questionnaire. The result of cluster analysis showed that the student of the target lecture was divided into three groups. Moreover, the result of principal component analysis showed the dispersion state for a student's understanding. Furthermore, it was observed that a student's understanding and an actual understanding have a difference from the figure in consideration of a report result. It is thought that the teacher can control speed of appropriate learning by knowing the student's achievement level. Since the teacher is required to make a high quality lecture in the future, we are going to propose the improved method of the effective lecture.

Acknowledgment. This work was supported by JSPS KAKENHI Grant Number 25330427.

References

1. Masako, K., Kunio, Y.: Characterization of Questionnaires of College Student Evaluation on Lectures with respect to Faculty Development. Japan Society of Education Information 24(2), 57–67 (2008)

2. Masahiko, N.: Statistical Analysis of a Feeling of Attainment for Students Taking a Class in Fundamental Physics and of its Correlation and Causal Relation with Student Evaluation of Instruction and Exam Scores. Japanese Society for Engineering Education 59(2), 3–10 (2011)
3. Rumiko, A., Shinya, N., Yoshinori, N.: Extraction of Priority of Questionnaire by using Principal Component Analysis and Application to Improvement of Lecture, IEICE Technical Report ET2010-108 (2011)
4. Masahiko, N.: Statistical Analysis of Evaluation Questionnaire on the Teaching of Fundamental Physics (Wave Mechanics, Thermodynamics, and Electricity) in an Introductory Course of Technology. Japanese Society for Engineering Education, 234–235 (2005)
5. Yoshitaka, M., Masahiro, N., Takeshi, K., Jun, U.: Some Factors Determining Students' Motivation to Study and Their Satisfaction in University Life. Japan Society for Education Technology 32(2), 189–196 (2008)
6. Mio, T., Hiroko, O., Risa, T., Noyuri, M.: Establishing a Sustainable Learning-support-organization and Practice of Peer Tutoring Centering on College Students(<Special Issue>Improvement of Higher Education and Faculty Development). Japanese Society for Engineering Education 36(3), 313–325 (2012)
7. Koshi, Y., Kenichi, K.: A Study on the Research Trends for Japan Journal of Educational Technology. Japanese Society for Engineering Education 34(suppl.), 1–4 (2010)
8. Hiromitsu, N., Yoshihiro, S.: Lecture Speech Analysis for Development of Support System for Improving Lecturers' Diction(<Special Issue>Development of Learning and Educational Technologies). Japan Society for Educational Technology 34(3), 171–179 (2010)
9. Light, G., Calkins, S.: The Experience of Faculty Development: Patterns of Variation in Conceptions of Teaching. International Journal of Academic Development 1(13), 27–40 (2008)

Challenging Robustness of Online Survey via Smartphones: A View from Utilizing Big Five Personal Traits Test

Shiro Uesugi

Graduate School of Business, Matsuyama University, 4-2 Bunkyo Matsuyama City,
Ehime, Japan
shiro.uesugi@nifty.com

Abstract. As smartphones are widely available, automatically generated data through smartphones gain more importance. They attract
large attention as the major source of big data. However, traditional
survey data do not lose the value because it is essential for obtaining
responses that reflect one's thinking process and responding behaviors.
Thus, online survey, which uses smartphones as input device, have become major method. Online survey is expected to have the benefits related to both size and thinking process. However, because of its nature of
easy-to-use, the data quality may not be as good as paper-based survey.
This paper looks into a case of the Big Five Personality Tests, which
were run on Japanese university students, comparing paper-based and
online-based/smartphone-based methods. Statistical analysis shows that
there are differences between them. This paper suggests that the Big
Five Personal Traits Test can be a yardstick to check the robustness of
smartphone-based online survey.

Keywords: Smartphone, Online survey, Big Five Personal Traits Test,
Robustness of survey.

1 Introduction

The diffusion of smartphones has become worldwide phenomena today. Consequently, the availability of the data which are generated through the smartphones
has boosted to a huge level of amount to make such a data called big data. While
big data attract large attention in both business and academic arena, data from
traditional survey do not lose its value. Big data which originate from smartphones are in many ways generated automatically, in other words, generated
without involvements of human's input.

In a survey, it is essential for obtaining responses that are provided as a
reflection of one's thought. Unlike many of big data, a survey is conducted with
respondent's consent. Therefore, the responses from surveys are considered as
one which comes from respondent's inner mind and trustworthy. Traditionally,
a survey is conducted in a manner of oral face-to-face interviews or paper-based
written forms. Recently, the wide availability of the internet brought web-based

L.S.-L. Wang et al. (Eds.): MISNC 2014, CCIS 473, pp. 345–354, 2014.

or online survey methodology. Then, as smartphones are increasingly used, online surveys which use smartphones as input device have become major method.

However, because of its nature of easy-to-use, the data quality may not be as good as paper based survey. The process of touching screen involves different physical efforts from the process of drawing by hands. Hence, the robustness of the responses in a survey may be compromised.

This paper emerged from the observations about the increase of the proportions of invalid responses among surveys which were performed yearly and changed methodology from paper-based survey to online-based to smartphone-based survey. This paper looks into a case of the Big Five Personality Traits Tests, which are conducted using either paper-base or online-base and smartphone-based methods. Statistical analysis shows that there are differences between paper-based and online-based or smartphone-based. This paper suggests that the Big Five Personal Traits Test can be a yardstick to check the robustness of smartphone-based online survey.

2   Related Works

Because this paper investigates the robustness of online-based survey and/or smartphone-based survey, the first category of the related works falls into the one which deals with survey methodology. On the very first step, literature review introduces a study about online-survey. It is followed by the study about the use of mobile devices.

Secondly, a brief explanation of the Big Five Personal Traits Test and its previous applications are introduced. Because the Big Five Personal Traits Test is so designed to ensure the robustness of the test results itself, there are numbers of responses which must have discarded because they did not satisfy the robustness testing process. However, in this paper, this is seen from the view point of whether it is appropriate to use such a threshold as for the mechanism to ensure the robustness of smartphone-based survey.

2.1   Literature Review

Evans and Mathur (2005) [1] conducted thorough survey about the value of online surveys. While they pick up 16 positive attributes of online surveys, they also pick up 9 "Major Potential Weakness." Among them, "Respondent lack of online experience/expertise" and "Low response rate" are worth revisited. They touch upon Fricker and Schonlau (2002) [2] in order to explain about "Low response rate" in online survey. It is said that the online survey cannot attain better results than other survey methods. However, the reason is not clear.

According to our survey, which will be described in the later part of this paper, response rate by using smartphones is about 50 per cent and is rather high even considering the classroom settings. Therefore, the use of smartphones itself may not be blamed of.

As regard to "Respondent lack of online experience/expertise," Evans and Mathur (2005) [1] state that there are difficulties "due to the lack of familiarity of possible respondents with internet protocols." Today, considering the fact that the use of smartphones is very popular, "lack of expertise" are not present.

Other literatures such as Bosnjak et al (2010)[3], de Bruijne and Wijnant (2013)[4] and Vicente et al (2009)[5] deal with issues related to mobile devices and web-based surveys. They are aware that the developments of mobile technology will affect the methodologies of survey. These literatures focus on the utilizations of mobile technologies and web-based surveys. They do not provide a yardstick to check the robustness of the responses. However, in this paper, the comparison between paper-based and online/smartphone-based survey is presented so as a candidate of a yardstick.

3 Big Five Personal Traits Test

The "Five Factor Model" or the "Big-Five" factors of personality, which is developed by Goldberg (1990)[6], is a widely used measurement in psychology to deal with human personality traits. In "Big-Five" personal traits, basic dimensions of personality are divided into five factors: Extroversion(E), Agreeableness(A), Conscientiousness(C), Emotional Stability/ Neuroticism(N), and Intelligence/ Openness to experience(O). There are variations of tests to obtain five factors scores. Usually, numbers of questions, which are called "Big-Five Inventory" (BFI) are given in the form of written statements to respondents. These may be in paper-based and/or online-based. The scores are calculated from the answers to see how one's overall response is deviated from the averages.

In Japan, the methodology to obtain these five dimensions was adopted and elaborated by Murakami and Murakami (1997)[8] and Murakami (2004)[7]. In Murakami and Murakami (1997)[8], the "Big-Five" traits were translated into Japanese terminology, then Murakami and Murakami (1999)[9] also constructed a Japanese version of the "Big-Five Inventory" (BFI) which consists of 70 questions. Additionally, Murakami and Murakami (1999)[9] provides a conversion table with standard deviations and means to calculate Z-score, taking into account the differences of age and gender. Murakami and Murakami (2008)[10] developed a handbook that illustrates entire process about "Big-Five Personal Traits Test" from the methodology of their development to the understanding of the test results.

Robustness of the BFI was utilized in Uesugi et al (2010)[11]. It examined the relevance between personal traits and privacy concerns. In Dell et al (2010)[12], the extended model of this, which included international comparison, was presented. Both of the studies indicated the Big Five personal traits have some explaining ability about the people's attitude in accepting specific kind of technologies, which are defined as "Potentially Privacy-Invading Technology, or PPIT" by them.

Following the design of Murakami and Murakami (2008)[10], Uesugi et al (2010)[11] used paper-based questionnaire sheets whereas Dell et al (2010)[12] used a combination of paper-based and web/computer-based survey methodology. Those researches brought meaningful results because samples were cleaned by the process of removing invalid responses as instructed in the handbook.

4   Framework and Illustration of Methods and Evaluation Results

4.1   Background

The threshold of validity of a response is set by Murakami and Murakami (2008)[10] as follows. The test asks a respondent to choose from three answers, namely "Applicable," "Not Applicable" and "Do not Know" for each of 70 questions. When the numbers of answers of "Do not Know" exceed 4 items (that is more than 5), appropriateness of this response is regarded doubtful and is discarded.

During the preparation for Uesugi and Okada (2014)[13], one of the authors noticed that the numbers of invalid answers are significant. It seemed that more valid responses are collected via surveys which used paper-based methods, more precisely to say, mark sheets. It evoked the memory about the other experiences of the authors trying to collect 2,000 samples via commercial web-based survey in 2010. It also used Big-Five Tests, and resulted to get only 40 per cent valid responses.

Before, they were not keen to looking at invalid responses because they were more interested in the analysis of valid responses. Looking back the past data, the proportions of invalid responses seem to be affected by the survey methods. Observed numbers are shown in Table 1.

Table 1. Numbers of Invalid Responses

Survey Method	Year	Total	Valid	Invalid	(Invalid Proportion)
Paper	2010	191	172	19	(10%)
	2011	221	186	35	(19%)
	2014	169	113	56	(33%)
Online	2010	183	69	114	(62%)
	2013	147	77	70	(48%)
	2014	2,000	774	1,226	(61%)

4.2   Framework and Illustrations of Methods

Reflecting the wide diffusion of smartphones, this paper's research only focuses on online surveys which use smartphones. The term **"smartphone-based"** is used hereinafter. The term **"paper-based"** means survey methods which use questionnaire written on paper and use mark sheets to respond.

A simple framework was developed in order to run a comparison of difference between "paper-based" and "smartphone-based" surveys. Samples were collected from students of Matsuyama University. Same questions were asked both in paper-based and smartphone-based forms. Comparisons of means between those responses are conducted in order to see if there are differences.

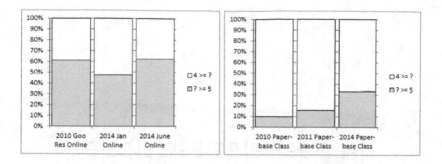

Fig. 1. Differences of Proportion: Online-based and Paper-based

About 340 students altogether in two classes on Matsuyama University were given "Big-Five" tests in two different occasions. The age group was between 18 to 25. Economics major and Business major students participated. They were told that, by participating this, they would be given 5 class points each time.

In early May 2014, the first tests, which were **"smartphone-based"** were conducted. As instructed in the Murakami and Murakami (2008)[10], students were given instructions such as "You need to respond spontaneously," and "You should select 'Yes' or 'No' and avoid selecting 'Do not Know' as possible as you can."

In late June 2014, students were given **"paper-based"** "Big-Five" test. One week after, a questionnaire about "Experience of Online Shopping" was conducted asking whether they use smartphones. This questionnaire is deployed to select responses only submitted from smartphones omitting the responses from PC. After conducted matching among these datasets, 169 valid samples were obtained.

Table 2. Descriptive Statistics of ONLY Valid Responses to Big Five Tests

Item	Total	Means	Var (n-1)	SD	Lower Limit	Upper Limit	NA	n
PBExtraversion	5471	48.416	131	11.439	67	32	0	113
SBExtraversion	5314	47.027	119	10.913	67	32	0	113
PBAgreeable	5527	48.912	74	8.611	60	20	0	113
SBAgreeable	5302	46.920	69	8.307	60	20	0	113
PBConscientious	5825	51.549	95	9.733	73	31	0	113
SBConscientious	5602	49.575	99	9.931	73	31	0	113
PBNeurotic	5503	48.699	101	10.052	68	32	0	113
SBNeurotic	5273	46.664	83	9.117	63	32	0	113
PBOpeness	5510	48.761	103	10.161	71	32	0	113
SBOpeness	5256	46.513	95	9.751	75	32	0	113
PBAttitude	5329	47.159	121	11.000	77	31	0	113
SBAttitude	5137	45.460	112	10.583	77	31	0	113
PB?	6551	57.973	150	12.242	79	49	0	113
SB?	8097	71.655	170	13.053	80	49	0	113
PBFrequency	5406	47.841	71	8.426	73	38	0	113
SBFrequency	5284	46.761	61	7.835	73	38	0	113

Table 3. Difference in Means between Corresponding Items

Item	n	Means	Differences	SD	Statistical Value	DF	p-Value	sig.	Lower Limit	Upper Limit	Value
PBExtraversion	113	48.416	1.389380531	4.5009	3.281	112	0.001	[**]	0.5505	2.2283	0.8389
SBExtraversion		47.027							0.2799	2.4989	1.1095
PBAgreeable	113	48.912	1.991150442	5.8102	3.643	112	0.000	[**]	0.9082	3.0741	1.083
SBAgreeable		46.92							0.5589	3.4234	1.4323
PBConscientious	113	51.549	1.973451327	6.2713	3.245	112	0.001	[**]	0.8045	3.1424	1.1689
SBConscientious		49.575							0.4275	3.5194	1.5459
PBNeurotic	113	48.699	2.03539823	6.1017	3.546	112	0.001	[**]	0.8981	3.1727	1.1373
SBNeurotic		46.664							0.5313	3.5395	1.5041
PBOpeness	113	48.761	2.247787611	6.5826	3.630	112	0.000	[**]	1.0208	3.4747	1.2269
SBOpeness		46.513							0.6251	3.8705	1.6227
PBAttitude	113	47.159	1.699115044	6.3399	2.849	112	0.005	[**]	0.5174	2.8808	1.1817
SBAttitude		45.46							0.1363	3.262	1.5629
PB?	113	57.973	−13.6814159	15.843	9.10	112	0.001	[**]	−16.63	−10.73	2.953
SB?		71.655							−17.59	−9.776	3.9055
PBFrequency	113	47.841	1.07964018	6.4988	1.766	112	0.080	[ ]	−0.132	2.291	1.2113
SBFrequency		46.761							−0.522	2.6817	1.602

4.3   Arguments

There are three arguments which derive from the statistical evaluations of the responses of the surveys as shown in the followings.

Firstly, as stated in the background section, it is apparent from Table 1 that there are difference between "paper-based" and "smartphone-based" responses when focused on the numbers of invalid responses. To enhance the data and its interpretation, **ONLY** those cases which passes the validation process was extracted. In Table2, a descriptive statistics are shown. Then, the **Difference in Means between Corresponding Items Tests** were conducted using MS-Excel and ad-on software *(Excel Antkeito Taiko)*. Table 3 shows the results. According to this result, it can be safely said that **there are statistically significant differences between "paper-based" responses and "smart phone-based" responses for each of the items of "Big Five" traits.**

Secondly, as is seen in Table 3, there is an item which indicates that there is not a significant difference in the means between paper-based and smartphone-based survey. The item is "Frequency." This item is one of three items that fall in the category of not to test the personal traits but to test the attitude of the respondents. These items are, namely, "Attitude," "?" and "Frequency." Firstly, "Attitude" is calculated to find whether the response is *"Tatemae"* or not. *"Tatemae"* is an attitude to pretend to be ideal person. According to Big Five Handbook[10], the threshold score is 65 and more to regard the response falling into this category. Secondly, "?" is calculated from the numbers of invalid answers in each response. When the numbers of invalid answers exceed 5, a score of 80 is automatically given, and the response is discarded. Finally, the item "Frequency" is calculated to look for degrees of "Inconsistent" responses. Interpretation of this item in Big Five Handbook [10] indicates desirable scores should be between 45 and 55. The Means of "Frequency" item are 47.841 and 46.761, respectively, in Table 3 for both paper-based and smartphone-based, and both of them fall in this interval. Therefore, even though there is no statistical significance about "Frequency" item, the argument which is mentioned above, firstly, can be regarded to hold safely.

Thirdly, looking at the difference between paper-based and smartphone-based responses in Table 3, we find all of them are positive, except for item "?". This can be interpreted as follows. Because paper-based response involves more physical reactions than smartphone-based response, the time required to answer each of questions becomes longer. Because it takes more time to respond, the answers are provided after a certain level of considerations. According to Murakami and Murakami (2008) [10], the Means of each factors of EACNO should be 50 and the distribution among each factors should follow normal distributions. From this view point, each scores of EACNO from "paper-based" is closer to 50 than that of "smartphone-based" survey. In all, "paper-based" survey scores better and, consequently, it accomplishes a better robustness than "smartphone-based" survey.

5   Conclusions and Future Work

This paper looked through a simple experimental comparison between paper-based survey method and smartphone-based survey method. It focused on the validity testing mechanism in the "Big Five Personal Traits Test" and, by using the mechanism, it tested the robustness of smartphone-based survey method. When compared to paper-based method, smartphone-based method indicated lesser scores in all of personal traits factors. The results are consistent either invalid responses are removed, or included. There is "Frequency" item which indicated that there is no statistical difference in the means between paper-based and smartphone-based methods.

Through a simple analysis, this research suggests that the robustness of the answers provided through smartphones maybe inferior to the one provided through paper-based methods, in general. As a matter of course, we should recognize that this is only a result from one case study and the statistical analysis is very simple. The limit of this study exists in several ways. The experimental setting is very limited. Statistical analysis is not in depth.

However, it may be fair to argue that "Big Five Personal Traits Test" can be utilized to detect respondents' intentional and/or unintentional mal-behaviors. Considering the "Frequency" item showed no significant differences in means, the questionnaires that are used to calculate "Frequency" score may be used to check whether a response is provided sincerely. There are 8 questionnaires for this item. Those questionnaires maybe added to original questionnaires in order to check the validity of the entire response. Because the threshold to discard a response is 5 invalid answers out of total 70, which is 7 per cent, 1 invalid answer to 8 questions may be used as a yardstick to maintain robustness of a smartphone-based survey method.

The diffusion and the use of smartphones are rapidly increasing. The applications that utilize the spontaneous responsiveness of smartphones have a large potential. To this day, there are numbers of applications that collect information through smartphones to generate big data. However, once human responses are involved, the robustness of the response maybe compromised. The "Big Five Personal Traits Test" can be utilized as a yardstick to detect the robustness of such a survey. More detailed development of the methodology is a part of the future work.

Acknowledgements.   This research is partly supported by the JSPS KAKEN24330127, 2012-2013, MEXT-Supported Program for the Strategic Research Foundation at Private Universities S1291006, 2012-2014, and Matsuyama University Research Grant. The author would like to express their thanks to students at Matsuyama University for participation of this research.

References

1. Evans, J.R., Mathur, A.: The value of online surveys. Internet Research 15(2), 195–219 (2005)
2. Fricker Jr., R.D., Schonlau, M.: Advantages and disadvantages of internet research surveys: evidence from the literature. Field Methods 14(4), 347–367 (2002)
3. Bosnjak, M., Metzger, G., Gräf, L.: Understanding the willingness to participate in mobile surveys: exploring the role of utilitarian, affective, hedonic, social, self-expressive, and trust-related factors. Social Science Computer Review 28(3), 350–370 (2010)
4. de Bruijne, M., Wijnant, A.: Comparing Survey Results Obtained via Mobile Devices and Computers: An Experiment With a Mobile Web Survey on a Heterogeneous Group of Mobile Devices Versus a Computer-Assisted Web Survey. Social Science Computer Review 31(4), 482–504 (2013)
5. Vicente, P., Reis, E., Santos, M.: Using mobile phones for survey research. International Journal of Market Research 51(5), 613–633 (2009)
6. Goldberg, L.R.: An alternative description of personality: The Big Five factor structure. Journal of Personality and Social Psychology 59(6), 1216–1229 (1990)
7. Murakami, Y.: The Big Five in the Japanese lexical approach. Memoirs of the Faculty of Education, Toyama University, University of Toyama, pp. 39–60 (2004)
8. Murakami, Y., Murakami, C.: Scale construction of a "Big Five" personality inventory. The Japanese Journal of Personality 6(1), 29–39 (1997)
9. Murakami, Y., Murakami, C.: The standardization of a Big Five personality inventory for separate generations. The Japanese Journal of Personality 8(1), 32–43 (1999)
10. Murakami, Y., Murakami, C.: Handbook of Shuyo 5 Inshi Seikaku Kensa Revised (Handbook of Big Five Personality Testing Revised), Gakugeitosho, Tokyo (2008) (in Japanese)
11. Uesugi, S., Okada, H., Sasaki, T.: The Impact of Personality on Acceptance of Privacy-sensitive Technologies: A Comparative Study of RFID and Finger Vein Authentication Systems. In: 2010 IEEE International Symposium on Technology and Society (ISTAS), pp. 111–122. IEEE Press, New York (2010)
12. Dell, P., Uesugi, S., Okada, H., Sasaki, T.: Problems Relating to the Adoption of Potentially Privacy–Invading Technologies. In: Pro. of 18th Biennial Conference of the International Telecommunications Society, Tokyo, June 27-30, 29–4–F–1, pp. 1–12 (2010)
13. Uesugi, S., Okada, H.: Relationship between Smartphone Diffusion and Personality: A Case of a Japanese University. Journal of Informatics and Regional Studies 6(1), 5–19 (2014)

Privacy in Sound-Based Social Networks

João Cordeiro[1,2] and Álvaro Barbosa[1,2]

[1] Faculty of Creative Industries, University of Saint Joseph,
London Street 16, NAPE, Macau
[2] CITAR – Research Center for Science and Technology of the Arts,
Portuguese Catholic University, Oporto
{joao.cordeiro,abarbosa}@usj.edu.mo

Abstract. In this paper we address the issue of privacy in Online Social Networks (OSN), focusing on those that use environmental sound as a contextual cue for users activity. Through the use of a costume-made research tool consisting of an Online Sound-Based Social Network (OSBSN), we undertook scientific experiments aiming to assess how users deal with the use of sound. Results show that contextual sound is regarded as important and useful for OSN but raises important privacy concerns. In order to deal with this constraint, we propose a system based on the automatic classification of sound environment rather than capturing and sharing actual audio.

Keywords: Acoustic Communication, Mobile Computing, Online Social Networks, Privacy, Online Sound-Based Social Networks, Soundscapes.

1   Introduction

The protection of personal data is one of the holly grails of the network society. In an era where data is a commodity and everything seems to be able to be digitized (thus turning into data), privacy gains a whole new meaning. The abundance of Social Media applications dedicated to the exchange of personal information among the peers of the network takes this concern even further, as their use has become the norm in modern societies.

Our study is focused on privacy issues raised by the discloser of personal media, with an emphasis on the acoustic information. How legitimate and safe is the broadcasting of personal audio messages throughout the network and what are the advantages and disadvantages of broadcasting this information? We approach these questions from a user-centric perspective, presenting technological and conceptual options for the use of sound in a novel and less-intrusive way.

We start by elaborating on the notion of privacy in OSN, proceeding with the presentation of the concept of Online Sound-Based Social Networks supported by a survey of software applications. After, we present a research tool designed and developed by us, aiming at gathering qualitative and quantitative information about the use of sound in social media. Finally, we present the methodology for data gathering followed by the results analysis and discussion.

L.S.-L. Wang et al. (Eds.): MISNC 2014, CCIS 473, pp. 355–367, 2014.

2   For a (Re)Definition of Privacy in Social Media

The Social Web of OSN and Social Media (SM) are constantly defying the boundaries of privacy, to a point where it is plausible to question if the distinction between private space and public space will still be valid in the future. That is the question raised by [1] in the article *La vie privée, un problème de vieux cons?*[1] where the author debates how the new generations of digital natives will understand privacy under the light of frivolous broadcasting on SM and OSN. Professor Ravi Sandhu compares the *data revolution* felt by the first generations of Web 2.0 users with the sexual revolution experienced by young couples during the 60s and 70s *"For a while, there were very few inhibitions and people did extremely risky things. That's where we are now with sharing information. (…) Over time, people learned, 'Hey, this is not risk-free.' At first the freedom was attractive, but then everybody realized the risks."* (Quoted in [2]). Personal data can assume various forms and natures, ranging from the banc account number to the private pictures exchanged between lovers over the web. In any case, the discussion in this scope does not focus on 'how safe is my information?' but rather 'how private is my information and how willing am I to make it public?'. That accounts for *self-preservation* and *self-discloser* in OSN. The first concept states that for any social interaction one always expects to control the impression caused on others about itself. *Self-discloser* regards the conscious or unconscious exposure of personal data. For example, communities of naturists[2] [3] do exist offline and are willing to expose their naked bodies online without great concern for privacy; they show a particular high degree of *self-discloser* when compared with other people. This is an extreme example that most people would consider confined to a particular group of people, but what happens is that such behavior may be more common than one expects. According to a studied conducted at Harvard University, one youngster in every five and one young adult in every three already shared online personal photographs or videos naked or semi-naked [3]. This is particularly worrying if we think that those kind of audiovisual contents usually represent the currency for blackmail in cyber bulling and online harassment.

A study conducted in 2008 showed that users in general are not aware, precisely, how much of their personal information is available for the others online. One of the reasons may be the fact that many social apps (add-ons) for OSN like Facebook, request access to personal information in order to install and run. The authors support that "OSNs must clearly indicate the bare minimum of private information needed for a particular set of interactions. If an external application requires access to list of friends and nothing else, then the default should be that bare minimum." [4].

---

[1] Private life. A fool old men's issue? (translation by the author). Due to the popularity of this article, the same was extended and publish in book one year later. Manach, J. M. (2010). La vie privée, un problème de vieux cons? (Vol. 10). Fyp.

[2] http://naturistcommunity.com

[3] http://perspective-numerique.net/wakka.php?wiki=ClubDesNaturistesNumeriques

Other important aspect of privacy is related with location-based services[4], which are one of the technologies that most alarm people regarding their online exposure. When combined to OSN it gets the name of Geo-Social Networks (GeoSN)[5] and two main issues emerge immediately: location privacy and absence privacy. "The former concerns the availability of information about the presence of users in specific locations at given times, while the later concerns the availability of information about the absence of an individual from specific locations during given periods of time." [5]. Besides location and privacy, there is also co-location and identity privacy issues, which result from the intersection of several bits of data gathered from different users, allowing the inference of a person's encounters and identification [6]. This is possible because data nowadays is highly interconnected and our data is aggregated with others data, allowing prediction algorithms to run over large datasets, decreasing the error deviation. "How machines see us depends on how our data connects to others. The tastes and interests of people who don't yet exist within systems can be easily predicted based on the patterns of others. And, when machines have access to a person's social network, the predictions are even stronger." [7]

According to a report from 2010, done by Forester Research, "just 4 percent of Americans have tried location-based services, and 1 percent use them weekly" (quoted in [8]). This technology, although raising privacy concerns, is of great interest for marketers, which soon realized its potential to deliver an immersive multichannel experience for in-store shoppers. As before, the success of this technology will probably be driven by marketing demands. Shopkick is an example of a geo-location application (with some OSN characteristics), dedicated to offer bonuses to costumers who walk into stores. On the company's website[6] one can read: *"You just do what you love doing anyway—walking into your favorite stores like Target, Macy's, Best Buy, Crate&Barrel, Old Navy, Exxon and Mobil convenience stores, and more—and you automatically rack up über-versatile points called "kicks." No applications to fill out, and no purchase required. That's right—you just walk in with your smartphone and instantly earn kicks."* While the market keeps pushing the boundaries of self-discloser; new generations of digital natives seem more opened to accept these technologies and use them in their own social advantage.

From this brief discussion on privacy issues regarding OSN, we agree with the opinion that the problem will not have a straight technological solution based on data control[7]. Developers determined to produce OSN have to have in mind that while the boundaries are being pushed, there are still blocking thresholds that may refrain people to use a particular technology due do privacy issues. In any case, once again, technology is shaping the way we understand the world and re-think established concepts. Solutions for privacy problems will probably redound to a redefinition of privacy itself.

---

[4] Mobile Location-based services can be achieved with several types of technologies, such as GPS, Cell ID, aGPS, Broadband Satellite Network, etc..

[5] Also named Geo-Aware Social Networks.

[6] http://www.shopkick.com/about

[7] As a consequent of the scandal evolving NSA spying practices, Vint Cerf, one of the fathers of the Internet, asserted that there is no technological cure for privacy issues. http://gigaom.com/2013/07/09/internet-inventor-no-technological-cure-for-privacy-ills/

3   Sound-Base Social Networks

In the scope of this project we define Sound-Based Social Networks (SBSN) as social networks where sound accounts for an important share on the establishment and maintenance of links between the nodes of a network. One may consider it as a conceptual framework to approach social networks rather than a classification category *per se*. This means that a social network identified as Sound-Based Social Network still affords other classifications and methodological approaches. Examples of SBSN are the community formed by the audience of a local radio broadcasting station or the online communities flocking around music (e.g. LastFM, MySpace). Despite such broad spectrum, our approach focus on a particular field of sound called Soundscapes, which defines the sound of a particular place from a phenomenological point of view. Soundscape studies have shown that Environmental Sound is a rich resource for understanding the social context of a place, its dynamics, problems and virtues, assuming sound as a resource rather than waste [9].

People, during their daily life, travel through a sea of sounds, eventually without even changing their geographic location. Passively or actively, Soundscapes vary along the hours, days, months and years, characterizing in each moment the sonic context of a place. By extending the analysis time span, sonic patterns about the place are unveiled, contributing to the characterization of its sonic profile. This data can then be extrapolated to other layers of significance, mainly when correlated with data gathered through multi-modal analysis (geo-location, time, weather conditions, etc.).

By shifting the *place* from static to dynamic, analysing the sound of a person along his or her day, we are characterizing not a geographical location but a node of a social network drifting in space, time and network position. The more data is collected, the more detailed and accurate is the sonic profile of the node. This information, when evaluated in context, can be of great relevance for daily social interactions, since the short-term and long-term analyses show different (but complementary) aspects of the social behaviour of the nodes of a network.

Narrowing the scope of our study towards online technologies, we define Online Sound-Based Social Networks (OSBSN) as Sound-Based Social Networks grounded on online systems, with a special focus in mobile platforms, which are regarded as more effective since they accompany the users at all times. Our research tool - Hurly-Burly – for which we give an overview of the design and implementation in section 4, falls under this category, along with other projects such as CenceMe [10] and Soundwalk [11], both incorporating sound-sensing mechanisms on mobile devices combined with a social network dimension.

4   Hurly-Burly: Online Sound-Based Social Network System

Hurly-Burly - a synonym for boisterous activity – is a sound-sensing software system, with characteristics of online social networks. Users can create links with other users, sharing and receiving information about each other and their environment on a mobile device. In this particular case, the information being shared regards the soundscapes of the places visited by each user during his/her daily life, completed with a rough description of his/her movements. The system is comprised of a mobile application, a web application and a webserver, combined in a typical client-server configuration (Fig. 1). The mobile application is responsible for input and output tasks, covering the visual and auditory display of real-time information as well as the soundscape sensing task. The sound is sensed while the application is working inconspicuously; feeding the social network with this information while keeping a personal record of the data. This way, each user acts as a terminal of a sensor network linked through the Internet.

The web application displays the sonic profile (corresponding to long-term information) and the webserver saves and retrieves information based on forms/queries (represents a functional piece of the system, potentially invisible for the user).

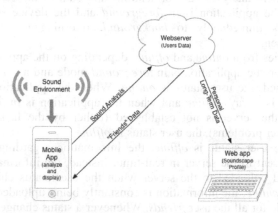

Fig. 1. Hurly-Burly configuration

The application was developed for the iOS platform, being compatible with iOS+4.3 systems. We have chosen the iOS platform for four main reasons:

- Processing Power: The nature of the software application demanded a powerful system in order to perform advanced tasks such as sound classification. iOS devices (iPhone, iPad and iPod) are among the most powerful mobile computing devices in the market.
- Programming Language: The official programing language for the iOS is the C-based language Objective C. However, we have used OpenFrameworks[8] - a popular C++ open source framework – which compiles C++ source code to run on iOS platforms. C++ is known for its efficiency and we were already familiarly with the framework, which is supported by an extensive and active community, developing Add-ons (libraries) that fulfilled most or our needs.
- Penetration Market: iOS devices are highly appreciated among our target audience (young adults, higher education, engaged in MOSNs), therefore a common device among our sample.
- Multitasking: Latest modes of iOS devices allow multitasking; including sound capture while the application is running in background mode. This was a key factor on our decision.

---

[8] www.openframeworks.cc (accessed on 15 June 2014).

The mobile application is comprised of four modules: 1) a soundscape-sensing module, 2) a movement detection module; 3) a client-server communication module and 4) a visualization module.

When users have the application running on their devices, it can either be in *background* or *foreground* mode. Working in *foreground* means that the application *graphic user interface* (GUI) is displayed on the screen and that Hurly-Burly is the main application being managed by the system. Working in *background* means that the application is represented on the screen only by a red bar on top, continuing the sound and movement analyses but not sending this data to the server and *friends*, consequently. If the application is in *foreground* and the device screen locks, the application passes immediately to *background*, returning to *foreground* when unlocking the screen.

User status varies from *online* and *offline*, depending on the application mode and other factors. When the application is in *foreground* mode and a connection with the server is established, the user status is *online*. When the application is working in *background*, user is always *offline* and when the application is in *foreground* but a connection with the server is not established (either by the lack of an Internet connection or server problems), the user status is *offline*.

Every time the user status is *offline*, the information regarding the sound and movement is not sent to the server in real-time. Instead, the information is recorded on the device and uploaded to the server when the user status changes to *online*. When the user is *online*, the information is constantly being uploaded onto the server and made available for all the user *friends*. Whenever a status change occurs, this bite of information is updated on the server.

The Soundscape Sensing Module is responsible for analysing the sound that gets into the device through its microphone. It executes sound analysis and machine audition tasks, using established techniques and algorithms. No sound or any information regarding audio descriptors is shared on the network. The only data being shared is the sound intensity of the soundscape and its classification according to three categories: music, speech and environmental sound.

The analysis is always performed on ±2,79 seconds mono audio samples (44.1kHz, 16bit), recorded every 3 minutes (or whenever the user double tap on the device' screen), using the built-in microphone of the mobile device. The soundscape-sensing module was built using visual programming language Pure Data[9] and embedded in the main C++ code using a software wrapper called LibPD[10]. The performance of the classification system was 87% of successful classifications. Since a mobile device has limited memory resources and processing power (when compared with high-end desktop units), the analysis task required a compromise between accuracy/precision and device's resources. The goal of this implementation was to achieve a satisfactory analysis and classification tasks that could provide a coarse approximation to the real acoustic environment. Producing a state-of-the-arte machine audition algorithm was out of the scope of this project. Details about the algorithm can be found [12].

The movement module detects if the device is steady or moving. This information is used as a control datum for crosschecking with the sound-based information. If

---

[9] Originally developed by Miller Puckette, http://puredata.info (accessed on 15 June 2014).

[10] Official webpage: http://libpd.cc (accessed on 15 June 2014)..

loud environmental noise is detected along with a consistent moving status, the user can infer that the sound analysis may correspond to the movement of the phone inside a bag, for example, rather than the sound environment where the user is placed.

The Graphic User Interface (GUI) of the mobile application is mostly based on the display of the information regarding the user and his friends' activity (Fig. 2). Each user (and his/her friends) is represented by an animated waveform, which has a specific shape and amplitude according to the type of sound, amplitude of the sound, movement of the device and user status. We used the analogy of waveforms so that the information was conveyed to the user in a fast, and meaningful way. The name of the user is displayed inside a grey box on the left end of the waveform, while his/her friends' names are inside grey boxes. Waveform's amplitude changes according to the intensity of the analysed sound: the louder the soundscape, the higher the wave's amplitude. When a waveform represents an offline user, its colour is always white, meaning that the information is not up-to-date. When the user is offline, all waveforms display in white. The GUI affords scrolling up and down through the list of users, passing one finger vertically across the screen. Other graphical elements of the GUI are: the bottom bar with a legend (explaining the meaning of the four types of waves and offline status), the top bar (with the name of the application and the slogan "your friends' soundscapes") and a white circle marking the analysis period.

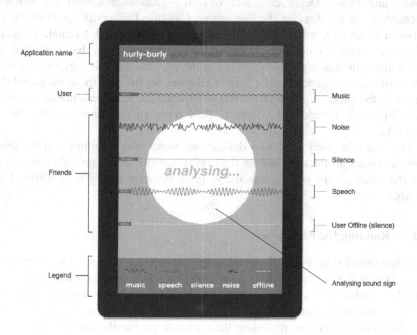

Fig. 2. Hurly-Burly Graphic User Interface

5   Methods

In order to assess privacy concerns about the use of sound in OSBSN, we undertook a scientific experiment using the research tool Hurly-Burly described in section 4.

The experiment involved running the application within several groups of participants, each group forming a social network with a cluster coefficient equal to one. The users were asked to run the application continually during 3 days and filling a questionnaire at the end.

5.1   Sample

The sample selection was particularly difficult as several requirements had to be attended, namely: owning a recent iOS device, be familiar with OSN, belonging to a highly clustered group, voluntary participation, be willing to share his/her personal data and be willing to install experimental software on their devices.

Considering these constraints, we were able to gather three groups (Fig. 4): one comprised of 9 students from Communication and Media Licentiate in University of Saint Joseph, Macau, 3 male and 6 female subjects, aged between 19 and 23. The second group, with 11 members, was constituted by teachers and researchers in the Sound and Image Department and CITAR – Research Center for Science and Technology of the Arts, in the Portuguese Catholic University, in Porto. It is aged between 27 and 43, with a gender distribution of 7 male and 4 female subjects. The third and smaller group was comprised of 3 people who have worked on the conceptualization and graphic interface of the application. It is a group specialized in interaction design, experts in the field of HCI but not necessarily in the sound domain. This was the smallest group of the whole experience, with three members only, aged from 24 to 32, 1 male and 2 female. They were all regular users of online social networks.

The research team used two devices as monitoring equipment. One iPod was carried around all the time and an iPad remained still at the group's research office. All the other groups were connected to these devices, seeing it as a *friend* in their devices.

5.2   Running the Experiment

The designated time frame for running the experiment within each group was three days. During this period, subjects were asked to run the application as exhaustively as possible, including during the night. The application affords limited input from the users, so the main request was that they kept the application running and regularly checked the information regarding their friends on the display. By the end of the experimental period, a questionnaire was delivered to the subjects in order to assess their experience using the application, and collect information about the users and their relation with the topic of mobile sound, online social networking, self-disclosure and privacy. The questionnaire was produced in digital format distributed by e-mail using the free service Google Drive (Form document). Accompanying the survey was a link for a webpage containing an interactive graph, with a history plot of users sonic and movement activity (Fig. 3). The graph afforded zooming in and out and timeline navigation. This webpage was only sent to the users showing a significant amount of data and was necessary for answering one question of the survey. After filling in the questionnaire the experimental procedure was over and users could uninstall the application from their devices.

5.3   The Questionnaire

A questionnaire was distributed in order to collect qualitative and quantitative data from the subjects regarding their use of the application. It was divided in 6 sections according to different aims and topics: 1) respondent identification; 2) subject relation with mobile devices; 3) subject relation with social media and online social networks; 4) user acknowledgment of the concept of soundscape; 5) assessment of the application usability and user experience and finally 6) the soundscape history graph.

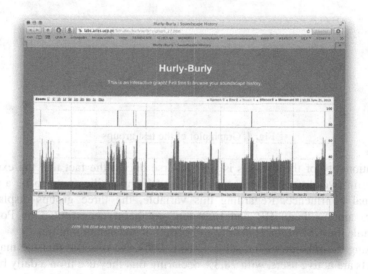

Fig. 3. Hurly-Burly Web Application

6   Results

Data collected through the questionnaires was organized into three logical parts that do not necessarily correspond to the original sections of the questionnaire. The parts are: 1) Characterization of the sample regarding its demographics, relation with social media, mobile sound and soundscapes, 2) The user experience regarding the application and 3) User's opinion about soundscapes and OSN. For the sake of the topic discussed in this paper – privacy in OSBSN - only data related to this topic will be presented and discussed. The three test groups are different in size; therefore we opted for using normalized graphic representations whenever we compare one against others.

The sample is comprised of 48% male respondents against 52% female respondents. While there is a balance in the overall sample, the group comparison shows that USJ and HB groups have a higher rate of female participants than UCP. Regarding the age of the participants, the groups are also distinct, in particular USJ and UCP. The former is comprised of undergraduate students with an average age of 20 years old while the later is comprised of teachers and researchers, showing an average age of 34. There is a gap of fourteen years between these two. HB group is situated approximately in this gap, presenting an average age of 27 years old.

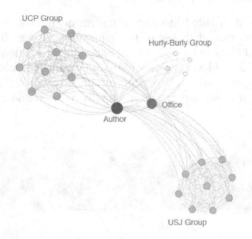

Fig. 4. Graph plot of the test groups

The nationality of the subjects is very diverse due to the fact that the experience took place in two different continents. Moreover, universities tend to be a place of international confluence. Regarding this variable, the three groups display some differences among them: HB and UCP tend to mainly constituted by Portuguese people, while the USJ group is mainly comprised of Asian nationalities.

Analyzing the relation of the subjects with OSN, we observe that the majority of the subjects are active users, with 78,3% declaring that they use it on a daily basis.

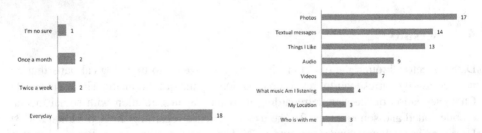

Fig. 5. Left: Characterization of the sample according to the usage frequency of OSN (absolute number of answers). Right: User's sharing preferences in SM (<u>absolute number of answers</u>).

When asked about the content they like to share and receive on SM (question 2.4), answers tend to be homogeneous among the groups, electing *photos* as the most popular media, followed by text and *likes* (Fig. 5).

When enquired about the use of sound in mobile devices, users elect "listening to music" as the principal activity. This appears ahead of calls in the rank (Fig. 6). However, when differentiating iPhone users from the other users, the former do elect calls as the main sound-related activity.

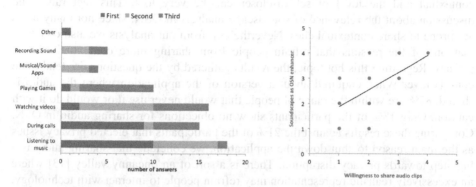

Fig. 6. Left: Usage of sound in mobile devices. Right: Correlation between users' recognition of soundscapes as OSN enhancer and their willingness to share audio clips.

When questioned about the first reason that would take them to quit the application (question 4.8), 72% of the users replied "low battery" issues, 21% evoked "privacy issues" and 7% mention "all the reasons" as valid to quit the application (this question was asked only to the group UCP and HB, as it was added to the questionnaire after the first round of experiments).

Regarding the use of sound in OSN, 96% of the subjects claim to have, at least, a notion of what soundscapes are and 70% recognized a medium to high potential in enhancing OSN.

Regarding the issue of privacy, we asked participants to imagine a new version of Hurly-Burly application where the actual soundscape (the sound recording) was shared within the network. The results show that 17% of the users would never use an application like that because it violates their privacy and 39% would only use it in very controlled situations. 31% said that they would use it with caution and 13% evaluated this feature as very appropriate and useful.

When correlating these results with those regarding the subjects' recognition of soundscapes as OSN enhancer, one observes a direct relation between the willingness to share sound and the acknowledgment of HB as an OSN enhancer (correlation = 0,6; Fig. 6).

7   Conclusions

After analyzing the results of the questionnaires we conclude that contextual cues are among the least data that users like to share/receive in OSN. Our study shows that only 8% of the sharing options contemplate information about the context of the user, with the two options being: *people accompanying the user* (4%) and *geo-localization* (4%). One may argue that text messages and photos - two highly regarded options – may also convey contextual information but the fact is that they are not strictly contextual and the level of self-discloser can be very low. This fact raises the discussion about the relevance of soundscape analysis in OSN, since not many users are prone to share contextual cues. Nevertheless, from our analysis we also conclude that one of the reasons that refrain people from sharing more contextual cues is privacy. Regarding this hot topic, the results gathered by the questionnaire tend to be conservative. When enquired about a version of the application where the audio is shared, 87% are within the range of people that would never use it or would do it with caution. Only 13% of the participants show no objections for sharing audio in OSN. Comparing these results against the 21% of the participants that elected privacy issues as the main reason to shut down the application, we conclude that sharing audio is a big step towards privacy disruption. There is a sort of an uncanny valley [13] where an excessively realistic representation may refrain people to interact with technology. From all the participants, those more willing to share sound are, in general, those who tend to acknowledge the potential of HB in OSN. This suggests that people showing a propensity for this type of applications are more open to share their actual soudscapes, revealing fewer privacy reservations.

A system like HB based on soundscape analysis instead of the actual sound may overcome these constraints by providing contextual cues without sharing sensitive information. Our study suggests that sharing a high-level description of context instead of the verbal information encoded in speech may be a successful approach to enhance the interaction within OSNs, bridging the gap between the virtual and physical worlds.

References

1. Manach, J.-M.: La vie privée, un problème de vieux cons? [Internet]. InternetACTU.net (2009), http://www.internetactu.net/2009/03/12/la-vie-privee-un-probleme-de-vieux-cons/ (cited July 10, 2013)
2. Peppers, D., Rogers, M.: The Social Benefits of Data Sharing [Internet] (2009), http://www.privacyassociation.org/publications/2009_01_the_social_benefits_of_data_sharing (cited July 10, 2013)
3. Palfrey, J., Sacco, D., Boyd, D. (eds.): Enhancing Child Safety & Online Technologies [Internet]. Final Rep. Internet Saf. Tech. Task Force to Multi-State Work. Gr. Soc. Netw. State Atty. Gen. United States. Carolina Academic Press, Durham, North Carolina, USA (December 2008), http://www.cap-press.com/pdf/1997.pdf

4. Krishnamurthy, B., Wills, C.: Characterizing privacy in online social networks. In: Proc. First Work. Online Soc. Networks [Internet], New York, NY, USA, pp. 37–42 (2008), http://dl.acm.org/citation.cfm?id=1397744 (cited November 2, 2012)
5. Freni, D., Vicente, C.R., Mascetti, S.: Preserving location and absence privacy in geo-social networks. In: Proc. 19th ACM Int. Conf. Inf. Knowl. Manag. ACM (2010)
6. Ruiz Vicente, C., Freni, D., Bettini, C., Jensen, C.S.: Location-Related Privacy in Geo-Social Networks. IEEE Internet Comput. [Internet] 15, 20–27 (1958), http://ieeexplore.ieee.org/lpdocs/epic03/wrapper.htm?arnumber=5719583
7. Boyd, D.M.: Networked Privacy. Surveill. Soc. Surveillance Studies Network 10, 348–350 (2012)
8. Miller, C., Wortham, J.: Technology Aside, Most People Still Decline to Be Located. New York Times [Internet]. New York (August 29, 2010), http://www.nytimes.com/2010/08/30/technology/30location.html?pagewanted=1&_r=2& (cited July 18, 2013)
9. Schafer, R.M.: The Soundscape: Our Sonic Environment and the Tuning of the World. Destiny Books, Rochester (1994)
10. Miluzzo, E., Lane, N.D., Eisenman, S.B., Campbell, A.T.: CenceMe – Injecting Sensing Presence into Social Networking Applications. In: Kortuem, G., Finney, J., Lea, R., Sundramoorthy, V. (eds.) EuroSSC 2007. LNCS, vol. 4793, pp. 1–28. Springer, Heidelberg (2007), http://dx.doi.org/10.1007/978-3-540-75696-5_1 (cited October 1, 2012)
11. Fink, A., Mechtley, B., Wichern, G., Liu, J., Thornburg, H.: Re-Sonification of Geographic Sound Activity Using Acoustic, Semantic, and Social Information. In: Proc. 16th Int. Conf. Audit. Display, Washington, D.C., June 9-15, pp. 305–312 (2010)
12. Cordeiro, J., Barbosa, Á., Afonso, B.: Soundscape-Sensing in Social Networks. In: Proc. AIA-DAGA 2013 Conf. Acoust. EAA Euroregio, Merano (2013)
13. Seyama, J., Nagayama, R.: The uncanny valley: Effect of realism on the impression of artificial human faces. Presence Teleoperators Virtual ... [Internet] (2007), http://www.mitpressjournals.org/doi/abs/10.1162/pres.16.4.337 (cited June 30, 2014)

College Student Performance Facilitated on Facebook: A Case Study

Pei-Lin Hsu[1,*] and Ying-Hung Yen[2]

[1] General Education Center, Oriental Institute of Technology,
[2] Department of Industrial and Commercial Design, Oriental Institute of Technology,
No.58, Sec. 2, Sih-Chuan Rd., Ban-Ciao District, New Taipei City 22061
Taiwan (R.O.C.)
peilingmz@gmail.com

Abstract. The purpose of this study was to explore the effect of using Facebook after class in college learning and whether it is a good teaching tool. The ten-week educational discussions among fifty Taiwanese college students on two Facebook discussion pages were recorded and analyzed. The researchers examined the relationship between Facebook Group participation and the academic achievement of college students. Results showed that there was a significant positive correlation between a student's online participation frequency and his GPA. The frequency of student participation in Facebook Groups could explain 15% to 19% of the variance in the grades they received for their exhibition projects. Furthermore, it was found that evaluation criteria designed by instructors can have an effect on the direction of online discussions. According to the results, we can conclude that educational discussions on Facebook can be beneficial to college students' learning. However, whether Facebook can serve as an effective teaching tool depends on how the educators use it in students' learning process.

Keywords: college student performance, evaluation criteria, Facebook, social network.

1 Introduction

Good teaching is an important contributor to student learning. Hatti [1] showed that teaching expertise accounts for about 30 percent of the variance in student achievement. Facing the changing world influenced by technology, the teachers have to know how to use it for increasing student learning.

Today the Internet has led to the formation of a whole new kind of community, the "virtual community," eliminating geography and time constraints. Parks [2] suggested that "being connected to others fosters a sense of purpose, belonging, and attachment that is central to the concept of community." Founded in 2004, Facebook is an online social networking service that has more than 1.28 billion active users and is highly

---

* Corresponding author.

L.S.-L. Wang et al. (Eds.): MISNC 2014, CCIS 473, pp. 368–382, 2014.

popular among college students [3]. Social networking sites serve as a genre of "networked publics", to transform how people think and communicate by providing a variety of technical features to their users [4]. These sites allow individuals to present themselves online, establish new connections, and maintain connections with existing contacts. As a social networking service provider, Facebook allows its users to share interests, interact with their own personal networks, and post comments on the profile page of other users. Therefore, Facebook was chosen as the example social networking service in this study.

Coleman [5] viewed communities and families as repositories of social capital and established the importance of these networks in fostering academic success. Semo and Karmel [6] argued that network associations and influences can increase educational engagement and achievement. Thus social networks in schools can be beneficial to students' learning and foster academic success.

The relationship between student engagement through the use of Facebook and student's academic performance was examined. It has been found that students who are more engaged during the learning process tend to perform better in their academic studies [7]. Therefore, an investigation on the relationship between student engagement and their academic performance by using Facebook as an educational enhancement tool is worthy of study.

The present study is to investigate how teachers can incorporate the use of Facebook in guiding and facilitating college student learning. Of Hew's 36 literature reviews, eight papers were based on content analysis of Facebook features [8]. However, the content generate on Facebook Group, a popular feature on Facebook has not been analysed yet. Also, few studies have been done on the educational use of Facebook. Therefore, in this paper, four specific research objectives are examined :

1. Do the students with different group perform differently?
2. Is there a significant correlation between students' participation frequency on Facebook and his academic achievement?
3. How does the instructor interact with the students on Facebook ?
4. How does the instructor motivate the students joining the Facebook discussion?

2   Literature Review

2.1   Social Networks on Campus

Many studies have been done to examine social networks and how people process information [9]. " Computer networks are inherently social networks, linking people, organizations, and knowledge" [10]. A social network may happen in Web pages, journal articles, countries, neighborhoods, and departments within organizations. It could include collaborations, friendship, trade ties, Web links, citations, resource flows, information flows, exchange of social support or any other possible connection between them [11].

Borgatti, Mehra, Brass, and Labianca [12] identified four broad categories of relations: similarities, social relations, interactions and flows. Internet data types include emails, discussion forums, and Web pages. Online forums are a good example of social networks. Its accumulated data can be used to conduct studies and experiments. It allows researchers to examine the collaborative process in online groups and communities [13].

Students accept social networking tools as a main platform of e-learning. They need more interactive learning environment and the role of instructors is to guide students and consult and advise students throughout their learning process [14]. There were a lot of studies about social network for the nomination number [15], the English teachers in-service development [16], and the teaching ESL writing [17]. It has been found that students seem to be open to using Facebook as a means to enrich their classroom learning [18]. "New tools must be developed to help people navigate and find knowledge in complex, fragmented, networked societies [10]." With the wide usage of Facebook among college students, would it be a good idea to incorporate this technology in teaching college students?

2.2   Facebook as a Social Network

Facebook's impact on college students has received a lot of attention and interests from researchers. It started out as a social networking site targeting the college-aged students [19], attracted users mainly through social motivation [20], and was used often for social interaction [8], [19], [21]. It has been found that Facebook usage improves a user's self-esteem by increasing his sense of belonging [22].

Furthermore, Facebook is also playing an important role in student academic performance. It can provide many pedagogical advantages to teachers and students. Facebook creates a comfortable learning environment for students, and allows teachers to provide students' guidance on group assignments and direct useful online educational resources to students [23]-[24]. Moreover, it was found that the use of Facebook as a communication tool can be beneficial in both social and academic settings [25].

Facebook was used most often for casual social interaction [26]. However, it has also become a valuable communication tool to support students' educational communications and collaborations with their instructors [18].

2.3   Facebook and Student Performance

Networking is important for students' learning [26]. Several studies have examined the educational use of Faceboook [27]. It was found that Facebook appeared not to play an effective role in students' academic achievement in Sweden. Rouis, Limayem, and Salehi-Sangari [28] found the proposed model including Facebook usage, satisfaction with life, and performance-goal orientation explained 15% of the impact on students' academic achievement. Using Facebook encourages students to catch up with friends. However, student's Facebook usage and discussions examined in this study was not guided by teachers.

Some researchers state that Facebook has positive effects on student academic performance [29]-[30] while others claim the opposite [31]-[32]. Hunley et al. [33]

found that there is no correlation between computer usage and academic achievement. Therefore, it is expected that Facebook, with users mostly being college students, can enhance interaction related to educational purpose.

Different methods are employed to investigate issues of using Facebook as an educational facilitation tool. In order to investigate the effect of Facebook usage in learning, online questionnaires were created to examine the relationship of frequency and duration of Facebook usage with the academic performance of Facebook users [7], [34]-[35]. A study done by Forkosh-Baruch & Hershkovitz suggested that social networking sites promote knowledge exchange through the facilitation of informal learning [29].

From the literature review, Facebook was recommended as a tool for quality educational act [34]. Bicena & Cavus suggested it might be worthwhile to integrate the use of Facebook into education and teaching [25]. Using Facebook for collecting and sharing information was positively correlated to academic performance [7].

Motivation is another important factor related to the academic performance. Motivation influences students' involvement and academic achievement [36]. Spending excessive time on the Internet can affect school performance [31]. For example, in a recent study, teachers of a high school as mentors may provide educational support on a Facebook Group discussion page, and promote a stronger relationship between the student members and the mentors. Most students believe that they learned more because of the use of Facebook group discussion page [37].

We can see the students using technology on campus all over the world. The teacher needs to think about how to integrate it into a curriculum. There are many ways of using technology in the classroom, such as some software programs, distance learning, multimedia and the Internet. Choosing the right tool for the classroom is important. Sturgeon and Walker [38] found that there is an indirect correlation between the use of Facebook in a classroom setting and student academic performance. They cited a professor who interviewed, "The more relaxed the student is, the more relaxed the faculty member is and the more learning can take place. If the relationship that exists is what makes for a better learning experience for the student, then Facebook most certainly has an indirect impact."

Evaluation criteria have an effect on students' learning [39]. The researchers suggested that adopting Facebook as an education enhancement tool could be beneficial to students, because it facilitates comfortable learning conditions and provides students a sense of community [40]. Furthermore, the research found that the faculty believed that grades were an important motivation for students [41].

So far studies on Facebook usage for educational purpose is inconclusive and still minimum, and the studies are done in various countries including America, Europe, and Asia [7], [29], [31], [34]-[35], [42]. It is believed that more studies on this subject can help find better ways to facilitate student learning, as well as improve student-educator interaction.

3   Methods

3.1   Context and Participants

Product Development is a required four-credit course for third-year students in the Industrial Design program at Oriental Institute of Technology. This course spans across two semesters. A student-centered curriculum was designed by the teacher to attain the following goals: (1) to help students to understand the technology and knowledge in product development process, (2) to help students gain more hands-on experience in the field of design, (3) to stimulate students to think and be creative, and to provide students with the opportunities to collaborate with their peers. The design exhibition took place at the end of the semester on June 17, 2011.

Fifty students (31 females, 19 males) participated in this study. Ten subgroups worked in the first semester. Because there were too many students in a class to discuss on Facebook, the students were merged into two groups in the second semester. They were B Group (13 males, 14 females) and F Group (6 males, 17 females). The researchers considered the balance of the abilities as well.

3.2   Measure

The researchers made some categories of the data so as to make them easy to read and analyze. Two faculty members were asked to review these categories. The agreement between the faculty members varied from 85% to 90%.

The data collected was categorized into three categories: (1) active participation : whereby students participate in online discussions and proactively post information (2) passive participation: whereby students respond to posts that have been posted proactively by other members (3) total participation counts in both Facebook discussion groups.

The pre-test scores are defined by the GPA received in the first semester in the same course before starting the two Facebook Groups and the post-test scores are defined by the grades received for this ten-week Facebook discussion group assignment.

Four evaluation criteria included in the study were to (1)come to class on time, (2) post news in the field of industrial and commercial design , (3) post events related to building professional skills and (4) write a reflection after each class.

There were four kinds of postings from the instructor. These categories were (1)answers to students' questions, (2)announcements, (3)social support (maintains and encourages students' discussions on the Facebook Group) and (4) question to stimulate students' learning.

3.3   Procedure

For the sake of design exhibition in June, the students were asked to create Facebook Groups to discuss how to make it successfully. Their interaction and conversations on the Facebook discussion pages are recorded and evaluated according to the evaluation criteria. It began from April 12 to June 17 for B Group and from April 15 to June 17 for F Group in 2011.

3.4   Analysis

The data collected were analyzed using SPSS Statistics 20.0. Descriptive statistical analyses were conducted to illustrate the demographic characteristics of the participants. Correlations were examined to evaluate the frequency of participation in Facebook discussions, the pre-test and post-test. The Independent sample T-test, the Chi-square test for independence and Goodness of fit were also used.

4   Results

4.1   Description of Participating Facebook Group

There are 3035 Facebook conversation records for B Group and 2279 records for F Group on Facebook Group. The frequency of active participation in B Group was 532 and that in F Group was 446. The frequency of passive participation in B Group was 2503 and that in F Group was 1833. The mean frequency of active participation in B Group was 19.00 (SD=42.68) and in F Group was 18.58 (SD=22.06). The mean frequency of passive participation in B Group was 89.39 (SD=116.39) and in F Group was 76.37 (SD=58.87).

4.2   Facebook Group Discussions and the Academic Performance

Before launching discussions on Facebook Group pages, we first checked whether there was a difference in the academic performance between the two student groups. The T-test was used to compare the academic achievement (pre-test) of the B group (M=67.00, SD=14.15, N=27) and that of the F group (M=66.00, SD=16.45, N=23). The result showed there was no significant difference (T-value=0.23) between B group and F group.

We then compare the academic performance between the two groups post FB discussions. The T- test was again used to compare the academic achievement of B group (M=67.93, SD=14.59, N=27) with that of F group (M=72.48, SD=7.93, N=23). The result indicated there was no significant difference (t=-1.40) between B group and F group after the ten-week discussions on Facebook Group pages.

The academic performance of the two groups in the pre- and post-Facebook discussions is shown in Table 1. As Table 1 indicates, the academic achievement of participants in B Group improved, but it was not significant (T=-0.317). The academic performance of the participants in the F Group improved and was statistical significant (T=-2.405). However, the total academic improvement of both groups was not statistically significant (T = -1.712).

Table 1. Pair-Sample T-test on the academic achievement

Group		M	N	SD	T-value
B	pre-test	67.00	27	14.15	-.317
	post-test	67.93	27	14.59	
F	pre-test	66.00	23	16.45	-2.405*
	post-test	72.48	23	7.93	
total	pre-test	66.54	50	15.10	-1.712
	post-test	70.02	50	12.10	

*$P<.05$

4.3   Participation Frequency of the Facebook Group Discussions and the Academic Performance

After the students were engaged in the Facebook Group discussions for 10 weeks, they had an exhibition and received grades. The correlation coefficients between variables are listed in Table 2. As shown in Table 2, coefficients of the frequency of active participation, passive participation, and total participation on Facebook Group discussion pages, as well as the academic achievement, including both the pre-test and the post-test, were all found to be significant ($p<.01$ & $p<.001$). The correlation between active participation and the pre-test was 0.367, and it was 0.434 between active participation and the post-test. The correlation between the total participation, the pre-test and the post-test varied from 0.406 to 0.410. These results showed that there is a significant positive relationship among the variables. The frequency of student's participation on Facebook discussion pages can explain 15% to 19% of the variance in the academic performance for the post-test /exhibition.

Table 2. Correlation matrix of variables

Variable	1	2	3	4	5
1.active participation		.896***	.943***	.367**	.434**
2.passive participation		–	.992***	.407**	.390**
3.total participation			–	.406**	.410**
4.pre-test				–	.459**
5.post-test					–

** $p<.01$ ***$p<.001$,   N=50

4.4   Teacher Participation in Facebook Group

Teacher participation was also examined in the study to clarify whether it was different between the groups. As shown in Table 3, there was no significant difference in teacher participation in the two groups ($\chi^2=1.852$, $P=.174$). However, the frequency of teacher passive participation (74.36%) was significantly higher than that of active participation (25.64%)($\chi^2=27.769$, $P<.01$).

Table 3. χ^2 Test for the teacher's participation

| Group | Participation | | Total |
	Active	Passive	
F	16(32.0%)	34(68%)	50(100%)
B	14(20.9%)	53(79.1%)	67(100%)
Total	30(25.64%)	87(74.36%)	117(100%)

$\chi^2=1.852$ $P=.174$

As shown in Fig. 1, it shows that the teacher participated more actively in both groups (FA, the teacher participated in Group F actively; BA, the teacher participated in Group B actively) after week 8. In F Group, the teacher participated passively / responded more in week 4, 6, 7 and 8 (FP, the teacher participated in Group F passively). In B Group, the teacher participated passively /responded more in week 5, 7, 8 and 9 (BP, the teacher participated in Group B passively).

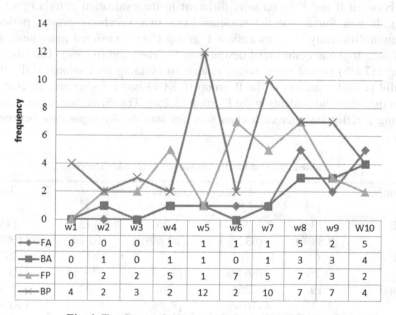

	w1	w2	w3	w4	w5	w6	w7	w8	w9	W10
FA	0	0	0	1	1	1	1	5	2	5
BA	0	1	0	1	1	0	1	3	3	4
FP	0	2	2	5	1	7	5	7	3	2
BP	4	2	3	2	12	2	10	7	7	4

Fig. 1. Teacher participation in Facebook in 10 weeks

We also examined topics the instructor posted on Facebook Groups. As shown in Table 4, fifty-six percent of the topics posted by the teacher were to answer the students' questions. And twenty-six percent of the topics were announcements to the students. Fifteen percent of the topics were to support students' discussion participation.

Table 4. Topics of the teacher posting on Facebook Group

Category	Messages	Group F		Group B		Total
		Active	Passive	Active	Passive	
1.Answers to students	1*	0	23	0	24	65(56%)
2.Announcement	2*	11	4	12	4	31(26%)
3.Social Support	3*	3	7	2	6	18(18%)
4.Question	4*	2	0	0	1	3(3%)
Total		16	34	14	53	17(100%)

1* "Thank you. Please inform other teammates.";"I'll check it."
2* "The Name of the exhibit is 2011 OIT EXPO."; 'There will be a conference tomorrow."
3* "You work hard." ;"Cheer up";"You're really smart."
4* "May I post it on the net?";"Is there the 3th works to be exhibited?"

4.5   Evaluation Criteria and Facebook Group Participation

In this study, four evaluation criteria were also investigated with the use of a Chi-Square Test. As shown in Table 5, the contents of active participation on Facebook Group between B and F group were different in the evaluation criteria (χ^2=37.303, $p<.001$). It was found that the students from two Facebook groups posted the information differently. The students in F group (86.6%) offered more news in the field of industrial and commercial design than B Group did (63.5%). The students in B group (17.6%) posted more events related to building professional skills than F group did (1.1%). Students in the B group (8.88%) had a higher rate of coming to class on time than did students in the F group (3.2%). The difference in the frequency of writing a reflection after each class was not statistically significant between the groups.

Table 5. χ2 Test for the evaluation criteria in Groups

Group	Evaluation criteria				Total
	1[a]	2[b]	3[c]	4[d]	
B	13	94	26	15	148
	(8.88%)	(63.5%)	(17.6%)	(10.1%)	(100%)
F	6	162	2	17	187
	(3.2%)	(86.6%)	(1.1%)	(9.1%)	(100%)
Total	19	256	28	32	335
	(95.7%)	(76.4%)	(8.4%)	(9.6%)	(100%)

χ^2=37.303[***] $p<.001$ Note: [a] Come to class on time; [b] Post news in the field of industrial and commercial design; [c] Post events related to building professional skills; [d] Write a reflection after each class

5 Discussion

5.1 Participants' Improvement in Academic Performance

Although the analyses showed that divided students from the same class into two Facebook discussion groups resulted in no difference in the grades they received for their exhibition projects, it was found that the frequency of student participation in Facebook Groups could explain 15% to 19% of the variance in their exhibition project grades. The result is similar to the findings by Rouis [28]. Encouraging the students to participate in the discussions in Facebook Group has some beneficial effect on their academic performance.

Students' motivation is an important factor contributing to the participation of Facebook Group and their academic achievement. A motivation to communicate was promotes Facebook usage [43]. Motivation influences students' involvement and academic achievement [36]. Critical thinking skills, interactions among students and interactions between students and faculty members are important variables related to motivation [44]. In the study, the students followed what the teachers demanded and had successful exhibits. They might care about the grades and devote their time to participating in the Facebook discussion then got good grades. Besides, there are other factors related to academic performance and worthy to be considered, including personality characteristics, self-efficacy, self-esteem, interest and competence.

5.2 Evaluation Criteria and Contents Discussed in Facebook Group

The researchers analyzed the contents discussed in the two Facebook Groups, and found that the discussion pages fostered collaboration among the students, and allowed them to share knowledge and exchange ideas with each other. Facebook Group also encouraged students to become active learners [45]. The students were stimulated to discuss and do well by the healthy competition resulted from having more than one discussion group. The factor affecting student's collaboration was how students use Facebook [46].

Furthermore, it was found that the students tailored their discussions around the evaluation criteria given by the teacher. This finding that the evaluation criteria lead students' discussions is consistent with the previous research [39]. Facebook Group promotes a small-group development. Tuckman suggested that a group goes through a five stage development process that includes forming, storming, norming, performing, and adjourning. Further research is recommended for understanding group development on Facebook [47].

The participation frequency in the Facebook discussions can show the students' interests in learning from each other about the topics. This study revealed that B Group participants did not improve significantly in the academic performance for the exhibit. However, the participants in F Group made great improvement in the academic achievement. It seemed that student interactions on Facebook Group were more beneficial to the students in F Group.

Furthermore, it was found that the groups posted different types of information on their discussion pages. F group students offered more news in the field of industrial and commercial design (86.6%) than B Group did (63.5%). This may explain in F Group make more improvement in their academic performance. The information they shared with Facebook Group helped F Group to gain better grades. However, looking from another perspective, it is possible that students in B group can make use of the learning opportunity and perform better in the future because they posted more discussions related to building professional skills (17.6%) than F group did (1.1%).

Returning to the second research question, we can definitively state that there is a positive relationship between the participation frequency in Facebook Group and the academic performance. Facebook group appears to play a role in the process by which students prepare for the design exhibit. It therefore can be considered as a teaching tool to motivate learning and facilitate student discussions.

5.3   How the Teacher Interacts with Students on Facebook

Interactions between the students and the teacher on Facebook are another factor considered in the Facebook usage for educational purpose. In the study, there was no significant difference in the teacher's participation frequency between the two groups. However, the teacher had a higher rate of passive participation than did active participation in both groups. Comments posted by the teacher supported and encouraged the students' discussions, and balanced his appearances for the two groups in the study. When the discussion happened at midnight, the teachers overloading occurred also.

Facebook creates positive interactions and collaborations between teachers and their students. As in the study, students communicated with each other and learned on the platform. The features of effective group work and cooperative learning are clear learning objectives directing the groups' activities, encourages learners to interact and help each other to reach objectives [48]. Whether the teacher's role is a facilitator in the online discussions needs to be further investigated.

5.4   Drawbacks of Using Facebook as a Tool in Education

Who can get the benefits in learning from Facebook group discussion? It might be the students who are shy and prefer not to express their thoughts and discussions in public. Facebook discussion group provides a good platform for those who are shy and prefer to communicate through text. Some students had low participation frequency on Facebook discussions and it may be that some of these students are already very knowledgeable about the field of design so they didn't need much of interactions and inputs from other students. Still there were also students who spent time on chatting or playing games online rather than taking the time to participate in the Facebook discussions for this course. Thus it's important that the instructor knows when to intervene in such case.

Students' motivation to participate in Facebook discussions is not necessary the same. We can't conclude that participating in Facebook discussions can benefit all students. Some students found social identity, and the opportunity to lead and exert influence on others during the Facebook discussions while other students might gained nothing from the discussions. Personalities can have an important effect on how people present themselves on Facebook. Open and extroverted students use social media more [28]. Participants with high frequency of computing results in the restricting time for thinking and other learning work. If we had allowed the Facebook group discussions to continue for a longer time, would it still give the same results? Before we use Facebook as a teaching tool, we need to think thoroughly. Which group of students benefit from using FB in academic learning, how to incorporate FB into curriculum design and what's the purpose of using FB in the curriculum design really need to be investigated.

6   Conclusion

Although there have been many research studies on the learning function of Facebook, too few are based on evidence from the classrooms, and even less consider the teaching design effects on student learning. This study is a report not a survey. The data collected were objective because they were obtained after the course was finished. The experimental bias might not be found. Four conclusions are drawn: (1)The students in one of the groups on Facebook made much improvement in academic performance after ten-week discussions.   (2) There is a significant positive correlation between a student's online participation frequency and his GPA. And the frequency of students' participation on Facebook Groups could explain 15% to 19% of the variance in his academic performance. (3)The teacher participated on the discussion more passively. (4) The evaluation criteria and teacher facilitation in discussions can have an effect on the direction of online discussion. The conclusions may give the hints to the instructors for a better curriculum design next time. If the instructors make good use of Facebook, it not only can play a positive role for the academic achievement in the discussion section after class, it's also a great tech tool in teaching.

The present results are yet to be generalized to other student populations. To determine if these results can be generalized to other students requires a wider sampling. Therefore, we suggest that there be a further large scale of research on larger and more representative samples and we recommend future research studies explore student interaction more extensively. In the future we would like to continue this research with a larger and more varies sample of students so that conclusions can be extended to macro level.

References

1. Hattie, J.: Teachers make a difference: what is the research evidence? Australian Council for Educational Research Annual Conference on Building Teacher Quality (2003), http://www.acer.edu.au/documents/RC2003_Hattie_TeachersMakeA Difference.pdf

2. Parks, M.R.: Social network sites as virtual communities. In: Papacharissi, Z. (ed.) A Networked Self: Identity, Community, and Culture on Social Network Sites, pp. 105–123. Routledge, New York (2011)
3. Facebook, https://newsroom.fb.com/company-info/
4. Boyd, D.: Social Network Sites as Networked Publics. In: Papacharissi, Z. (ed.) A Networked Self: Identity, Community, and Culture on Social Network Sites, pp. 39–58. Routledge, NY (2011)
5. Coleman, J.S.: Social capital in the creation of human capital. Am. J. Sociol. 94, S95–S120 (1988)
6. Semo, R., Karmel, T.: Social capital and youth transitions: do Young people's networks improve their participation in education and training? Occasional Paper, National Centre for NCVER, pp. 1–40, ED524430 (2011)
7. Junco, R.: Too much face and not enough books: the relationship between multiple indices of Facebook use and academic performance. Comput. Hum. Behav. 28, 187–198 (2012)
8. Hew, K.F.: Review: students' and teachers' use of Facebook. Comput. Hum. Behav. 27, 662–676 (2011)
9. Friedkin, N.E., Johnsen, E.C.: Social Influence Network Theory, Cambridge, New York (2011)
10. Wellman, B.: Computer networks as social networks. Science 293, 2031–2034 (2001)
11. Marin, A., Wellman, B.: Social network analysis: an introduction. In: Scott, J., Carrington, J. (eds.) The SAGE Handbook of Social Network Analysis, pp. 11–25. SAGE, CA (2011)
12. Borgatti, S., Mehra, A., Brass, D., Labianca, G.: Network analysis in the social sciences. Science 323, 892–895 (2009)
13. Gruzd, A., Haythornthwaite, C.: Networking online: cybercommunities. In: Scott, J., Carrington, J. (eds.) The SAGE Handbook of Social Network Analysis, pp. 167–179. SAGE, CA (2011)
14. Tasir, Z., Al-Dheleai, Y.M.H., Harun, J., Shukor, N.A.: Students' perception towards the use of social networking as an e-learning platform. Recent Researches in Education, 70–75 (2011), http://www.wseas.us/e-library/conferences/2011/Penang/EDU/EDU-10.pdf
15. Yang, H.H., Wu, C.I., Lei, M.K., Yang, H.J.: The nomination number of adolescent social networks. In: Proc. of the 8th WSEAS International Conference on Telecommunications and Informatics, pp. 55–61 (2009), http://www.wseas.us/e-library/conferences/2009/istanbul/../TELE-INFO-07.pdf
16. Kuo, L.H., Yang, H.H., Yu, J.C., Yang, H.J., Sue, S.M.: The social network structure of in-service advancement education for English literature teachers in high school, pp. 158–163 (2010), http://www.wseas.us/e-library/conferences/2010/Taipei/AIBE/AIBE-25.pdf
17. Yunus, M.M., Salehi, H., Sun, C.H., Yen, J.Y.P.: Using facebook groups in teaching ESL writing, pp. 75–80 (2011), http://www.wseas.us/e-library/conferences/2011/Montreux/COMICICBIO/COMICICBIO-11.pdf
18. Roblyer, M.D., McDaniel, M., Webb, M., Herman, J., Witty, J.V.: Findings on facebook in higher education: a comparison of college faculty and student uses and perceptions of social networking sites. Internet High. Educ. 13(3), 134–140 (2010)
19. Pempek, T.A., Yermolayeva, Y.A., Calvert, S.L.: College students' social networking experiences on Facebook. J. Appl. Dev. Psychol. 30, 227–238 (2009)
20. Tsai, C.-M., Huang, Y.-T., Hsieh, J.-L.: Taiwanese facebook users' motivation and the access of information technology. In: Stephanidis, C. (ed.) Posters, Part I, HCII 2011. CCIS, vol. 173, pp. 469–473. Springer, Heidelberg (2011)

21. Tu, B.M., Wu, H.C., Hsieh, C., Chen, P.H.: Establishing new friendships—from Face-to-Face to Facebook: a case study of college students. In: Proc. of the 44th Hawaii International Conference on System Sciences, pp. 1–10 (2011)
22. Nadkarni, A., Hofmann, S.G.: Why do people use Facebook? Pers. Indiv. Differ. 52, 243–249 (2012)
23. Muñoz, C.L., Towner, T.L.: Opening Facebook: how to use Facebook in the college classroom. In: Society for Information Technology and Teacher Education Conference, Charleston, SC (2009)
24. Baran, B.: Facebook as a formal instructional environment. Brit. J. Educ. Technol. 41(6), E146–E149 (2010)
25. Bicen, H., Cavus, N.: Social network sites usage habits of undergraduate students: case study of Facebook. Procedia Social and Behavior Sciences 28, 943–947 (2011)
26. Coromina, L., Capo, A., Guia, J., Coenders, G.: Effect of background, attitudinal and social network variables on PhD students' academic performance, a multimethod approach. Estudios Sobre Educacion 20, 233–253 (2011)
27. Mazman, S.G., Usluel, Y.K.: Modeling Educational Usage of Facebook. Comput. Educ. 55, 444–453 (2010)
28. Rouis, S., Limayem, M., Salehi-Sangari, E.: Impact of facebook usage on students' academic achievement: role of self-regulation and trust. Electronic Journal of Research in Educational Psychology 9(3), 961–994 (2011)
29. Forkosh-Baruch, A., Hershkovitz, A.: A case study of Israeli higher-education institutes sharing scholarly information with the community via social networks. Internet High. Educ. 15, 58–68 (2012)
30. LaRue, E.M.: Using Facebook as course management software: a case study. Teaching and Learning in Nursing 7, 17–22 (2012)
31. Kirschner, P.A., Karpinski, A.C.: Facebook and academic performance. Comput. Hum. Behav. 26, 1237–1245 (2010)
32. Junco, R., Cotten, S.R.: The relationship between multitasking and academic performance. Comput. Educ. (2012), doi:10.1016/j.compedu.2011.12.023
33. Hunley, S., Evans, J., Delgado-Hachey, M., Krise, J., Rich, T., Schell, C.: Adolescent computer use and academic achievement. Adolescence 40(158), 307–318 (2005)
34. Grossecka, G., Bran, R., Tiru, L.: Dear teacher, what should I write on my wall? a case study on academic uses of Facebook. Procedia Social and Behavioral Sciences 15, 1425–1430 (2011)
35. Junco, R.: The relationship between frequency of Facebook use, participation in Facebook activities, and student engagement. Comput. Educ. 58, 162–171 (2012)
36. Gambrell, L.B.: What we know about motivation to read. In: Flippo, R.F. (ed.) Reading Researchers in Search of Common Ground, pp. 129–143. International Reading Association, Newark (2001)
37. Pollara, P., Zhu, J.: Social networking and education: using Facebook as an edusocial space. In: Proc. of Society for Information Technology & Teacher Education International Conference, pp. 3330–3338. ACCE, Chesapeake (2011)
38. Sturgeon, C.M., Walker, C.: Faculty on Facebook: confirm or deny? In: 14th Annual Instructional Technology Conference, Middle Tennessee State University, pp. 29–31 (March 2009)
39. Dahlgren, L.O., Fejes, A., Abrandt-Dahlgr, M., Trowald, N.: Grading systems, features of assessment and students' approaches to learning. Teaching in Higher Education 14(2), 185–194 (2009)

40. Mazman, S.G., Usluel, Y.K.: Modeling educational usage of Facebook. Comput. Educ. 55, 444–453 (2010)
41. Haughey, E.J.: An Analysis of Student and Faculty Attitudes Toward Grading Practices in Relation to Student Characteristics of Entering Students and Attrition. A Research Project in Nova University, Fort Lauderdale, FL, ED150922 (1977)
42. Kabilan, M.K., Ahmad, N., Abidin, M.J.: Facebook: an online environment for learning of English in institutions of higher education? Internet High. Educ. 13(4), 179–187 (2010)
43. Ross, C., Orr, E.S., Sisic, M., Arseneault, J.M., Simmering, M.G., Orr, R.R.: Personality and motivations associated with facebook use. Comput. Hum. Behav. 25, 578–586 (2009)
44. Rugutt, J., Chemosit, C.C.: What motivates students to learn? contribution of student-to-Student Relations, student-faculty interaction and critical thinking skills. Educational Research Quarterly 11, 16–28 (2009)
45. Ajjan, H., Hartshorne, R.: Investigating faculty decisions to adopt Web 2.0 technologies: Theory and empirical tests. Internet High. Educ. 11, 71–80 (2008)
46. Lampe, C., Wohn, D.Y., Vitak, J., Ellison, N.B., Wash, R.: Student use of facebook for organizing collaborative classroom activities. International Journal of Computer-Supported Collaborative Learning 6(3), 329–347 (2011)
47. Tuckman, B.W., Jensen, M.A.: Stages of small-group development revisited. Group & Organization Studies 2(4), 419–427 (1977)
48. Johnson, D.W., Johnson, R.: Learning Together and Alone: Cooperation, Competition and Individualization, 8th edn. Allyn & Bacon, Needham Heights (2006)

Extraction of Indirect Effect
among Sectors in Industrial Network
Based on Input-Output Data

Koji Okuhara[1], Hiroshi Tsuda[2], and Hiroe Tsubaki[3]

[1] Osaka University, 2-1, Yamadaoka, Suita, 565-0871 Japan
okuhara@ist.osaka-u.ac.jp
[2] Department of Mathematical Sciences, Doshisha University,
Kyotanabe, Kyoto, 610-0321, Japan
[3] The Institute of Statistical Mathematics,
10-3, Midori-cho, Tachikawa, Tokyo, 190-8562, Japan

Abstract. Input-Output table is important for analyzing the relationship among industrial sectors when we decide a policy making. In this paper we propose to consider the indirect effect between sectors in sales and sectors in purchases from Input-Output table. We adopt a economical interaction model permitting consideration of several path selection for each origin-destination pair. The economical interaction model is considered as a general model because it has been developed by entropy maximization principle. From observed data about total output in sales and purchases, our model can estimate interdependencies between different industrial sectors. The parameter estimation procedure for the proposed model is developed.

Keywords: input-output table, indirect effect, industrial policy making, economical interaction model.

1 Introduction

In this paper we explain the concept of power of sector in industry and describe the conditions that such index should satisfy. It is understood that the eigen value and eigen vector play important role in derivation of power of sector in industry. Next we develop a economical interaction model which permits consideration of several path selection for each origin-destination (OD) pair. Based on the entropy maximization principle, we derive theorem about relationship between the power of sector and interdependency in economical interaction model.

Furthermore in order to obtain the input volume from origin to destination depending on observed data about input flow, algorithm to estimate model parameters must be derived. For the purpose we give useful theorem about derivation of model parameters. The relationship between the power of sector and interdependency of the economical interaction model is revealed based on entropy maximization principle.

L.S.-L. Wang et al. (Eds.): MISNC 2014, CCIS 473, pp. 383–392, 2014.

This paper is organized as follows. The next section devotes to introduce the explanation of concept of Input-Output table used for industrial policy making. We formulate indirect effect by using extended spatial interaction model in which path selection can be considered from origin to destination. The relationship between Input-Output table and graph of Decision making trial and evaluation laboratory (DEMATEL) [1] [2] are discussed.

In section 3, we introduce extended spatial interaction model in which path selection can be considered from origin to destination and propose interdependency based on an entropy maximization principle[3]. We apply its concept to estimate the power of sector in industry.

In section 4, a relation among a power of sector and interdependency is shown, and an estimation algorithm to derive flow volume from observed data will be derived. The availability of application of properties to data mining of industrial interdependency is also shown in section4.

2   Extraction of Network Structure and Influence Rate

Input-output table is given by matrix structure whose elements measure represent the interdependencies between different industrial sectors for particular national economy or different regional economies. Input-output analysis seeks to explain how one industry sector affects others. The analysis is represented as a matrix, where different rows and columns are filled with values representing the inputs and outputs of various sectors. It can show that the output of one sector can in turn become an input for another sector, which results in an interlinked economic system. We denote an origin sector i by o_i and a destination sector j by d_j.

In economics, an input-output table presents a quantitative economic technique like input-output analysis. It was one of the major conceptual models for a socialist planned economy and is used for industrial policy making. For example, the table about output of gross illustrates the relationship between total gross output, value added, and gross domestic product. Commodities consumed by industries gives intermediate inputs and by value added gives final use. Total gross output is the sum of intermediate inputs and value added. The total gross domestic product is equal to the sum of value added.

It is generally accepted that input-output tables are useful and important for economic and industrial structural analyses, and economic projections. Table 1 shows a sample of input-output table for both supply sectors and demand sectors.

Now we introduce a new concept like a background effect among intermediate inputs, value added and final demand. Background effect consists of direct and indirect impacts. For example, it is assumed that sectors of input-output table are given by a set $\Omega = \{1, 2, 3, 4, 5, 6, 7\}$. It satisfies $\Omega = \Omega_o \cup \Omega_d$ where $o_i \in \Omega_o$ ($\forall i$) and $d_j \in \Omega_d$ ($\forall j$) as follows; 1: Agriculture, 2: Mining, 3: Manufacturing, 4: $\cdots$, 5: Services, 6: Value added, 7: Final Demand. In this study we represent the relationship between those economical sectors by using graph network.

Table 1. Illustrative matrix of input-output table

Sales	Purchases					Final Demand	Total Output
	Agriculture	Mining	Manufacturing	$\cdots$	Services		
Agriculture	$r_{o_1 d_1}$	$r_{o_1 d_2}$	$r_{o_1 d_3}$	$\cdots$	$r_{o_1 d_{m-1}}$	$r_{o_1 d_m}$	V_{o_1}
Mining	$r_{o_2 d_1}$	$r_{o_2 d_2}$	$r_{o_2 d_3}$	$\cdots$	$r_{o_2 d_{m-1}}$	$r_{o_2 d_m}$	V_{o_2}
Manufacturing	$r_{o_3 d_1}$	$r_{o_3 d_2}$	$r_{o_3 d_3}$	$\cdots$	$r_{o_3 d_{m-1}}$	$r_{o_3 d_m}$	V_{o_3}
$\vdots$	$\vdots$	$\vdots$	$\vdots$	$\vdots$	$\vdots$	$\vdots$	$\vdots$
Services	$r_{o_{n-1} d_1}$	$r_{o_{n-1} d_2}$	$r_{o_{n-1} d_3}$	$\cdots$	$r_{o_{n-1} d_{m-1}}$	$r_{o_{n-1} d_m}$	V_{o_4}
Value Added	$r_{o_n d_1}$	$r_{o_n d_2}$	$r_{o_n d_3}$	$\cdots$	$r_{o_n d_{m-1}}$	$-$	$-$
Total Output	U_{d_1}	U_{d_2}	U_{d_3}	$\cdots$	$U_{d_{m-1}}$	$-$	T

Effect from one sector to another sector is total sum of effects which derived by direct and indirect path selection. Our model is important in identifying such background effect theoretically. We apply DEMATEL for extracting network structure and influence rate among sectors in industry. DEMATEL as a tool to support decision-makers is used for evaluating experimental attempts. It was developed in Battelle Institute of America to face worldwide complex problems. Its application area lies in the analysis of complex and uncertain factors within a problem. Its feature is trying to take full advantage from the institution information and as well people related to the problem under study.

Having this in mind, we used this approach in order to understand the overall impact of the relationship between the items related to industrial information. First, the information regarding ratio from input-output data is arranged vertically and horizontally. Then, the items are represented into seven criteria and through a paired comparison the influence of an item i over j is analyzed given a direct relation matrix $\mathbf{A}$: $[a_{ij}]$.

Table 2. Illustrative direct relation matrix $\mathbf{A}$

	1	2	3	4	5	6	7
1	0	a_{12}	a_{13}	a_{14}	a_{15}	a_{16}	a_{17}
2	a_{21}	0	a_{23}	a_{24}	a_{25}	a_{26}	a_{27}
3	a_{31}	a_{32}	0	a_{34}	a_{35}	a_{36}	a_{37}
4	a_{41}	a_{42}	a_{43}	0	a_{45}	a_{46}	a_{47}
5	a_{51}	a_{52}	a_{53}	a_{54}	0	a_{56}	a_{57}
6	a_{61}	a_{62}	a_{63}	a_{64}	a_{65}	0	a_{67}
7	a_{71}	a_{72}	a_{73}	a_{74}	a_{75}	a_{176}	0

Based on this, the initial direct influence $\mathbf{X}$:$[x_{ij}]$ which relatively expressed strength was calculated. The initial direct influence matrix $\mathbf{X}$ computed the sum of each line of the direct relation matrix $\mathbf{A}$ by breaking each element of $\mathbf{A}$ and maximizing them.

$$\mathbf{X} = \frac{1}{\min_{1 \leq j \leq n} \sum_{i=1}^{n} |a_{ij}|} \mathbf{A} \tag{1}$$

The indirect influences were computed by introducing an identity matrix $\mathbf{I}$ in the total relation matrix $\mathbf{F}{:}[f_{ij}]$ as presented next.

$$\mathbf{F} = \mathbf{X} + \mathbf{X}^2 + \mathbf{X}^3 + \mathbf{X}^4 + \cdots = \mathbf{X}\,(\mathbf{I} - \mathbf{X})^{-1} \tag{2}$$

Consider a direct impact matrix as shown in Table 2 referring to the data in Table 1 herein. Table 2, from which we based our analysis, shows an example where are evaluated 7 relative strengths of the causal relationship leading to Services from Manufacturing.

The total relation matrix $\mathbf{F}$ obtained by the DEMATEL method. Row sum d of the total relation matrix $\mathbf{F}$ represents the sum of the effects that certain items give to other items. Column sum r of the total relation matrix $\mathbf{F}$ represents the sum of the effects that certain items receives from other items. $d+r$ corresponds to the sum of non-impact and impact, the item indicates whether $\mathbf{I}$ has a central role in the extent to which the issue structure.

Figure 1 shows a sample when one sector effects to another sector each other which depends on economical network structure. It shows the causal relationship between the items through a directed graph.

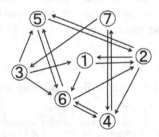

Fig. 1. Sample of causal relationship between the items through directed graph

In addition, ij element of the square of the direct relation matrix $\mathbf{A}$ represents the number of paths to reach item j from item i. Similarly, ij element of the cube of the $\mathbf{A}$ represents the number of lengths of the path from i to j. The sum $A, A^2, A^3, \cdots$ represents the number of paths in order to represent the possibility of arrival, i.e., the presence of the path to obtain a reachability matrix $\mathbf{R}$.

$$\mathbf{R} = \frac{\mathbf{I}}{\mathbf{I} - \mathbf{A}} = \mathbf{I} + \mathbf{A} + \mathbf{A}^2 + \mathbf{A}^3 + \cdots \tag{3}$$

The formulation of problem in this paper is described here. We develop an extraction method of indirect effect among sectors under the following two conditions. First condition is that we can not grasp accurate data about values $r_{o_i d_j}$

in input-output table are given in advance by collection, however, we can obtain the row sum U_{d_j} and column sum V_{o_i}. Second condition is that we can obtain the graph of causal relationship between sectors as items and value r_{ij} which is calculated by the element in the following matrix $\tilde{\mathbf{R}}$ derived from reachability matrix $\mathbf{R}$.

$$\tilde{\mathbf{R}} = \mathbf{R} - \mathbf{I} = \mathbf{A} + \mathbf{A}^2 + \mathbf{A}^3 + \cdots \tag{4}$$

3 Interdependency in Economical Interaction Model

An OD matrix usually represents the transportation on a network. If the origins are listed as the rows and the destinations as the columns a from/to or $r_{o_i d_i}$ matrix is produced like Table 1. The row sums, V_i, will be the total origin traffic, the column sums, U_j, the total destination traffic and T denotes total traffic counts. Especially traffic counts at the OD pair is a required input to estimate the elements of future OD matrix, estimation of zone traffic volume from base OD matrix is a fundamental problem for the planning and evaluation of transportation systems. The OD matrix information is important for not only traffic volume analysis but also information network model, if we regard the interaction $r_{o_i d_i}$ reflects data packet [4].

A spatial interaction model which permits several paths for each OD pair can be derived from an entropy maximization principle as follows:

$$\text{Maximize} \quad \frac{T!}{\prod_{i=1}^{n} \prod_{j=1}^{m} \prod_{k=1}^{K_{ij}} r_{o_i d_j}^k!} \tag{5}$$

$$\text{Subject to} \quad E_{V_i} = V_{o_i} - \sum_{j=1}^{m} \sum_{k=1}^{K_{ij}} r_{o_i d_j}^k = 0 \quad (\forall i) \tag{6}$$

$$E_{U_j} = U_{d_j} - \sum_{i=1}^{n} \sum_{k=1}^{K_{ij}} r_{o_i d_j}^k = 0 \quad (\forall j) \tag{7}$$

$$E_C = TC - \sum_{i=1}^{n} \sum_{j=1}^{m} \sum_{k=1}^{K_{ij}} c_{o_i d_j}^k r_{o_i d_j}^k = 0 \tag{8}$$

where $c_{o_i d_j}$ denotes measure reflecting distance concept from an origin node o_i to a destination node d_j and C is constant [5].

Consider an application of Stirling's approximation, $\ln T! \approx T \ln T - T$, Lagrange undetermined multipliers γ_i, μ_j, ρ for each constraint and the condition of total sum of $r_{o_i d_j}^k$ must be equivalent to T, the interaction can be derived as

$$r_{o_i d_j}^k = \varphi \exp(-\gamma_i - \mu_j - \rho c_{o_i d_j}^k) \tag{9}$$

where coefficient φ is defined by

$$\varphi = \frac{T}{\sum_{i=1}^{n} \sum_{j=1}^{m} \sum_{k=1}^{K_{ij}} \exp(-\gamma_i - \mu_j - \rho c_{o_i d_j}^k)} \tag{10}$$

Now introduce the following variables:

$$\alpha_{o_i} = \frac{\exp(-\gamma_i)\sqrt{\varphi}}{V_{o_i}} \tag{11}$$

$$\beta_{d_j} = \frac{\exp(-\mu_j)\sqrt{\varphi}}{U_{d_j}} \tag{12}$$

so that

$$r_{o_i d_j}^k = \alpha_{o_i} V_{o_i} \beta_{d_j} U_{d_j} \exp(-\rho c_{o_i d_j}^k) \tag{13}$$

In a maximum entropy model with a negative exponential distribution as an attenuation function, consider the following translation:

$$c_{o_i d_j}^k = \log \phi_{o_i d_j}^k \tag{14}$$

therefore we obtain

$$r_{o_i d_j} = \sum_{k=1}^{K_{ij}} \alpha_{o_i} V_{o_i} \beta_{d_j} U_{d_j} \left(\phi_{o_i d_j}^k \right)^{-\rho} \tag{15}$$

This means a gravity model with negative polynomial function as an attenuation function can be regarded as one kind of the maximum entropy model.

From constraints about sums of row and column in OD matrix, we can define

$$\alpha_{o_i d_j}^{-1} = \sum_{k=1}^{K_{ij}} \beta_{d_j} U_{d_j} \exp(-\rho c_{o_i d_j}^k) \tag{16}$$

$$\beta_{o_i d_j}^{-1} = \sum_{k=1}^{K_{ij}} \alpha_{o_i} V_{o_i} \exp(-\rho c_{o_i d_j}^k) \tag{17}$$

In order to understand what interdependency means, suppose that there is two side like decision maker side and facility side. An inverse value of $\beta_{o_i d_j}$ represents sum of accessibilities for all decision makers, each interdependency for the sector in sales side is calculated by damping with respect to distance concept of decision maker at zone o_i to facility at zone d_j. That is, it means the interdependency (or attraction and so on) of each facility from decision maker's viewpoint. Similarly, an inverse value of $\alpha_{o_i d_j}$ represents an interdependency of each sector in sales side from industrial viewpoint. Note that absolute of it value has no meaning, it is only possible to compare relatively.

4   Derivation of Power of Sector from Interdependency

Here we will show the relationship between the power of sector and the interdependency in economical interaction model. The power of sector has been given for each node, thus we derive the same one about the interdependency as follows;

$$\alpha_{o_i}^{-1} = \sum_{j=1}^{m} \alpha_{o_i d_j}^{-1} \tag{18}$$

$$\beta_{d_j}^{-1} = \sum_{i=1}^{n} \beta_{o_i d_j}^{-1} \tag{19}$$

$\alpha_{o_i}^{-1}$ is total interdependency of sector o_i in sales side from all sector in purchases' viewpoint, also $\beta_{d_j}^{-1}$ is total interdependency of sector d_j in purchases side from all sector in sales' viewpoint. We can show the following theorem to recognize the relationship between the power of sector and the interdependency.

Theorem 1. *The power of sector* $\mathbf{g}_s \in \Re^n \times \Re^1$ *is derived as the eigen vector for the matrix* $\mathbf{M} \in \Re^n \times \Re^n$ *whose element consists of the interdependency* $\alpha_{o_i d_j}^{-1}$ *of each sales from sector* d_j *in purchases side and the total interdependency* π_{o_i} *in sector* o_i *from all purchases' viewpoint.*

Proof. We can obtain the transition probability from o_i to d_j as follows;

$$m_{ji} = \frac{r_{o_i d_j}}{V_{o_i}}$$

$$= \alpha_{o_i} \sum_{k=1}^{K_{ij}} \beta_{d_j} U_{d_j} \exp(-\rho c_{o_i d_j}^k) \tag{20}$$

This relationship implies that the modified matrix $\mathbf{M}$ can be reconstructed by using interdependency as follows;

$$m_{ij} = \frac{\alpha_{o_j d_i}^{-1}}{\alpha_{o_i}^{-1}} \tag{21}$$

From these results, the power of sector in industry can be regarded as the eigen vector for the matrix whose element consists of the total interdependency from all purchases' viewpoint and the interdependency of each sector of sales from purchases side. □

Therefore derivation of power of sector in industry from observed data based on the input-output flow at node is extension model for usual one. In order to derive the index of power of sector in industry the parameters must be estimated, for the purpose we can apply the nonparametric approach like a mixture of expert model [6]. However a prospective convergence time of parameter estimation may be long, thus we derive the following useful theorem and an estimation algorithm based on the theorem is developed.

Theorem 2. *Assuming* $\gamma_i = -\log g_i^{(\gamma)}$, *then the derivation of power of sector in industry* $\mathbf{g}^{(\gamma)} = [g_1^{(\gamma)}, g_2^{(\gamma)}, g_3^{(\gamma)}, \cdots, g_n^{(\gamma)}]^{\mathrm{T}} \in \Re^n \times \Re^1$ *is the same of finding eigen vector for matrix* $\mathbf{M}^{(\gamma)} \in \Re^n \times \Re^n$.

Proof. From Eqs. (6) and (9) we can find

$$g_i^{(\gamma)} = \frac{V_{o_i}}{T} \sum_{j=1}^{n} \frac{\sum_{s=1}^{m} \sum_{k=1}^{K_{js}} \exp(-\mu_s - \rho c_{o_j d_s}^k)}{\sum_{s'=1}^{m} \sum_{k'=1}^{K_{is'}} \exp(-\mu_{s'} - \rho c_{o_i d_{s'}}^{k'})} g_j^{(\gamma)} \tag{22}$$

Also we consider

$$a_{ij}^{(\gamma)} = V_{o_i} \sum_{s=1}^{m} \sum_{k=1}^{K_{js}} \exp(-\mu_s - \rho c_{o_j d_s}^k) \tag{23}$$

then we can obtain

$$g_i^{(\gamma)} = \sum_{j=1}^{n} \frac{a_{ij}^{(\gamma)}}{\sum_{t'=1}^{n} a_{t'i}^{(\gamma)}} g_j^{(\gamma)} \tag{24}$$

Let define an element of matrix $\mathbf{M}^{(\gamma)}$ by

$$m_{ij}^{(\gamma)} = \frac{a_{ij}^{(\gamma)}}{\sum_{t'=1}^{n} a_{t'i}^{(\gamma)}} \tag{25}$$

then the following formulation can be considered

$$\mathbf{Mg} = \lambda \mathbf{g} \tag{26}$$

We can find that the sum of row in matrix $\mathbf{M}^{(\gamma)}$ becomes as follows;

$$\sum_{i=1}^{n} m_{ij}^{(\gamma)} = 1 \tag{27}$$

This means that the maximum eigen value is $\lambda_{max} = 1$ and the vector $\mathbf{g}^{(\gamma)}$ is an eigen vector for the maximium eigen value of matrix $\mathbf{M}^{(\gamma)}$.   □

Similarly for the parameter μ_j we have

$$a_{ij}^{(\mu)} = U_{d_i} \sum_{s=1}^{n} \sum_{k=1}^{K_{sj}} \exp(-\gamma_s - \rho c_{o_s d_j}^k) \tag{28}$$

and

$$m_{ij}^{(\mu)} = \frac{a_{ij}^{(\mu)}}{\sum_{t'=1}^{n} a_{t'i}^{(\mu)}} \tag{29}$$

The eigen vector for the maximum eigen value of matrices $\mathbf{M}^{(\gamma)}$ and $\mathbf{M}^{(\mu)}$ can be obtained by using power method.

[**Algorithm of Parameter Estimation**]

Step 1. Give the initial values of $\rho^{(0)}$ and $\mu_i^{(0)}$.

Step 2. Calculate $\gamma_i^{(t+1)}$ as an ememt of eigen vector for maximum eigen value based on Eqs. (26) and (25) by using parameters $\rho^{(t)}$ and $\mu_i^{(t)}$. t denotes iteration number.

Step 3. Calculate $\mu_i^{(t+1)}$ as an ememt of eigen vector for maximum eigen value based on Eqs. (26) and (29) by using parameters $\rho^{(t)}$ and $\gamma_i^{(t+1)}$.

Step 4. If $|\mu_i^{(t+1)} - \mu_i^{(t)}| < \epsilon$ and $|\gamma_i^{(t+1)} - \gamma_i^{(t)}| < \epsilon$ are satisfied then go to next, otherwise return to **Step 2**. ϵ is small value given to judge convergence for $t > 0$.

Step 5. Check the condition of Eq. (8) if the condition is satisfied then stop procedure, otherwise calculate $\rho^{(t+1)} = \rho^{(t)} + \delta(E_C)$ where $\delta(E_C)$ takes a positive small value ϵ^+ if E_C is negative otherwise a negative small value ϵ^- if E_C is positive, then return to **Step 2**.

5 Conclusion

In this paper, we explained the concept of power of sector in industry and proposed an extended spatial interaction model permitting consideration of several path selection for each origin-destination pair. Based on the entropy maximization principle, we derived theorem about relationship between the power of sector in industry and interdependency in economical interaction model. From observed data about traffic flow to obtain the traffic volume from origin to destination, algorithm to estimate model parameters has been derived by using eigenvalue problem. It can be regarded as a merged procedure between the power of sector in industry and interdependency of the economical interaction model in unified way. In numerical experiment, the availability of our model which can consider not only a network structure but also a traffic flow has been shown. As the further work, project management have a number of common points with traffic assignment [7], thus we think that such research area is one of applicable scope.

Acknowledgment. This work was supported by JSPS KAKENHI Grant Number 22240030.

References

1. Toyoda, T., Horii, H.: Application of Structural Modeling Analysis to Social Problems. For Falsification Issues at Tepco's BWR Plant. Sociotechnology Research Network 1, 16–24 (2003)

2. Toyota, N., Saegusa, T.: Importance Measure of Instructional Elements in Teaching Strategies. IEICE Technical Report. Educational Technology 96(148), 121–128 (1996)
3. Okuhara, K., Yeh, K.Y., Shibata, J., Ishii, H., Hsia, H.C.: Evaluation and Assignment of Traffic Volume for Urban Planning Based on Planner and User Stance. International Journal of Innovative Computing, Information and Control 4(5), 1151–1160 (2008)
4. Okuhara, K., Tanaka, T., Ishii, H.: Routing and flow control by genetic algorithm for a flow model. Systems and Computers in Japan 34(1), 11–20 (2003)
5. Wilson, A.G.: Entropy in urban and regional modelling. Pion, London (1970)
6. Okuhara, K., Ishii, H., Uchida, M.: Support of decision making by data mining using neural system. Systems and Computers in Japan 36(11), 102–110 (2005)
7. Okuhara, K., Shibata, J., Ishii, H.: Adaptive Worker's Arrangement and Workload Control for Project Management by Genetic Algorithm. International Journal of Innovative Computing, Information and Control 3(1), 175–188 (2007)

Load Optimization of E-government System Based on Hiphop-PHP

Bingyu Ge[1], Zhiyi Fang[1], Ce Han[2], Lei Pu[2], Quan Zhang[2], and Hong Pei[1]

[1] College of Computer Science and Technology, Jilin University,
Changchun, 130012, P.R. China
[2] School of Information and Software Engineering,
Northeast Normal University, Changchun, 130117, P.R. China

Abstract. In this paper, we study the load capacity of e-government system and try to find the parts that can be optimized. In order to increase the load capacity of the server, we introduce Hiphop-PHP technology into the e-government system that was developed by php language. By using Hiphop-PHP program, PHP language can be converted into compiled C++ code. It can significantly reduce the computing time of server and the load pressure of the web server. At the same time, this paper also proposes other ways to improve server performance based on Hiphop-PHP technology. It combines the advanced cache reading technology, memory management technology and Hiphop-PHP technology together to improve the load capacity of the server. The simulation uses apache http server benchmarking tool that was developed by the apache software foundation to test the server. The result shows that the system can face up to the worst case of user requests. The introduction of Hiphop-PHP technology can significantly improve the load capacity of the server.

Keywords: load capacity, e-government, server, Hiphop-PHP.

1 Introduction

With the development of national economy, all kinds of industries begin the process of information technology. Also social networks have a rapid development. The main role of social networking service is to create an online community for a group of people that have the same interests and like to do the same activities. Such services are often based on Internet to provide users with a various ways to contact with each other such as email, instant messaging services and so on. With the evolution of social networks, the image of a person on the network tends to be more completely. The social networks appear at this time. Dating is just the beginning of social networks, as Google is just the backlinks of each web page the beginning. Social networks are just to get your personal information and friends list at the beginning. Social networks have undergoned such a process of development: Early conceptualization stage: SixDegrees represents for Six Degrees of Separation; Meeting strangers stages :The theory that Friendster helps you to build a weak relationship in order to get higher social capital; Entertainment stages: The theory that MySpace creates multimedia personalized space and attracts the

L.S.-L. Wang et al. (Eds.): MISNC 2014, CCIS 473, pp. 393–403, 2014.
© Springer-Verlag Berlin Heidelberg 2014

attention of people; Social graph stage: The theory that Facebook copied real social networks to online world for low-cost management. The whole process of SNS is to gradually move the lives of people in real world to the online world for low-cost management. This allows the virtual socializing cross with the real world socializing.

The government office is a part of the social life. So the process of e-government is also imperative. The overall implication of e-government is: Accelerate government office rate by the use of modern information technology and computer technology; Optimize and streamline government services; Optimize and balance the resource allocation of different regions and departments; Integrate data resources of different sectors; Avoid the situation that all kinds of information is provided separately so that it forms a separate information silos. E-government is an important part of national economic construction. The realization of making the e-government business electronic and intelligent has great significance[1,2]. Many e-government systems are developed by PHP. Thus, in this structure, improving the operating speed of the server and reducing the load pressure of the web server has great significance in making the system maintain stable when there are a large number of user requests[3,4].

2   Popular Server System Architectures

2.1   ASP / ASP.net + IIS

In early 1996, Microsoft company released WEB service framework called ASP. By using this service framework, programmers can easily embed VB script scripting language or JAVA script scripting language into the WEB page without using complicated CGI structure. In 2002, Microsoft company released ASP.net. From the name you can clearly see the relationship between this new service framework and ASP. This framework is an upgraded version of ASP. It is a kind of ASP that can support services framework of .net. The new service framework brings obvious advantages. ASP can only support VB script, JAVA Script and other weakly typed programming language. But ASP.net introduces C # programming language so it can support strong type programming language and makes the programmers can make good use of object-oriented methods in WEB development.

2.2   PHP+Apache

PHP is a kind of web programming language. The syntax and style of this language absorb the syntax of C language and Perl. The developers who are familiar to C language and Perl can quickly adapt to PHP and use it to develop system.

PHP also belongs to an interpreted language, which uses a kind of tag that can be identified by the server to separate itself from the html language. The structure of tags is as followed:

\<p> the content that server requires to explain. </ p>

When the server software recognizes this kind of tag structure in html codes, it will call the PHP interpreter to explain the contents that the label contains.

The server software that supports PHP language we commonly use is the software named Apache which is developed by the apache software foundation. Currently, Apache server software is very popular. On the one hand, it provides a very reliable security. On the other hand, Apache supports cross-platform service. The customers who use Apache no longer limited to windows operating system. They can choose a more secure operating system. In the process of constructing many e-government systems, it is very popular using PHP + Apache structure to build e-government systems as PHP has the advantage of openness and flexibility.

2.3   JSP + Apache + Tomcat

The full name of JSP is Java Server Pages. It is developed by Sun Microsystems and it is similar to WEB programming architecture. JSP is also composed by the label that can be identified by the server and the html language that can be rendered by the browser.

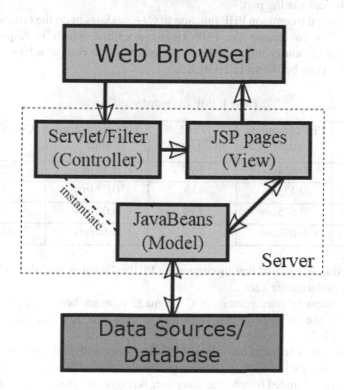

Fig. 1. The flowchart of JSP deals with user requests

Similar to PHP, JSP uses <% ...%> tags. But the difference between ASP and PHP is: JSP is not entirely an interpreted language. When JSP is called for, Java attachments on the server will convert the content in<% ....%> in the JSP web page into a java file called Servlet. Then by using the Java compiler, it becomes the binary code that can be running in the java virtual machine. Finally, pass the JSP pages components combined with html language to the client browser. Process is shown in Fig. 1.

3 The Design of Server Backend Based on Hiphop-PHP

3.1 Principle of Improving Load Capacity by Using Hiphop-PHP

Through the analysis of e-government system, the complex general calculation section(Encryption, decryption, other algorithms) in PHP is the section that occupies the computing performance of the server most. As PHP is a scripting interpreted language, including Asp, etc. The utilization of memory and CPU for this type language is very low. So the best way is to use C / C++ to replace PHP section which is responsible for calculating part.

By converting the common PHP function to C++ modules to do the test, we find that using C / C++ to replace the PHP language section which is responsible for calculating can obviously improve the performance in the case of a large number of users[5,6] .As it can be shown in TABLE I.

Table 1. PHP compared with C++

Test function	PHPcomputing time	PHP CPUload	C++computing time	C++ CPUload
MD5	0.624ms	76%	0.163ms	21%
Base64	0.153ms	43%	0.074ms	13%
AES	0.525ms	73%	0.087ms	15%
sha1_file()	0.692ms	77%	0.107ms	19%

In TABLE I, we can find that by converting PHP calculation module to C++ we can improve the performance a lot.

Since the memory management of C language is much better than that of PHP, therefore, the use of Hiphop-PHP technology uses less resource than the direct application of PHP.

Every time a user requests the server, PHP interpreter will explain every request for every time. When we use Hiphop technology, the key part of the explanation for WEB program has been handed over to the compiler. Background process uses binaries that have been compiled to run and to respond for each user request [7]. We can save the explanation time by this method. The ordinary running process is shown in Fig .2. The running process using Hiphop is shown in Fig. 3.

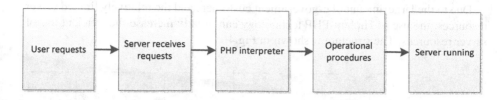

Fig. 2. The ordinary running process

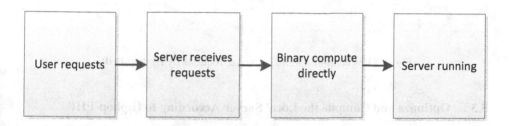

Fig. 3. The running process using Hiphop

3.2   E-government Systems with the Use of Hiphop-PHP Technology

Hiphop-PHP was originally a secret project of the global SNS giant Facebook. The project's purpose is to significantly enhance the utilization of server. The majority of pages and handlers of Facebook use PHP. With the growing number of users, the load of Facebook server is increasing greatly. Although companies can relieve the pressure by buying servers, the newly added servers also brought the same problem.

On the one hand, the servers brought a lot of power consumption. On the other hand, the servers also brought a variety of hardware failures. Although Facebook is currently the world's largest SNS websites, the team scale of engineers is relatively small. Facing up to of the rapid growth of user pressure, Facebook developed Hiphop-PHP technology in this situation. By using this technology, developers can quickly convert PHP critical computing section to C++ code. So it can avoid a large number of manual operations. On the other hand, it can also increase the error possibility of the converted process.

Currently, Facebook has made this technology known to public. The open source code of this technology is published on Github website. Any developer that needs them can obtain them easily. However, the technology has just appeared. Few companies use it.

Like most technologies, Hiphop-PHP technology also has a long evolution. Hiphop-PHP was originally composed by an interpreter called HPHPi syntax interpreter. This syntax interpreter is based on AST (Abstract Syntax Tree) abstract syntax tree. Through the static analysis of original PHP syntax, it can generate the C language intermediate code for the computer architecture of X86/X64. We can get the code that can be directly runned after C language intermediate code being compiled by the GCC.

Due to the large amount of e-government customers and the relatively limited server resources, the use of Hiphop-PHP technology can greatly increase the efficient use of server resources. The structure is shown in Fig.4.

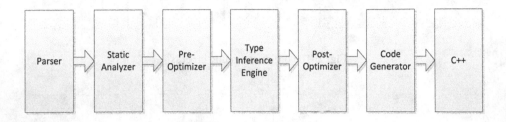

Fig. 4. Flow chart of optimizing server requests by Hiphop-PHP

3.3   Optimize and Compile the Local Server According to Hiphop-PHP

3.3.1   The Configuration of Server System

Our server system configuration is as followed:

Operating system: CentOS 6.3;

Processor: Intel Xeon E7 X4, 16 cores;

Memory: 32G

Nic: Intel Pro 1 G

3.3.2   The Step of Optimizing and Compiling the Local Server According to Hiphop-PHP

Step 1: install the necessary depending packages

We make a list of the required package files and realize automatic configuration by script:

```
#!/bin/bash
sudo yum install gitsvncpp make \
autoconfautomakelibtool patch \
memcachedgcc-c++ cmakewget \
boost-develmysql-develpcre-devel \
gd-devel libxml2-devel expat-devel \
libicu-devel bzip2-devel oniguruma-devel\
openldap-develreadline-devel \
libc-client-devellibcap-devel \
```

binutils-devel pam-devel \

elfutils-libelf-devel

In addition to the scripts that include some packages, Hiphop-PHP also needs libmcrypt package. This package contains some commonly encryption algorithms used in PHP, which include GOST, RIJNDAEL, DES, 3DES, Twofish, IDEA, CAST-256, ARCFOUR, SERPENT, SAFER +, etc. But libmcrypt encryption extension algorithm library is not including in Centos program library. In order to make Hiphop-PHP use the algorithm above normally, we need to use wget to manually download the extensions, then install the package manuall.

Step Two: Use Git Synchronizes Hiphop-PHP source

In fact,There is another important reason why we use the source code to compile and install. The reason is that the precompiled package is relatively old in general and can not be updated in time according to the need. But by using Git distributed source code management, we can easily get the latest hiphop source code and then do personalized configuration and installation.

first of all, we need to use Github to establish a new local hiphop source code repository:

'git clone git://github.com/facebook/Hiphop-PHP.git'

we also need to set environment variables for Hiphop-PHP after using git command to synchronize source code. In this way, Hiphop-PHP can know the location in which we want to install it. The environment path is set as follows:

CMAKE_PREFIX_PATH=/usr

HPHP_HOME=`/bin/pwd`

HPHP_LIB=`/bin/pwd`/bin

USE_HHVM=1

Step 3: Use third-party libraries to optimize server performance

With the use of Hiphop-PHP technology in e-government system, we also use third-party libraries to optimize server performance. The list of third-party libraries are as follows: libevent, libCurl, libunwind, Google glog, libmemcached, JEMalloc, Tbb, libdwarf[8].

Libevent: Libevent provides a network library functions that based on messages transferring. Using this library we can optimize some event handling mechanism of Hiphop-PHP.

libCurl: The function of libCurl is to add server that supports PHP to the recompiled C language. In order to make the php page compiled into binary code by using HPHPc directly, we need to rely on this library.

libunwind: The purpose of this library is mainly used to achieve parsing function based on AST (abstract syntax tree). Computer can automatically expand and merge the function and analyse its functions by using libunwind. This library is the core library when Hiphop-PHP converts php syntax to C++ automatically. We are unable to achieve automatic conversion function without this library.

Google glog: The library can provide the ability of printing log debug information. After the PHP is automatically converted into C++ language, the library can provide the ability to print log debug information .With this library, Hiphop-PHP can automatically handle some exceptions and reduce the probability of arising bugs.

libmemcached: The library provides a distributed computing capabilities. The class objects analysed by libunwind can be cached in memories by using this library, making distributed computing nodes can communicate with the access class objects.

JeMalloc :JeMalloc is an advanced memory management library, which uses technology called Arena. Generally speaking, this technique separates the shared memory of different threads logically, so that each thread is independent to read and write the memory it belongs to. By using this method, the memory fragmentation that generated in the runtime can be reduced to a very high degree. This is also an important method to optimize server computing. The working method of Jemalloc is shown in Fig. 5.

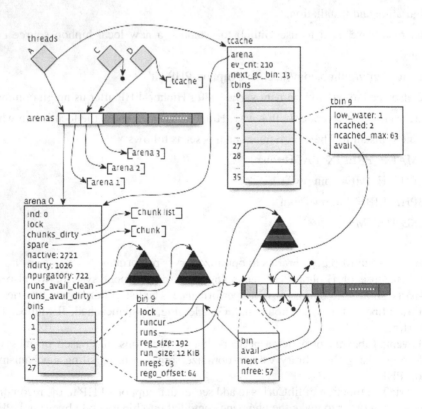

Fig. 5. The schematic of Memory allocation

As can be seen in Fig. 5, the region that memory threads occupy is divided into a number of arenas, and each arena has its own lock mechanism and id. Through this measure, thread can read and write in their own memory area instead of caring the whole allocation of memory.

Comparing with other memory managing mechanism, we can see the jemalloc have a very good throughput, as is shown in Fig. 6.

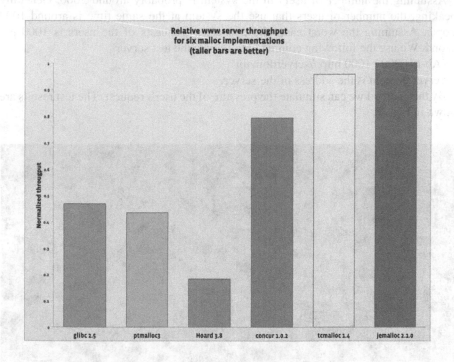

Fig. 6. The comparison

Tbb: Tbb is a parallel computing framework developed by intel, called thread building blocks. With this module, we can create threads in C / C++ language freely without worrying about the distribution of the thread.

libdwarf: The library record information and store debug information through property.

4   Simulation and Experiment

4.1   Experimental Environment Configuration

The operating system is CentOS 6.3. Processor is Intel Xeon E7 X4 with16 core. RAM is 32G. NIC is Intel Pro 1G. We use Apache HTTP server benchmarking tool that is developed by the Apache Software Foundation to test the server.

4.2   Test Methods and Results

Generally speaking, about the stress test we concernning most are the number of failed requests and the the ability of concurrent requests. To enable users to have a good web experience, we also concern about the average response time.

The test method is as followed:

Assuming the number of users in the system is probably around 5000. Generally speaking, the number of users that use the system at the same time is around 1000 people. Assuming the worst case, the concurrent requests of the users is 1000 per second. We use the following command to initiate the test server

"Ab-n10000-c1000 http://serverdomain"

Server domain is the address of the server.

By this method we can simulate the pressure of the user's request. The test results are shown in Fig . 7.

```
Server Software:        Apache
Server Hostname:        serverdomain
Server Port:            80

Document Path:          /
Document Length:        2017 bytes

Concurrency Level:      000
Time taken for tests:   4.876 seconds
Complete requests:      10000
Failed requests:        0
Write errors:           0
Non-2xx responses:      10000
Total transferred:      36510000 bytes
HTML transferred:       20130000 bytes
Requests per second:    2050.66 [#/sec] (mean)
Time per request:       487.648 [ms] (mean)
Time per request:       0.488 [ms] (mean, across all concurrent requests)
Transfer rate:          730.95 [Kbytes/sec] received

Connection Times (ms)
              min  mean[+/-sd] median   max
Connect:        1    6  81.1      2     301
Processing:     0    2  68.0     20     482
Waiting:        0    1  68.0     19     482
Total:         13   18  81.0     22     487
```

Fig. 7. Analysis of test results

Through the test results we can know that the number of failed requests we mostly concerned is 0. The average waiting time is only 48.648ms.The concurrent capacity of this system is fully meets the design requirements.

5   Conclusion

This article uses Hiphop-PHP technology to improve the load capacity of the e-government system. By using this technology, the user can face up to a large number of user requests. Simulation results show that the introduction of Hiphop-PHP technology can significantly improve the load capacity of the server. However, the realization in this paper still has room for improvement in some degree. For example, we can not conduct compiler optimization for the entire web source code seamlessly. The compilation process is too cumbersome,etc. Therefore, about the issue of load optimization of e-government system, we will do further research.

Acknowledgements. The authors of this paper would like to thank the reviewers and editors for their helpful comments and suggestions.

References

[1] Wang, F.: Analysis of China's E-government Development. Wuhan Maritime Vocational and Technical College 2(4), 12–17 (2007)

[2] Zhang, R., Qiao, L.: The Influence and Demand of E-government Construction for Government Function Transformation in Our Country. Henan University Technology Journal 6(3), 182–186 (2005)

[3] Shen, X.: The Analysis and Optimization of WEB Server Performance, pp. 50–54. East China Normal University Press, Shanghai (2009)

[4] Zhang, Q., Riska, A., Sun, W., et al.: Workload Aware Load Balancing for Clustered Web Servers. IEEE Transactions on Parallel and Distributed Systems 16(3), 219–233 (2005)

[5] Gao, K.: Analysis and Research about Technologies Related to Web Performance. 11 volumes of Modern Telecommunications Technology 11(2), 46–58 (2003)

[6] Zhang, G., Zheng, M., Ju, J.: WebMark: A test of Web server performance. Software Journal 14(7), 1318–1323 (2003)

[7] Abdelzaher, T.F., Shin, K.G., Bhatti, N.: Performance Guarantees for Web Server End-Systems: A control-Theoretical Approach. IEEE Transactions on Parallel and Distributed Systems 13(1), 80–96 (2002)

[8] Surhone, L.M., Timpledon, M.T., Marseken, S.F.: Hiphop for PHP. Betascript Publishing, Mauritius (2010)

A Hybrid Algorithm by Combining Swarm Intelligence Methods and Neural Network for Gold Price Prediction

Zhen-Yao Chen[*]

Department of Business Administration, DE LIN Institute of Technology
No. 1, Ln 380, Qingyun Rd., New Taipei City, 23654, Taiwan
keyzyc@gmail.com

Abstract. This paper attempts to enhance the learning performance of radial basis function neural network (RBFN) through swarm intelligence methods and self-organizing map (SOM) neural network (SOMnet). Further, the particle swarm optimization (PSO) and genetic algorithm (GA)-based method (i.e., PG approach) is employed to train RBFN. The proposed SOMnet + PG approach (called: SPG) algorithm combines the automatically clustering ability of SOMnet with PG approach. The simulation results revealed that SOMnet, PSO, and GA methods can be integrated ingeniously and redeveloped into a hybrid algorithm which aims for obtaining the best accurate learning performance among other algorithms in this study. On the other hand, method evaluation results for two benchmark problems and a gold price prediction case showed that the proposed SPG algorithm outperforms other algorithms and the auto-regressive integrated moving average (ARIMA) models in accuracy.

Keywords: Self-organizing map neural network, swarm intelligence, particle swarm optimization, genetic algorithm, radial basis function neural network.

1 Introduction

Box and Jenkins [3] in 1976 developed the auto-regressive integrated moving average (ARIMA) methodology for forecasting time series events. Time series data are often examined in hopes of discovering a historical pattern that can be exploited in the forecast [3]. The widely used time series models for forecasting purpose especially ARIMA model is generally applicable to linear modeling and it hardly captures the non-linearity inherent in time series data [19].

For forecasting purpose, neural networks (NNs) do not assume linearity, they are capable of fitting a nonlinear function to the given data and do not need the data to be made stationary [4]. Next, radial basis function (RBF) neural network (RBFN) was proposed by Duda and Hart [14] in 1973, it has a number of advantages over other types of NNs and these include better approximation capabilities, simpler network structures and faster learning algorithms [38]. The previous researchers have adopted the RBFN structure along with other single method, such as particle swarm

---

[*] Corresponding author.

L.S.-L. Wang et al. (Eds.): MISNC 2014, CCIS 473, pp. 404–416, 2014.

optimization (PSO) [16] and genetic algorithm (GA) [41] to implement the learning of the RBFN.

In the literature, different combination techniques have been proposed in order to overcome the deficiencies of single models and yield more accurate hybrid models. In recent years, more hybrid forecasting models have been proposed and applied in many areas with good prediction performance [22]. As every single technique always exists with some drawbacks, hybridizing is a reasonable way to take strengths and avoid weakness. Hence, the hybrid methods become very popular for the combinatorial optimization problem [39]. For example, Kuo et al. [26] in 2009 supplement a practical application on the historical sales forecasting data of papaya milk to expound the superiority of the hybrid of PSO and GA (HPG) algorithm for training RBFN. In addition, facing forecasting problems, most of this research was able to use self-organizing map (SOM) to cluster the original data set into the different groups. Forecast procedures could then be applied to each data set to make an accurate forecasting result [34]. However, the construction of a quality RBFN for generalization error and classification accuracy can be a time-consuming process as the modeler must select both a suitable set of inputs and a suitable RBFN structure [38].

As such, this study intends to propose an algorithm by combining SOM neural network (SOMnet) with PSO and GA methods (i.e., PG approach) for training RBFN and make suitable performance verification and comparison. The proposed SOMnet + PG approach (called: SPG) algorithm combines the automatically clustering ability of SOMnet with PG approach. Next, the SPG algorithm applies to two benchmark problems, which are frequently used in the literature to be the comparative benchmark of algorithm performance in the experiment. Furthermore, it can utilize this verified SPG algorithm, in terms of forecasting accuracy, to make predictions in the gold price prediction case.

2   Literature Review

This section will present general backgrounds regarding SOMnet, RBFN, and some swarm intelligence methods.

The SOM algorithm, first introduced by Kohonen in 1990 [24], is one of the most popular NN models based on the unsupervised competitive learning paradigm [49]. SOMnets' ability to associate new data with similar previously learnt data can be applied to forecasting applications [33]. Thus, a plausible explanation is that while time series prediction is considered a function approximation problem, the SOM has usually been seen as architecture suitable for vector quantization, clustering and visualization [48]. Additionally, Yu et al. [51] in 2010 proposes an RBFN-ensemble forecasting model to obtain accurate prediction results and improve prediction quality further. Also, an RBFN model was developed to forecast the total ecological footprint (TEF) from 2006 to 2015 [29] as well. Moreover, the self-organizing RBF (SORBF) network was proposed by Moradkhani et al. [35] in 2004 for constructing a daily stream flow forecasting model. The SOM is used to approximate the distribution of input data with a small number of synaptic weights. In SORBF, the synaptic weights of the SOM are all used as hidden neuron centers of the RBFN [31]. Next, Chang et al. [5] in 2006 developed a hybrid model by integrating SOMnet, GAs and fuzzy rule base to forecast the future sales of a printed circuit board factory.

Subsequently, swarm intelligence (SI) method has drawn the attention of researchers because of its advantages such as scalability, fault tolerance, adaptation, speed, modularity, autonomy, and parallelism [37]. In the past decades, some SI algorithms, inspired by the social behaviors of birds, fish or insects, have been proposed to solve NP-complete optimization problems [10], such as PSO [21], ant colony optimization (ACO) [13], artificial bee colony (ABC) [20], cat swarm optimization (CSO) [8], firefly algorithm (FA) [50], etc. [45]. For example, PSO is a population based algorithm where each particle represents a potential solution. This algorithm is first invented by Kennedy & Eberhart [21] in 1995. Next, Feng [16] in 2006 proposed an evolutional PSO learning-based RBFN system to solve non-linear control and modelling problems. On the other hand, the GA is an iterative random search algorithm for nonlinear problems based on mechanics of natural selection and natural genetics [17]. Further, Sarimveis et al. [41] in 2004 proposed a GA-based algorithm, the objective is the minimization of an error function with respect to the structure of the RBFN, the hidden node centers and the weights between the hidden layer and the output layer.

Moreover, the difference between PSO and GA is that PSO lacks crossover and mutation and more easily to falls into local optimal solutions. But PSO can memorize the global best and affect the movement of other particles, resulting in quick convergence and falling into a local optimal solution [27]. Thus, the algorithms related to PSO and GA have been studied for its manner of combination extensively with different hybrid algorithms and proved to have better performance. For example, Kuo et al. [26] in 2009 proposed the HPG algorithm for training RBFN. In addition, to avoid the particle to be stuck in the local minimum, Kuo & Han [25] in 2011 integrated the mutation mechanism of GA with PSO. Also, Valdez et al. [44] in 2011 combined GA and PSO using fuzzy logic to integrate the results of both methods and for parameter tuning.

Therefore, the prediction accuracy can be improved if two different models are applied to the same data rather than a single model [4]. Especially, a NN combined with pre-processed input feature data will achieve better prediction accuracy [2]. As such, there are still spaces for improvement in terms of the fitting accuracy of the function approximation and prediction. Next, this paper presents a hybrid algorithm to enhance its fitting accuracy.

3   Methodology

Combining the automatically clustering ability of SOMnet with PG (i.e., PSO and GA-based method) approach, this paper proposed the SPG (i.e., SOMnet + PG approach) algorithm to improve the accuracy of function approximation by RBFN. The algorithm provides the settings of some parameters, such as the RBFN hidden node neuron, width, and weight.

The traditional SOM formulation includes a decaying neighborhood width over time to produce a more finely tuned output mapping [40]. Thus, during the process of the SPG algorithm, SOMnet determines the number of center and its position values at first through its automatically clustering ability. The results are used as the number of neuron in RBFN. The algorithm for the PG approach provides the settings of some parameters, such as the width and weight in RBFN. Then, the framework for the proposed SPG algorithm is illustrated in Fig. 1.

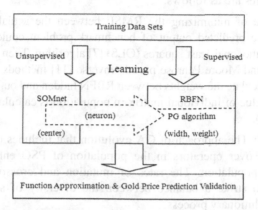

Fig. 1. The framework for the proposed SPG algorithm

3.1 The Analysis of the SPG Algorithm

The proposed SPG algorithm, which combines SOMnet with PG approach, was designed to resolve the problem of RBFN parameters training and solving. Next, through the PG approach of the SPG algorithm, it intends to solve proper values of the parameters from the setting domain in the experiment.

The SPG algorithm combines SOMnet with virtues of PG approach to enhance learning efficiency of RBFN. The optimal values of parameters solution can be obtained and used in the SPG algorithm with RBFN to solve the problem for function approximation. In addition, a typical hidden node in a RBFN is characterized by its center, which is a vector with dimension equals to the number of inputs to the node. Next, the nonlinear function that the RBFN hidden layer adopted is the Gaussian function, and the fitness value of individuals in population is calculated by formula (1). The fitness value for all algorithms in the experiment is computed by maximizing the inverse of root mean squared error (RMSE^{-1}) [28] defined as:

$$Fitness = RMSE^{-1} = \sqrt{\frac{N}{\sum_{j=1}^{N}(y_j - \hat{y}_j)^2}}, \tag{1}$$

where y_j is the actual output, $\hat{y}_j$ is the predicted output of the learned RBFN model for the jth training pattern, and N is the number of the testing set.

3.2   The Detailed Description of the SPG Algorithm

The evolutionary procedures for PG approach of the SPG algorithm was performed and summarized as follows.

Step1. Initialization: The initialization corresponding to nature random selection ensures the diversity among individuals and benefits the evolutionary procedure afterwards. An initial population with a number of individuals is generated and the initializing procedures are as follows.

(1)   The procedure of minimizing the RMSE between the actual output of trained RBFN and the predicted output of benchmark problem could be accomplished through an orthogonal least squares (OLS) [7] algorithm. Then the Gram-Schmidt scheme [18] and Moore Penrose pseudo-inverse [11] methods of the basis matrix are used to calculate the weights between RBFN hidden and output layers.

(2)   The fitness value of individual matrix in population is calculated by formula (1) (i.e., $RMSE^{-1}$).

Step2. GA method: The approach of GA evolution that includes one-point mutation and one-point crossover operators in the population of PSO enhanced learning is called [PSO+GA] population. The rates of mutation and crossover are decreased linearly in order to strike a balance between the diversity of solution space and stability during evolutionary process.

Step3. Duplication: The population enhanced by the GA method is duplicated and thus called [GA-only] population.

Step4. PSO method: PSO starts with an initial population of the possible solutions. Each solution is called a 'particle' in PSO where on the contrary to the former new solutions are not created from the parents within the evolution process. In PSO, the minimum value of Max. selection type PSO learning method [16] (i.e., PSO method) will be considered the active number of RBFs for all particles and ensure that the same vector length is achieved. The solution of RBFN correlated values of parameters that are included in the individuals of particle population, is equivalent to a set of the RBFN solution.

The PSO method is one step which will be executed in one generation and continue in the following process with PG method of the SPG algorithm. This step can update the values of velocity and the embedded values of all particle matrices to record the local best (Lbest) values. Finally, the optimal global best (Gbest) will be obtained in the evolutionary process. The related procedures of the PSO method are as follows.

(1)   The number of the RBFN hidden node neurons that use individual matrix of initialize population is regarded as the number of the neurons for each particle with PSO learning population, and thus is called particle matrix to progress the evolutionary process afterwards.

(2)   Initially, the particles in population don't move toward any particular direction until the Lbest and Gbest of the present particle are calculated. Thus, the velocity value of all particles initialized as zero to reduce entropy in later evolutionary processes.

(3)   The fitness value of each particle is calculated by formula (1) (i.e., $RMSE^{-1}$).

Step5. Reproduction: The [GA-only] and [PSO+GA] populations are combined after evolutionary process. Same amount of individuals from the initial population are randomly selected by the roulette wheel (proportional) selection [17] for the evolution afterwards.

Step6. Termination: The PG approach of the SPG algorithm will not stop returning to ***Step 2*** unless a specific number of iteration has been achieved.

As the SPG algorithm proceeds, the members of the population improve gradually. For population in PSO method, the particle figures out the best solution after consulting itself and other particles, and decides the proceeding direction and velocity. Through the memory mechanism [47], the obtained parameter solution in the population will be more advanced than the initial ones to facilitate the evolutionary process afterwards. In addition, due to the property of global search with GA method, no matter what the fitness values of the individuals in population are, they all have the chances to proceed with some genetic operators and enter the next generation of population to evolve. In this way, the PG method of the SPG algorithm meets the spirit of GA method and ensures the genetic diversity in the future evolution process, and proceeds to obtain a new enhanced population. Thus, the solution space in population could be changed gradually and converge toward the optimal solution.

In the latter experiment, the SPG algorithm stops and the RBFN corresponding to the maximum fitness value is selected. After those critical parameter values are set, RBFN can initiate the training of approximation and learning through two benchmark problems.

4   Simulation Results

Section 3 has presented the proposed SPG algorithm. Two benchmark problems were used in experimental analysis and the results from the SPG algorithm are compared than RBFN [6] and HPG [26] algorithms to verify the accuracy of the SPG algorithm.

4.1   Two Benchmark Problems

The SPG algorithm has better performance among other algorithms through the experiment in two benchmark problems, including Mackey-Glass time series [46] and B2 [42] continuous test functions, which are defined as follows.

The first experiment, the Mackey-Glass time series [46] is expressed as formula (2):

$$\frac{dx(t)}{d(t)} = 0.1x(t) + 0.2 \cdot \frac{x(t-17)}{1 + x(t-17)^{10}} \cdot \qquad (2)$$

The research for the retrieved time step was in the range from 118 to 1118 with the Mackey-Glass time series function, from which 1000 samples were generated randomly. In the second experiment, B2 function [42] is expressed as formula (3):

$$B2(x_j, x_{j+1}) = x_j^2 + 2x_{j+1}^2 - 0.3\cos(3\pi x_j) - 0.4\cos(4\pi x_{j+1}) + 0.7 \qquad (3)$$

a)   search domain: $-100 \leqq x_j \leqq 100, j = 1$;

b)   one global minima: $(x_1, x_2) = (0, 0)$; $B2(x_1, x_2) = 0$.

In addition, all algorithms start with the selection of the parameters setting for two benchmark problems shown in Table 1.

Table 1. Parameter setup for two benchmark problems

Description	Continuous Test Function:	
	Mackey-Glass time series	B2
Search domain	[0.4, 1.6]	[-100, 100]
The widths of RBFN hidden layer	[0.1, 0.2]	[24000, 25000]

4.2   Performance Analysis of Experimental Results

The SPG algorithm will carry out learning on the several RBFN parameters solutions that are generated by the population during the operation of the evolutionary procedure in the experiment. Next, 1000 randomly generated data sets are divided into three parts to train RBFN: Looney [32] in 1996 recommends 65% of the parent database to be used for training, 25% for testing, and 10% for validation.

This paper used the SPG algorithm to solve the optimal RBFN parameters solution, and it randomly generates unrepeated 65% training set from 1000 generated data and input the set to network for learning. Then, with the same procedure, it randomly generates unrepeated 25% testing set to verify individual parameters solution in population and calculates the fitness value. So far, RBFN has used 90% dataset in learning stage. After 1000 iterations in the evolutionary process had been progressed, the optimal RBFN parameters solution had been obtained. Finally, it randomly generates unrepeated 10% validation set to prove how the individual parameters solution approximate two benchmark problems and record the RMSE values to confirm the learning situation of RBFN.

Table 2. Comparison results among all algorithms in the experiment

Algorithm	Experiment Mackey-Glass time series		B2 function	
	Training set	Validation set	Training set	Validation set
RBFN [6]	6.82E2 ± 4.05E1	6.94E2 ± 5.36E1	4.05E3 ± 6.47E2	4.33E3 ± 7.29E2
HPG [26]	3.54E-3 ± 2.76E-4	3.49E-3 ± 2.97E-4	24.17E-2 ± 3.28E-3	28.06E-2 ± 8.25E-3
SPG	**2.37E-3 ± 1.03E-4**	**2.12E-3 ± 2.35E-4**	**19.64E-2 ± 2.47E-3**	**23.83E-2 ± 6.68E-3**

The above mentioned learning and validation stages were implements for 50 runs, and then the average RMSE values were calculated. Thus, the values of the average RMSE ± standard deviation (SD) are shown in Table 2.

The results indicate that SPG algorithm acquires the smallest values with stable performance during the whole training process in the experiment. After the training of RBFN by the SPG algorithm is finished, the individual with the optimal solution of values of parameters, such as neuron, width, and weight, is the exact network setting.

5   Model Evaluation Results

In this section, in order to ensure the predictions of ARIMA models to be complete, the case study for gold price prediction was adopted to verify the models. Next, the case includes gold price trend data in approximately one year. Additionally, the London Afternoon (PM) Gold Price is adopted as observations in this case. The detailed data distribution of this case is shown in Table 3. In addition, in the prediction and verification of gold price, it directly priced by US dollar. Moreover, the analysis had assumed that the influence of external experimental factors did not exist. The trend of gold price was not interfered by any special events. Further, there are several values of parameters within RBFN that must be set up in advance to perform training for the case of prediction analysis.

Table 3. The observations data distribution of gold price prediction case

Case study	The observations: month-day-year (number of samples)	
	Learning set (90%)	Forecasting set (10%)
Gold price trend	02/01/2008~12/23/2008 (228)	12/29/2008~02/02/2009 (24)

Table 4. Parameters setting for the SPG algorithm in the gold price prediction case

Parameter	Description	Value
S	Population size	50
E	The maximum number of iterations	1000
G	The maximum number of generations of SOMnet	100000
C	The number of centers of SOMnet	[1, 100]
σ	The radius of SOMnet	10
ε	The learning rate of SOMnet	0.7
wd_i^s	The width of RBFN hidden layer	[1000, 40000]
c_1, c_2	Acceleration coefficients	1.5
k	Inertia weight	0.75
P_c	Crossover rate (One-point crossover)	[0.5, 0.6]
P_m	Mutation rate (One-point mutation)	[0.1, 0.3]

Subsequently, the Taguchi (robust design) method [43] is a powerful experimental design tool [36] for solving the problems of optimizing the performance, quality and cost of a product or process in a simpler, more efficient and systematic manner than traditional trial-and-error processes [30]. In this case study, the parameters' setting for the SPG algorithm is obtained according to the literatures and Taguchi method. Meanwhile, the statistical software MINITAB 14 was used in the analysis of parameter design. Consequently, the Taguchi trials were configured in an L_9 (3^4) orthogonal array for the SPG algorithm after the experiment was implemented thirty times. Finally, the SPG algorithm starts with parameters setting shown in Table 4 and are meant to ensure consistent basis in this case.

5.1   Input Data and RBFN Learning

The application example with gold price prediction is based on time series data distribution and applied to forecasting analysis. For confidential reasons, the data are linearly normalized between [0, 1] and formula (4) was used [52] for trend data of gold price:

$$X_i' = \frac{(X_i - X_{min})}{(X_{max} - X_{min})} \quad ,$$
(4)

where X_i', X_i, X_{max}, and X_{min} are the normalized value of the observed gold price data, actual value of the observed trend data, maximum observation value of the trend dataset, and the minimum observation value of the gold price dataset respectively.

Most studies in the literatures use convenient ratio of splitting for in- and out-of-samples such as 70:30%, 80:20%, or 90:10% [52]. It uses the ratio of 90:10% here as the basis of division. In addition, Looney [32] in 1996 recommends 65% of the parent database to be used for training, 25% for testing, and 10% for validation. Thus, the learning stage of RBFN will be based on daily gold price data; it contains 65% training set and 25% testing set. The training was conducted by entering in turn four normalized vector data pulled from 65% training set to RBFN. In the process, the individual parameters solution within the population will verify along with the entire evolution procedure, and then the fitness values of all individuals within the population could be calculated with the 25% testing set. At this point, 90% of the gold price data had been used to the learning stage of RBFN, which eventually generated an individual parameters solution with the most precise prediction. Also, the approximation performance of the RBFN prediction shall be examined with the 10% validation set.

Furthermore, the following predicted values were generated in turn by the moving window method. The first 90% of the observations were used for model estimation while the remaining 10% were used for validation and one-step-ahead forecasting. This study elaborates how data is input to RBFN for prediction through all algorithms, and comparison with ARIMA models.

5.2   Building ARIMA Models

The study carries out price prediction based on ARIMA models. The ARIMA (p, d, q) modeling procedure has three steps: (a) identifying the model order, i.e., identifying p

and q; (b) estimating the model coefficients; and (c) forecasting the data [4]. On the other hand, EViews™ 6.0 software was also used the analysis of Box-Jenkins models to calculate the numerical results. This study precedes the data identification of ARIMA models through augmented Dickey-Fuller (ADF) [12] testing. Akaike information criterion (AIC) [1] criteria were employed to sift the optimal model out [15]. The results of model diagnosis reveal that the values of Q-statistic (i.e., Ljung-Box statistic) [23] are greater than 0.05 in result of ARIMA models, in which the results are serial non-correlation (i.e., white noise) and it had been suitable fitted (i.e., ARIMA (2, 1, 2) model).

5.3   Error Measure of the Prediction Performance in This Case

The mean absolute error (MAE) and mean absolute percentage error (MAPE) are the most commonly used error measures in business, and were used to evaluate the forecast models [9]. Thus, the prediction performances of above mentioned algorithms with the case data is presented in Table 5.

Table 5. The prediction errors comparison for all algorithms using gold price trend

Algorithm Error	ARIMA (2, 1, 2) model	RBFN [6]	HPG [26]	SPG
MAE	12.227	6.381	6.31E-3	**7.59E-4**
MAPE	1.409	1.928	19.24E-1	**10.61E-2**

Among these algorithms, the results derived from MAE and MAPE of the proposed SPG algorithm were the smallest ones. Therefore, the SPG algorithm can substantially improve the accuracy of gold price prediction among others algorithm.

6   Conclusions

Inspired by the evolutionary learning of automatically clustering (i.e., SOMnet) and swarm intelligence methods (i.e., PG approach), a SPG (i.e., SOMnet + PG approach) algorithm was proposed in this research. The complementation of some evolutionary procedures that improves the diversity of populations also increases the precision of the results. Next, the proposed SPG algorithm provides the settings of RBFN parameters and has better parameter setting of network and consequently enables RBFN to perform better learning and approximation in two benchmark problems and application in the gold price prediction case. This analytical outcome will be beneficial in practice to allow lower the investment risk and determine appropriate marketing strategy.

References

1. Akaike, H.: A new look at the statistical model identification. IEEE Trans. on Automat. Control AC 19, 716–723 (1974)
2. Anbazhagan, S., Kumarappan, N.: Day-ahead deregulated electricity market price forecasting using neural network input featured by DCT. Energy Conversion and Management 78, 711–719 (2014)
3. Box, G.E.P., Jenkins, G.: Time Series Analysis, Forecasting and Control. Holden-Day, San Francisco (1976)
4. Babu, C.N., Reddy, B.E.: A moving-average-filter-based hybrid ARIMA-ANN model for forecasting time series data. Applied Soft Computing (in press, 2014)
5. Chang, P., Liu, C., Wang, Y.: A hybrid model by clustering and evolving fuzzy rules for sales decision supports in printed circuit board industry. Decision Support Systems 42, 1254–1269 (2006)
6. Chen, S., Cowan, C.F.N., Grant, P.M.: Orthogonal least squares learning algorithm for radial basis function networks. IEEE Trans. Neural Networks 2(2), 302–309 (1991)
7. Chen, S., Wu, Y., Luk, G.L.: Combined Genetic Algorithm Optimization and Regularized Orthogonal Least Squares Learning for Radial Basis Function Networks. IEEE Transactions on Neural Networks 10(5), 1239–1243 (1999)
8. Chu, S.C., Tsai, P.W.: Computational intelligence based on behaviors of cats. Int. J. Innov. Comput., Inform. Control 3(1), 163–173 (2007)
9. Co, H.C., Boosarawongse, R.: Forecasting Thailand's rice export: Statistical techniques vs. artificial neural networks. Computers & Industrial Engineering 53, 610–627 (2007)
10. Cui, Z., Gao, X.: Theory and applications of swarm intelligence. Neur. Comput. Appl. 21(2), 205–206 (2012)
11. Denker, J.S.: Neural network models of learning and adaptation. Physica D 22, 216–232 (1986)
12. Dickey, D.A., Fuller, W.A.: Likelihood Ration Statistics for Autoregressive Time Series with A Unit Root. Econometrica 49(4), 1057–1072 (1981)
13. Dorigo, M., Maniezzo, V., Colorni, A.: The ant system: optimization by a colony of cooperating agents. IEEE Trans. Syst., Man Cybernet. – Part B: Cybernet. 26, 29–41 (1996)
14. Duda, R.O., Hart, P.E.: Pattern Classification and Scene Analysis. John Wiley & Sons, New York (1973)
15. Engle, R.F., Robert, F., Yoo, B.S.: Forecasting and Testing in Cointegrated Systems. Journal of Econometrics 35, 588–589 (1987)
16. Feng, H.M.: Self-generation RBFNs using evolutional PSO learning. Neurocomputing 70, 241–251 (2006)
17. Goldberg, D.E.: Genetic Algorithms in Search, Optimization & Machine Learning. Addison-Wesley, Reading (1989)
18. Golub, G.H., Loan, C.F.V.: Matrix Computations, 3rd edn. Johns Hopkins Univ. Press, Baltimore (1996)
19. Jaipuria, S., Mahapatra, S.S.: An improved demand forecasting method to reduce bullwhip effect in supply chains. Expert Systems with Applications 41, 2395–2408 (2014)
20. Karaboga, D.: An Idea Based on Honey Bee Swarm for Numerical Optimization. Technical Report-TR06, Erciyes University, Engineering Faculty, Computer engineering Department (2005)

21. Kennedy, J., Eberhart, R.C.: Particle swarm optimization. In: Proceedings of IEEE International Conference on Neural Networks, Perth, Australia, IEEE Service Center, pp. 1942–1948 (1995)
22. Khashei, M., Bijari, M.: A new class of hybrid models for time series forecasting. Expert Systems with Applications 39, 4344–4357 (2012)
23. Kmenta, J.: Elements of Econometrics, 2nd edn., p. 332. Macmillan Publishing Co., New York (1986)
24. Kohonen, T.: The Self-Organizing Map. Proc. IEEE 78(9), 1464–1480 (1990)
25. Kuo, R.J., Han, Y.S.: A hybrid of genetic algorithm and particle swarm optimization for solving bi-level linear programming problem-a case study on supply chain mode. Applied Mathematical Modelling 35(8), 3905–3917 (2011)
26. Kuo, R.J., Hu, T.-L., Chen, Z.-Y.: Sales Forecasting Using an Evolutionary Algorithm Based Radial Basis Function Neural Network. In: Yang, J., Ginige, A., Mayr, H.C., Kutsche, R.-D. (eds.) Information Systems: Modeling, Development, and Integration. LNBIP, vol. 20, pp. 65–74. Springer, Heidelberg (2009)
27. Kuo, R.J., Syu, Y.J., Chen, Z.Y., Tien, F.C.: Integration of Particle Swarm Optimization and Genetic Algorithm for Dynamic Clustering. Information Sciences 195, 124–140 (2012)
28. Lee, Z.J.: A novel hybrid algorithm for function approximation. Expert Systems with Applications 34, 384–390 (2008)
29. Li, X.M., Xiao, R.B., Yuan, S.H., Chen, J.A., Zhou, J.X.: Urban total ecological footprint forecasting by using radial basis function neural network: A case study of Wuhan city, China. Ecological Indicators 10, 241–248 (2010)
30. Lin, C.F., Wu, C.C., Yang, P.H., Kuo, T.Y.: Application of Taguchi method in light-emitting diode backlight design for wide color gamut displays. Journal of Display Technology 5(8), 323–330 (2009)
31. Lin, G.F., Wu, M.C.: An RBF network with a two-step learning algorithm for developing a reservoir inflow forecasting model. Journal of Hydrology 405, 439–450 (2011)
32. Looney, C.G.: Advances in feed-forward neural networks: demystifying knowledge acquiring black boxes. IEEE Trans. Knowledge Data Eng. 8(2), 211–226 (1996)
33. Lopez, M., Valero, S., Senabre, C., Aparicio, J., Gabaldon, A.: Application of SOM neural networks to short-term load forecasting: The Spanish electricity market case study. Electric Power Systems Research 91, 18–27 (2012)
34. Lu, C.J., Wang, Y.W.: Combining independent component analysis and growing hierarchical self-organizing maps with support vector regression in product demand forecasting. Int. J. Production Economics 128, 603–613 (2010)
35. Moradkhani, H., Hsu, K.L., Gupta, H.V., Sorooshian, S.: Improved streamflow forecasting using self-organizing radial basis function artificial neural networks. Journal of Hydrology 295, 246–262 (2004)
36. Olabi, A.G.: Using Taguchi method to optimize welding pool of dissimilar laser-welded components. Opt. Laser Technol. 40, 379–388 (2008)
37. Pan, Q.K., Wang, L., Mao, K., Zhao, J.H., Zhang, M.: An Effective Artificial Bee Colony Algorithm for a Real-World Hybrid Flowshop Problem in Steelmaking Process. IEEE Trans. on Automation Science and Engineering 10(2), 307–322 (2013)
38. Qasem, S.N., Shamsuddin, S.M., Zain, A.M.: Multi-objective hybrid evolutionary algorithms for radial basis function neural network design. Knowledge-Based Systems 27, 475–497 (2012)
39. Qiu, X., Lau, H.Y.K.: An AIS-based hybrid algorithm for static job shop scheduling problem. Journal of Intelligent Manufacturing 25, 489–503 (2014)

40. Rumbell, T., Denham, S.L., Wennekers, T.: A Spiking Self-Organizing Map Combining STDP, Oscillations, and Continuous Learning. IEEE Trans. on Neural, Networks and Learning Systems 25(5), 894–907 (2014)
41. Sarimveis, H., Alexandridis, A., Mazarakis, S., Bafas, G.: A new algorithm for developing dynamic radial basis function neural network models based on genetic algorithms. Computers and Chemical Engineering 28, 209–217 (2004)
42. Shelokar, P.S., Siarry, P., Jayaraman, V.K., Kulkarni, B.D.: Particle swarm and colony algorithms hybridized for improved continuous optimization. Applied Mathematics and Computation 188, 129–142 (2007)
43. Taguchi, G., Yokoyama, T.: Taguchi Methods: Design of Experiments. ASI Press, Dearbon (1993)
44. Valdez, F., Melin, P., Castillo, O.: An improved evolutionary method with fuzzy logic for combining particle swarm optimization and genetic algorithms. Applied Soft Computing 11(2), 2625–2632 (2011)
45. Wang, H., Wu, Z., Rahnamayan, S., Sun, H., Liu, Y., Pan, J.S.: Multi-strategy ensemble artificial bee colony algorithm. Information Sciences 279, 587–603 (2014)
46. Whitehead, B.A., Choate, T.D.: Cooperative-competitive genetic evolution of radial basis function centers and widths for time series prediction. IEEE Trans. Neural Networks 7(4), 869–880 (1996)
47. Xu, R., Venayagamoorthy, G.K., Wunsch, D.C.: Modeling of gene regulatory networks with hybrid differential evolution and particle swarm optimization. Neural Networks 20, 917–927 (2007)
48. Xu, R., Wunsch, D.: Survey of clustering algorithms. IEEE Transactions on Neural Networks 16(3), 645–678 (2005)
49. Yadav, V., Srinivasan, D.: A SOM-based hybrid linear-neural model for short-term load forecasting. Neurocomputing 74, 2874–2885 (2011)
50. Yang, X.S.: Firefly algorithm, stochastic test functions and design optimization. Int. J. Bio-Insp. Comput. 2(2), 78–84 (2010)
51. Yu, L., Wang, S., Lai, K.K., Wen, F.: A multiscale neural network learning paradigm for financial crisis forecasting. Neurocomputing 73, 716–725 (2010)
52. Zou, H.F., Xia, G.P., Yang, F.T., Wang, H.Y.: An investigation and comparison of artificial neural network and time series models for Chinese food grain price forecasting. Neurocomputing 70, 2913–2923 (2007)

Updating the Built FUSP Trees with Sequence Deletion Based on Prelarge Concept

Chun-Wei Lin[1,2], Wensheng Gan[1], Tzung-Pei Hong[3,4], and Jeng-Shyang Pan[1,2]

[1] Innovative Information Industry Research Center (IIIRC)
[2] Shenzhen Key Laboratory of Internet Information Collaboration
School of Computer Science and Technology
Harbin Institute of Technology Shenzhen Graduate School
HIT Campus Shenzhen University Town, Xili, Shenzhen, P.R. China
[3] Department of Computer Science and Information Engineering
National University of Kaohsiung, Kaohsiung, Taiwan, R.O.C.
[4] Department of Computer Science and Engineering
National Sun Yat-sen University, Kaohsiung, Taiwan, R.O.C.
jerrylin@ieee.org, {wsgan001,jengshyangpan}@gmail.com,
tphong@nuk.edu.tw

Abstract. Among various data mining techniques, sequential-pattern mining is used to discover the frequent subsequences from a sequence database. Most research handles the static database in batch mode to discover the desired sequential patterns. Transactions or customer sequences are, however, dynamically changed in real-world applications. In the past, the FUSP tree was designed to maintain and update the discovered information based on Fast UPdated (FUP) approach with sequence insertion and sequence deletion. The original customer sequences is still required to be rescanned if it is necessary. In this paper, the prelarge concept is adopted to maintain and update the built FUSP tree with sequence deletion. When the number of deleted customers is smaller than the safety bound of the prelarge concept, the original database is unnecessary to be rescanned but the sequential patterns can still be actually maintained and updated. Experiments are also conducted to show the performance of the proposed algorithm in terms of execution time and number of tree nodes.

Keywords: prelarge concept, sequential patterns, deletion, dynamic databases, FUSP tree.

1   Introduction

Sequential-pattern mining (SPM) is extended from association-rule mining [2, 5, 6] to concern the ordered sequence data such as DNA sequences, usage of Web log, Web-click streams, or the logs of network flow. Agrawal et al. first designed AprioriAll algorithm [4] to level-wisely generate-and-test the candidates for deriving the sequential patterns (large sequences). Many algorithms have been proposed to

L.S.-L. Wang et al. (Eds.): MISNC 2014, CCIS 473, pp. 417–426, 2014.
© Springer-Verlag Berlin Heidelberg 2014

discover the sequential patterns from the static database in batch mode [9, 14, 18, 19, 20, 22]. When customer sequences are changed whether sequence insertion [15] or sequence deletion [16], the discovered sequential patterns may become invalid or some sequential patterns may arise. An intuitive way to handle the dynamic databases whether sequence insertion or sequence deletion is to re-scan the original database and re-mine the sequential patterns, which is not suitable in real-world applications.

In the past, the Fast UPdated (FUP) [7] and FUP2 [8] concepts were respectively proposed to maintain and update the frequent itemsets of association-rule mining with transaction insertion and transaction deletion. Lin et al. then extended the FUP concept and designed an incremental FASTUP algorithm [17] to maintain and update the discovered sequential patterns. Lin et al. designed a fast updated sequential pattern (FUSP)-tree structure and algorithms to respectively maintain and update the FUSP tree with sequence insertion [15] and sequence deletion [16]. Based on the FUSP tree with an index Header_Table, it is easier to maintain and update the built FUSP tree in the dynamic databases. Original database is, however, required to be rescanned if the small sequences are necessary to be maintained in the updated database.

In the past, Hong et al. extended the prelarge concept [11] of association-rule mining to respectively maintain and update the discovered sequential patterns with sequence insertion [13] and sequence deletion [21]. In this paper, a maintenance algorithm for sequence deletion based on the prelarge concept is proposed to efficiently maintain and update the built FUSP and the prelarge 1-sequences for later mining process. The original database is unnecessary to be rescanned until the cumulative number of deleted customers achieves the designed safety bound of the prelarge concept. From the experimental results, the proposed algorithm has better performance than the batch-mode SPM algorithms or other maintenance algorithms with sequence deletion.

2   Review of Related Work

In this section, the sequential-pattern mining (SPM) algorithms and the prelarge concept are respectively reviewed.

2.1   SPM Algorithm

In the past, Agrawal et al. proposed the AprioriAll algorithm [4] to generate-and-test the candidates for mining the sequential patterns from a static database. Pei et al. designed the PrefixSpan algorithm to efficiently mine the sequential patterns based on the projection mechanism [19]. A sequence database is recursively projected into several smaller sets of projected database to speed up the computations for mining sequential patterns. Zaki et al. proposed a SPADE algorithm to fast mine the sequential patterns [22]. Based on SPADE algorithm, the sequential patterns can be derived with three database scans. Many algorithms have been proposed to mine the sequential patterns, but most of them are performed to handle the static database. When the sequences are changed whether sequence insertion [15] or deletion [16] in the original database, the discovered sequential patterns may become invalid or new sequential patterns may arise. An intuitive way to update the sequential patterns is to re-process the entire updated database in batch mode, which is inefficient in real-world applications.

Lin et al. first proposed a FASTUP algorithm [17] to incrementally maintain and update the discovered sequential patterns with sequence insertion. The original database is still, however, required to be rescanned if the discovered sequential pattern is large in the added sequences but small in the original database based on the FASTUP concept. Hong et al. then extended the prelarge concept [11] of association-rule mining to respectively maintain and update the discovered sequential patterns with sequence insertion [13] and sequence deletion [21]. The prelarge concept is also based on the Apriori-like approach [3] to generate-and-test candidates in a level-wise way for mining sequential patterns, which requires more computations of database rescan. Lin first extended the fast updated frequent pattern (FUFP)-tree [12] to design a fast updated sequential pattern (FUSP)-tree and developed the algorithms for respectively maintaining and updating the built FUSP tree with sequence insertion [15] and sequence deletion [16]. The original database is, however, required to be rescanned if it is necessary to maintain a sequence which is small in the original database but large in the inserted sequences with sequence insertion or a sequence is small both in the original database and in the deleted sequences with sequence deletion.

2.2   Prelarge Concept

A prelarge sequence [13] is not truly large, but has highly probability to be large when the database is updated. The upper (S_u) and lower support (S_l) thresholds are used to respectively set the large and prelarge sequences. When the support ratio of a sequence is larger than or equal to S_u (the same as minimum support threshold in traditional data mining approach), it is considered as the large sequence; otherwise, if the support ratio of a sequence lies between the S_u and S_l, it is considered as the prelarge sequence. The prelarge sequences can be concerned as the buffer to reduce the movement of sequences directly from large to small and vice-versa in the maintenance process. When some customer sequences are deleted from a database, two situations are then arisen with sequence deletion as:

1.   All sequences of an old customer are completely deleted. This situation will also modify the number of customers in the original database.
2.   Partial sequences from an old customer are deleted. This situation will not change the number of customers in the original database.

Considering the customer sequences in the original database and in the deleted parts, nine cases of the prelarge concept are arisen and shown in Figure 1. Based on the prelarge sequences, the sequence cannot possible be large for the updated database as long as q is smaller than the safety bound as [21]:

$$q \leq f = \frac{(S_u - S_l) \times |D|}{S_u},$$

where q is the number of deleted customers of the deleted sequences, $|D|$ is the number of customers from the original database, S_u is the upper threshold, and S_l is the lower threshold.

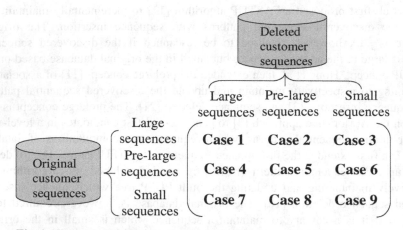

Fig. 1. Nine cases arising from the original database and the deleted sequences

3   Proposed Maintenance Algorithm for Sequence Deletion

Before the maintenance approach of sequence deletion, a fast updated sequential pattern (FUSP)-tree [15, 16] must built in advance to keep large 1-sequences. A set is also created to keep the prelarge 1-sequences for speeding up the maintenance approach. When sequences are deleted from the original database, the proposed maintenance algorithm is then performed to divide the discovered information into three parts with nine cases. Each case is then processed by the designed procedure to maintain and update the built FUSP tree and prelarge 1-sequences. Based on the designed algorithm, the original database is unnecessary to be rescanned until the cumulative number of deleted customers achieves the safety bound. The proposed algorithm is described below.

Proposed maintenance algorithm:

INPUT: The built FUSP tree for large 1-sequences from D, a set of P_Seqs to keep prelarge 1-sequences, a set of deleted sequences of T, an upper support threshold S_u, a lower support threshold S_l, and the number of deleted customers last time c.

OUTPUT: An updated FUSP tree with the discovered sequential patterns from $D - T$.

STEP 1: Calculate the safety bound f as follows:

$$f = \frac{(S_u - S_l) \times |D|}{S_u}.$$

STEP 2: Count the number q of deleted customers (with all sequences are deleted) from T.

STEP 3: Scan the deleted customer sequences to get all 1-sequences and their counts.

STEP 4: Divide the discovered 1-sequences from T into three parts according to whether they are large, prelarge or small in the original database D.

STEP 5: For each s which is large in the original database (existing in Header_Table), do the following substeps (**Cases 1, 2 and 3**):

Substep 5-1:   Set the updated $S^U(s)$ of s as:
$$S^U(s) = S^D(s) - S^T(s),$$
where $S^D(s)$ is the count of s in the Header_Table of the built FUSP tree and $S^T(s)$ is the count of s in the deleted sequences.

Substep 5-2:   If $S_u \leq S^U(s)/(|D| - c - q)$, update the count of s in the Header_Table as $S^U(s)$; put s into the set of *Reduced_Seqs*, which will be further processed in STEP 7. Otherwise, if $S_l \leq S^U(s)/(|D| - c - q) < S_u$, remove s from the Header_Table, put s with its updated $S^U(s)$ in the set of *P_Seq*. Otherwise, s becomes small after the database is updated; remove s from the Header_Table and connect each parent node of s directly to its child node in the built FUSP tree.

STEP 6: For each s which is prelarge from *P_Seqs*, do the following substeps (**Cases 4, 5 and 6**):

Substep 6-1:   Set the updated count $S^U(s)$ of s as:
$$S^U(s) = S^D(s) - S^T(s).$$

Substep 6-2:   If $S_u \leq S^U(s)/(|D| - c - q)$, s will become large after the database is updated; remove s from *P_Seqs*, and put s with its updated count $S^D(s)$ in the end of Header_Table; besides, put s into the set of *Branch_Seqs*. Ootherwise, if $S_l \leq S^U(s)/(|D| - c - q) < S_u$, s remains prelarge after the database is updated; update s with its new count $S^D(s)$ in the set of *P_Seqs*. Otherwise, remove s from *P_Seqs*.

STEP 7: For each deleted sequence with a 1-sequence J existing in the *Reduced_Seqs*, subtract 1 from the count of J node at the corresponding branch of the built FUSP tree.

STEP 8: For each 1-seqeunce s which is neither large nor prelarge in the original customer sequences but small in the deleted sequences (**Cases 9**), put s in the set of *Rescan_Seqs*, which is used when rescanning the original database in STEP 9 is necessary.

STEP 9: If $c + q \leq f$ or *Rescan_ Seqs* is **null**, then do nothing; otherwise, do the following substeps for each s in the set of *Rescan_Seqs*:

Substep 9-1:   Rescan the original database to find the count $S^D(s)$ of s.

Substep 9-2:   Set the updated count $S^U(s)$ of s as:
$$S^U(s) = S^D(s) - S^T(s).$$

Substep 9-3:   If $S_u \leq S^U(s)/(|D| - c - q)$, s will become large after the database is updated; put s in the *Branch_Seqs* and insert s to the end of the Header_Table according to the descending order of their updated counts. Otherwise, if $S_l \leq S^U(s)/(|D| -$

$c - q) < S_u$, s will become prelarge after the database is updated; put s with its updated count $S^U(s)$ into the set of P_Seqs. Otherwise, do nothing.

STEP 10: For each original sequence with a 1-sequence J existing in the $Branch\_Seqs$, if J has not been at the corresponding branch of the built FUSP tree, insert J at its correct position in the original database and set its count at 1; otherwise, add 1 to the count of the node J.

STEP 11: If $c + q > f$, then set $/D/ = /D/ - c - q$ and $c = 0$; otherwise, set $c = c + q$.

Note that in STEP 7, a *corresponding branch* is the branch generated from the large 1-sequences and corresponding to the order of sequences in the databases. Based on the proposed algorithm, the desired large sequences after sequence deletion can also be determined by using the FP-growth-like [10] mining approach.

4   An Illustrated Example

An example is given to illustrate the proposed algorithm to maintain and update the built FUSP tree for later mining process. An original database is shown in Table 1, which consists of 10 customer sequences with nine purchased items. For the given example, the S_u and S_l are respectively set at 50% and 30%. The built FUSP tree from the original database is shown in Figure 2.

Table 1. Original customer sequences

CID	Customer sequence
1	(A)(B)
2	(A)(E, F, G)(C, D)
3	(A, H, G)
4	(A)(E, G)(B)
5	(B)(C)
6	(A)(B, C)
7	(A)(B, C, D)
8	(E, G)
9	(F)(I)
10	(H, I)

Suppose four sequences are deleted from the original database shown in Table 2. The proposed maintenance algorithm is then processed by the proposed maintenance algorithm step-by-step. The variable c is initially set at 0 since no sequences are deleted last time.

Table 2. Four deleted sequences

CID	Customer sequence
4	(A)(E, G)(B)
5	(B)
6	(B)
10	(H, I)

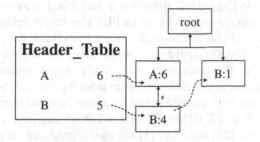

Fig. 2. Initial constructed FUSP tree with its Header_Table

The formula [13] of prelarge concept to determine whether the original database is required to be rescanned. It is calculated as:

$$f = \frac{(S_u - S_l) \times |D|}{S_u} = \frac{(0.5 - 0.3) \times 10}{0.5} = 4.$$

For the given example, when the cumulative number of deleted customers achieves 4, the original database is required to be rescanned. Otherwise, the proposed algorithm is then processed to maintain and update the built FUSP tree and the prelarge 1-sequences in the set of *P_Seqs*. From Table 2, it can be found that the number of deleted customers is 2 since the customers 4 and 10 are entirely deleted from the original database, and the sequence (B) is partial deleted from customers 5 and 6. The variable *q* is thus set at 2. The built FUSP tree is then maintained and updated by the proposed algorithm. After all procedures of the proposed algorithm are performed, the updated FUSP tree is shown in Figure 3.

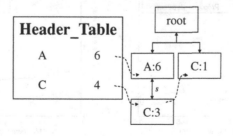

Fig. 3. Final updated FUSP tree with its Header_Table

5   Experimental Evaluation

Several experiments were conducted to compare the performance of the SPADE [22], PrefixSpan [19], prelarge AprioriAll algorithm with sequence deletion (PRE-AprioriAll-DEL) [13], the fast updated approach for maintaining the FUSP tree with sequence deletion (FUP-FUSP-TREE-DEL) [16], and proposed algorithm. When sequences are deleted, the SPADE and PrefixSpan algorithms are performed in batch mode. The FUP-FUSP-TREE-DEL algorithm is performed to maintain and update the built FUSP for later mining process based on FUP concept. A deletion ratio (DR) and a partial deletion ratio (PDR) are respectively set to evaluate the performance of the proposed algorithm. The DR is the percentage to delete sequences of the original database and PDR is the percentage to partially delete sequences of DR. The sequences for deletion were selected bottom-up from the original database.

The experiments are implemented in the Java language and executed on a PC with an Intel-Core i5-3470 at 3.2 GHz CPU and 4 GB of memory. Two real databases called BMSWebview-1 [23] and retail [1] are used to evaluate the performance of the proposed algorithm. The execution times of five algorithms are thus compared in two databases for various minimum support thresholds. When sequences are deleted, the SPADE and the PrefixSpan algorithms are required to process the updated database in batch mode for rescanning the original database to mine the sequential patterns. The PRE-AprioriAll-DEL, FUP-FUSP-TREE-DEL and proposed algorithm are performed to maintain and update the built FUSP tree. Each iteration in the experiments processes six times of sequence deletion. To evaluate the performance of the proposed algorithm at different minimum support thresholds in two databases, the S_u is the same as the user predefined threshold like the common sequential pattern-mining algorithms, and the S_l is set lower to the S_u in any value. For the BMSWebview-1 database, S_l values are set at S_u values minus 0.001%. For the retail database, S_u values minus 0.1%. The results of running time for two databases are shown in Figure 4.

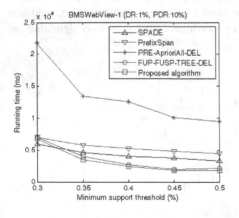
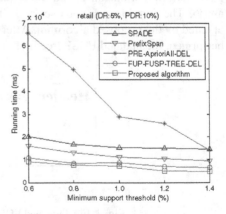

Fig. 4. Running time for different minimum support thresholds in two databases

From Figure 4, it is obvious to see that the proposed algorithm has better performance than the others whether the batch-mode algorithms or the maintenance ones with sequence deletion. The number of tree nodes are also compared to show the space complexity of the proposed algorithm. The FUSP-TREE-BATCH approach [16] is used as a baseline to show the complexity of tree node. The results are shown in Figure 5.

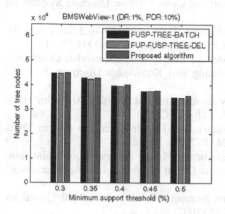 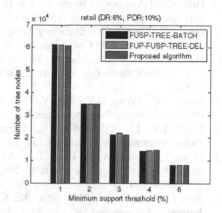

Fig. 5. Number of tree nodes for different minimum support thresholds in two databases

It can be observed from Figure 5 that the proposed algorithm has slightly generated more tree nodes than the other two tree-based algorithms since the sorted order of 1-sequences in the Header_Table is different than the FUSP-TREE-DEL and FUSP-TRE-BATCH algorithms. From the observation from Figures 4 and 5, the results of the proposed algorithm are acceptable.

Acknowledgement. This research was partially supported by the Shenzhen Peacock Project, China, under grant KQC201109020055A, by the Natural Scientific Research Innovation Foundation in Harbin Institute of Technology under grant HIT.NSRIF.2014100, and by the Shenzhen Strategic Emerging Industries Program under grant ZDSY20120613125016389.

References

1. A Sequential Pattern Mining Framework (2010),
   http://www.philippe-fournier-viger.com/spmf/index.php
2. Agrawal, R., Imielinski, T., Swami, A.: Database Mining: A Performance Perspective. IEEE Transactions on Knowledge and Data Engineering 5, 914–925 (1993)
3. Agrawal, R., Srikant, R.: Fast Algorithms for Mining Association Rules in Large Databases. In: The International Conference on Very Large Data Bases, pp. 487–499 (1994)
4. Agrawal, R., Srikant, R.: Mining Sequential Patterns. In: The International Conference on Data Engineering, pp. 3–14 (1995)

5. Bodon, F.: A Fast Apriori Implementation. In: IEEE ICDM Workshop on Frequent Itemset Mining Implementations (2003)
6. Chen, M.S., Han, J., Yu, P.S.: Data Mining: An Overview from a Database Perspective. IEEE Transactions on Knowledge and Data Engineering 8, 866–883 (1996)
7. Cheung, D.W., Han, J., Ng, V., Wong, C.Y.: Maintenance of Discovered Association Rules in Large Databases: An Incremental Updating Technique. In: International Conference on Data Engineering, pp. 106–114 (1996)
8. Cheung, D.W., Lee, S.D., Kao, B.: A General Incremental Technique for Maintaining Discovered Association Rules. In: The International Conference on Database Systems for Advanced Applications, pp. 185–194 (1997)
9. Guyet, T., Quiniou, R.: Extracting Temporal Patterns from Interval-Based Sequences. In: The International Joint Conference on Artificial Intelligence, pp. 1306–1311 (2011)
10. Han, J., Pei, J., Yin, Y., Mao, R.: Mining Frequent Patterns without Candidate Generation: A Frequent-Pattern Tree Approach. Data Mining and Knowledge Discovery 8, 53–87 (2004)
11. Hong, T.P., Wang, C.Y., Tao, Y.H.: A New Incremental Data Mining Algorithm using Pre-large Itemsets. Intelligent Data Analysis 5, 111–129 (2001)
12. Hong, T.P., Lin, C.W., Wu, Y.L.: Incrementally Fast Updated Frequent Pattern Trees. Expert Systems with Applications 34, 2424–2435 (2008)
13. Hong, T.P., Wang, C.Y., Tseng, S.S.: An Incremental Mining Algorithm for Maintaining Sequential Patterns using Pre-large Sequences. Expert Systems with Applications 38, 7051–7058 (2011)
14. Kim, C., Lim, J.H., Ng, R.T., Shim, K.: Squire: Sequential Pattern Mining with Quantities. Journal of Systems and Software 80, 1726–1745 (2007)
15. Lin, C.W., Hong, T.P., Lu, W.H., Lin, W.Y.: An Incremental FUSP-Tree Maintenance Algorithm. In: The International Conference on Intelligent Systems Design and Applications, pp. 445–449 (2008)
16. Lin, C.W., Hong, T.P., Lu, W.H.: An Efficient FUSP-Tree Update Algorithm for Deleted Data in Customer Sequences. In: International Conference on Innovative Computing, Information and Control, pp. 1491–1494 (2009)
17. Lin, M.Y., Lee, S.Y.: Incremental Update on Sequential Patterns in Large Databases. In: IEEE International Conference on Tools with Artificial Intelligence, pp. 24–31 (1998)
18. Mooney, C.H., Roddick, J.F.: Sequential Pattern Mining - Approaches and Algorithms. ACM Computing Surveys 45, 1–39 (2013)
19. Pei, J., Han, J., Mortazavi-Asl, B., Wang, J., Pinto, H., Chen, Q., Dayal, U., Hsu, M.C.: Mining Sequential Patterns by Pattern-Growth: the PrefixSpan Approach. IEEE Transactions on Knowledge and Data Engineering 16, 1424–1440 (2004)
20. Srikant, R., Agrawal, R.: Mining Sequential Patterns: Generalizations and Performance Improvements. In: Apers, P.M.G., Bouzeghoub, M., Gardarin, G. (eds.) EDBT 1996. LNCS, vol. 1057, pp. 3–17. Springer, Heidelberg (1996)
21. Wang, C.Y., Hong, T.P., Tseng, S.S.: Maintenance of Sequential Patterns for Record Deletion. In: IEEE International Conference on Data Mining, pp. 536–541 (2001)
22. Zaki, M.J.: SPADE: An Efficient Algorithm for Mining Frequent Sequences. Machine Learning 42, 31–60 (2001)
23. Zheng, Z., Kohavi, R., Mason, L.: Real World Performance of Association Rule Algorithms. In: ACM International Conference on Knowledge Discovery and Data Mining, pp. 401–406 (2001)

Novel Reversible Data Hiding Scheme
for AMBTC-Compressed Images
by Reference Matrix

Jeng-Shyang Pan[1], Wei Li[1,*], and Chia-Chen Lin[2]

[1] School of Computer Science & Technology, Harbin Institute of Technology
Shenzhen Graduate School, Shenzhen, China
{jengshyangpan,weil0819}@gmail.com
[2] Department of Computer Science and Information Management,
Providence University, Taichung City, Taiwan
ally.cclin@gmail.com

Abstract. This paper proposes a novel reversible data hidng scheme in images compressed by absolute moment block truncation coding (AMBTC). In this scheme, the secret data is embedded into the quantization levels of each AMBTC-compressed image block based on a reference matrix. Original quantization levels are transformed into another watermarking message which will combine with bitmap in each image block. The reconstructed image quality is exactly the same as the original AMBTC-compressed version due to the reversibility. Extensive experimental results demonstrate the effectiveness of the proposed scheme and good image quality of the embedded image.

Keywords: reversible data hiding, absolute moment block truncation coding, image compression, reference matrix.

1 Introduction

With the rapid and continuous development of the networking technologies, data communication over the Internet has become more and more popular. The security problems such as interception, modification, and duplication become an important issue in recent years. Steganography [1] and cryptography [2] are two typical technologies to fulfill the security and secrecy of the transmitted information. In this paper, we focus on techniques for hiding information in images.

Data hiding is a technique to embed copyright information or secret information into images, audio, video without perceivable degradation which can be broadly classified into two categories, i.e., irreversible data hiding [3]-[4] and reversible data hiding [5]-[6]. In the irreversible data hiding schemes, secret data is first embedded into the original cover images distorted and irreversible. On the contrary, the reversible data hiding schemes can extract the secret data, and

---

* Corresponding author.

L.S.-L. Wang et al. (Eds.): MISNC 2014, CCIS 473, pp. 427–436, 2014.
© Springer-Verlag Berlin Heidelberg 2014

recover the original cover images simultaneously. Nowadays, most reversible data hiding methods are derived from two principles, histogram shifting [7]-[8] and difference expansion [9].

To reduce the transmission volume while transmitting images over the Internet, most digital images are stored in compressed forms, such as JPEG and JPEG2000, or transmitted using VQ (vector quantization) [10] or BTC (block truncation coding) [11] techniques. BTC is a popular used lossy image compression prototype with low computation complexity, high efficiency and an acceptable compression rate. Absolute moment block truncation coding (AMBTC) [12] is a variant of BTC that can improve the performance of conventional BTC technique.

In this paper, we design a reversible data hiding scheme for the AMBTC-compressed images. Each block image is compressed into two quantization levels and a bitmap, and the secret data are embedded into the quantization levels of the compressed image blocks. The recovery results will still satisfy the original AMBTC-compressed version. The rest of this paper is organized as follows. In Section 2, we briefly describe the basic concept of AMBTC. Section 3 will contains a detailed introduction of the proposed algorithm. The experimental results and analysis will be presented in Section 4. Finally, some conclusions will be given in Section 5.

2   Related Works

2.1   Absolute Moment Block Truncation Coding (AMBTC)

To improve the compression performance, a variation of BTC called Absolute Moment Block Truncation Coding (AMBTC) is proposed [12]. The AMBTC consists of the image encoding and the image decoding procedures. In the image encoding procedure, a grayscale image to be compressed is first decomposed into a set of non-overlapping $n \times n$-sized blocks. It is noted that n is set to be 4 during the conventional AMBTC encoding phase. For each image block, its mean pixel value AVG and its standard deviation var are then computed by Eqs. (1) and (2), respectively.

$$AVG = \frac{\sum_{i=1}^{n \times n} x_i}{n \times n}. \tag{1}$$

$$var = \frac{\sum_{i=1}^{n \times n} \|x_i - AVG\|}{n \times n}, \tag{2}$$

where x_i represents the ith pixel value in each image block, AVG represents the pixel mean value, var denotes the standard deviation, and n denotes the size of the image block.

For each block, each pixel value x_i is compared with its AVG in such a way that a bitmap BM that is composed of two groups is generated according to the following two rules: If the pixel value x_i is less than its corresponding AVG, then the corresponding bit in the bitmap of the ith block is determined as group-0 and is denoted as 0 in the bitmap BM; otherwise, the bit belongs to group-1 and is denoted as 1 in the bitmap BM. Pixels in the same group will be encoded by the same quantization level. In other words, two quantization levels are be generated to represent the pixels in these two groups. Let L and H denotes the quantization levels in the group-0 and group-1, respectively. They can be derived by using the following equations:

$$L = AVG - \frac{n \times n \times var}{2(n \times n - q)} .$$ (3)

$$H = AVG + \frac{n \times n \times var}{2q} ,$$ (4)

where q is the number of pixels whose values are greater than or equal to AVG derived by using Eqs.(1). Each compressed image block generates a trio (L, H, BM), where L and H are the two quantization levels, and BM stands for the bitmap. By sequentially compressing each image block in the same way, the whole image is then compressed.

The image decoding procedure of AMBTC is the same as that of BTC. To reconstruct each compressed image block using the received trio (L, H, BM), the corresponding pixel is reconstructed by quantization level L if a corresponding bit valued 0 is found in the bitmap. Otherwise, it is recovered by quantization level H. When each image block is sequentially recovered by using the above-mentioned steps, the whole decoded image of AMBTC can be reconstructed.

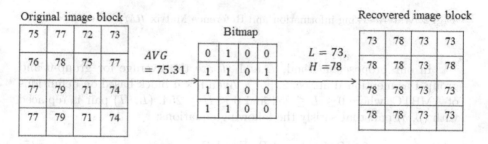

Fig. 1. An example of AMBTC image compression procedures with block size 4×4

3   The Proposed Scheme

3.1   Embedding Procedure

Suppose a $W \times H$ grayscale image is to be processed. It is divided into a set of non-overlapping $n \times n$ blocks, where W and H are the image size and n is the block size. A total of $w \times h$ compressed image blocks of AMBTC are stored where $w = \frac{W}{n}$ and $h = \frac{H}{n}$. That is to say, the $w \times h$ trios of (L, H, BM). Here, L and H are high and low quantization levels for these two groups. Let $mark$ denote the secret bits that will be embedded into the value of the quantization levels generated from a binary image which is shown in Fig. 2(a).

Phase 1: Embed secret data in quantization levels

To embed the secret data $mark$ into the compressed trio (L, H, BM), we utilize a Reference Matrix RM_B which is used for embedding $mark_B$ and $0 \leq mark_B \leq B-1$. In other words, bitstream of secret data are transformed into digits in base B. Each Reference Matrix contains B values ranging from 0 to $B-1$ and the number of each value is B. We choose $B = 3$ in our scheme. Fig. 2(b) depicts the Reference Matrix RM_B when $B = 3$.

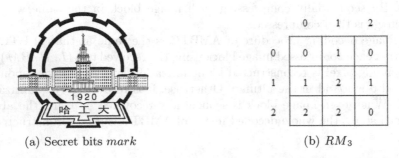

(a) Secret bits $mark$ (b) RM_3

Fig. 2. Watermarking information and Reference Matrix RM_B when $B = 3$

In our proposed method, L and H are the average for group-0 and group-1 as mentioned in Sec. 2. For a given $n \times n$ block based on definitions of AMBTC, where $0 \leq L \leq 254$ and $0 \leq H \leq 254$. (L, H) pair is replaced with (p, q) pair that satisfy the following equations:

$$RM_B (p \bmod B, q \bmod B) = mark_B . \tag{5}$$

Candidate for

$$(L, H) = \{(p, q) \mid |p - L| \leq B \cap |q - H| \leq B\} . \tag{6}$$

In our scheme, 1-bit $mark$ is hidden in each compressed image block. We can choose to embed secret data $mark$ into $mark_B$. Then to select the best candidate in all eligible (p, q)pair which satisfies Eq. (6), the incurred squared Euclidean distance for each candidate is computed. The candidate (p, q)pair with the least distortion is selected as the new (L', H')pair.

An example of the secret data embedding process is described in the following. In this example, the AMBTC compressed trio (73, 78, (010011011100

1100)$_2$) of the 4×4 image block as shown in Fig. 1 is used to embed the first secret data of 1-bit $mark$=0. According to reference matrix RM_3, 1-bit $mark$=0 can transformed into 3 base$mark_B = 0$. Here, we can choose to embed $mark$ into $mark_B$. Suppose using Eq. (6) as embedding equation, $RM_3\,(p \bmod 3,\, q \bmod 3) = 0$. Using Fig. 2(b) as RM_3, we find there are three options including $RM_3\,(0,0) = 0$, $RM_3\,(0,2) = 0$ and $RM_3\,(2,2) = 0$. Therefore, we have a table which lists all candidates as follows:

According to Eq. (6), $(avg0, avg1) = \{(p,q) |\, |p - 73| \le 3 \cap |q - 78| \le 3\}$, $70 \le p \le 76$ and $75 \le q \le 81$.

For Option **00**, $(p \bmod 3, q \bmod 3) = (0,0)$, there are four qualified candidates including $(p, q) = (72, 75)$, $(72, 78)$, $(72, 81)$, $(75, 75)$, $(75, 78)$, $(75, 81)$.

For Option **02**, $(p \bmod 3, q \bmod 3) = (0,2)$, there are four qualified candidates including $(p, q) = (72, 77)$, $(72, 78)$, $(75, 77)$, $(75, 78)$.

For Option **22**, $(p \bmod 3, q \bmod 3) = (2,2)$, there are four qualified candidates including $(p, q) = (71, 77)$, $(71, 80)$, $(74, 77)$, $(74, 80)$.

To select the best candidate in three options, the incurred squared Euclidean distance for each candidate is computed according to the following equation:

$$dist = (p - avg0)^2 + (q - avg1)^2 . \qquad (7)$$

After computing the incurred distortions of these candidates, the candidate with the least distortion is selected and the corresponding information of this candidate replace the original $(avg0, avg1)$pair. Therefore, the modified two quantization levels $(L', H') = (72, 78)$.

Phase 2: Embed quantization levels in bitmap

In order to satisfy the reversible in AMBTC-compressed images, we must remain the original two quantization levels that will be transmitted to receiver along with the compressed trio (L, H, BM).

As the low and high quantization level in each compressed image block are both integer falling into [0, 255] which can be represented by 8 bits. Denote the bits of L as L_0, L_1, L_2, L_3, L_4, L_5, L_6, L_7. Thus

$$L_k = \left\lfloor \frac{L}{2^k} \right\rfloor \bmod 2, k = 1, 2, ..., 7 . \qquad (8)$$

Do the same with H and obtain the sequence H_0, H_1, H_2, H_3, H_4, H_5, H_6, H_7. These two bit sequence are combined with $n \times n$ bits of the bitmap one-by-one. Then we joint the two bit into a new integer whose range is 0 to 3. There are four different cases for each position of new bitmap. Fig. 3 shows an example of how to generate a marked bitmap BM' after combining the bit sequence of quantization levels and original bitmap.

The detailed descriptions of the proposed embedding strategies for each case are listed as follows:

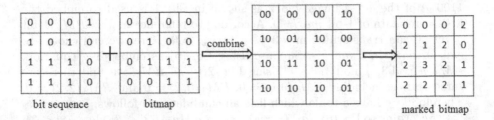

Fig. 3. Example of the generating a marked bitmap BM'

Case 00: If the quantization level bit is 0 and the corresponding bit in the original bitmap is 0, that belongs to Case 00. And the corresponding values of new bitmap is set to 0. The new quantization level L' is not changed.

Case 01: If the quantization level bit is 0 and the corresponding bit in the original bitmap is 1, that belongs to Case 01. And the corresponding values of new bitmap is set to 1. The new quantization level H' is not changed.

Case 10: If the quantization level bit is 1 and the corresponding bit in the original bitmap is 0, that belongs to Case 10. And the corresponding values of new bitmap is set to 2. The new quantization level L' is replaced by $L' + 1$.

Case 11: If the quantization level bit is 1 and the corresponding bit in the original bitmap is 1, that belongs to Case 11. And the corresponding values of new bitmap is set to 3. The new quantization level H' is replaced by H' - 1.

By successively modifying each (L, H) pair to embed secret data and quantization level bit, the embedding process is done.

3.2   Extracting Procedure

In the extracting phase, the secret data can be extracted and the original AMBTC-compressed image is recovered.

Input: A embedded AMBTC-compressed image and the marked bitmap BM'

Output: A reconstructed AMBTC-compressed image and extracted secret data

Phase 1: Extract secret data from the quantization levels

The extracted set of the secret data will be generated by using the marked bitmap BM' and reference matrix RM_B. Traversing each bitmap of $n \times n$ bits, if the bit is equal 0, the pixel value in the corresponding position of AMBTC-compressed image is regarded as L'. If the bit is equal 1, the value is regarded as H'. If the bit is equal 2, the value is regarded as $L' + 1$. If the bit is equal 3, the value is regarded as H' - 1. The extracted (L', H')pair replace (p, q)pair. According to Eqs. (6), we can get extracted secret data $emark_B$ by looking through the reference matrix RM_B.

Phase 2: Reconstruct AMBTC-compressed image based on the marked bitmap

Step 1. Reconstruct the quantization levels bits sequence. If the bit in marked bitmap BM' is 0 or 1, the corresponding quantization levels bit is 0. If the bit in marked bitmap BM' is 2 or 3, the corresponding quantization levels bit is 1.

Step 2. Rearrange the quantization levels bits sequence. The extracted bits sequence can be divided into two groups. Then, we transform each group into a integer number L and H which are the original quantization levels in each AMBTC-compressed image block.

Step 3. Reconstruct the AMBTC-compressed image by using the marked bitmap BM' and obtained quantization levels L and H. If the bit in marked bitmap BM' is 0 or 2, the reconstructed pixel value of this position is equal L. Otherwise, the reconstructed pixel value of this position is set as H.

4 Experimental Results and Discussion

To verify the efficiency of the proposed scheme, several simulations are performed in this section. All of the experiments were performed with six commonly used grayscale images,"Lena", "Airplane", "Girl" of the same size 512×512. In the experiments, the block size for AMBTC compression method was set as 4×4.

The peak signal-to-noise rate (PSNR) is used to measure the image quality. The computation of PSNR is defined as

$$PSNR = 10 \times \log_{10} \frac{255^2}{MSE} dB \ . \tag{9}$$

$$MSE = \frac{1}{(W \times H)} \sum_{i=1}^{W} \sum_{j=1}^{H} \left(x\left(i,j\right) - x'\left(i,j\right) \right)^2 \ . \tag{10}$$

In Eq. (10), $x\left(i,j\right)$ and $x'\left(i,j\right)$ indicate the pixel values for the position (i,j) of the original image and of the embedded AMBTC-compressed image, respectively, and W and H represent the width and height of the image.

Image qualities of the embedded images of the proposed scheme are listed in Table 1 when the block size is set to 4×4. From Table 1, average embedded image qualities of 33.97dB. In other words, the image quality losses of 0.15dB are obtained by the proposed scheme. Fig. 4-6 shows the embedded images and the corresponding PSNR values.

(a)

Fig. 4. Experimental results on Lena image, (a) the reconstructed image with AMBTC scheme, PSNR=34.40dB, (b) the embedded image with the proposed scheme, PSNR=34.21dB, (c) the reconstructed image with the proposed scheme, PSNR=34.40dB

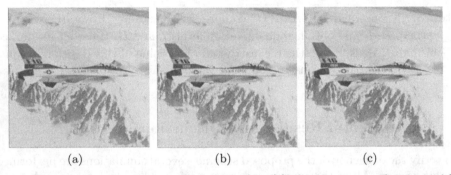

(a)   (b)   (c)

Fig. 5. Experimental results on Airplane image, (a) the reconstructed image with AMBTC scheme, PSNR=33.54dB, (b) the embedded image with the proposed scheme, PSNR=33.44dB, (c) the reconstructed image with the proposed scheme, PSNR=33.56dB

(a)   (b)   (c)

Fig. 6. Experimental results on Girl image, (a) the reconstructed image with AMBTC scheme, PSNR=34.40dB, (b) the embedded image with the proposed scheme, PSNR=34.25dB, (c) the reconstructed image with the proposed scheme, PSNR=34.40dB

Table 1. PSNR(dB) of traditional AMBTC image compression and proposed scheme

Images	AMBTC	Embedded image	Recovery image
Lena	34.40	34.21	34.40
Airplane	33.56	33.44	33.56
Girl	34.40	34.25	34.40
AVG	34.12	33.97	34.12

Experimental results show that the confidential data is embedded into the AMBTC-compressed filed, and can be extracted without distortion, results demonstrate that the proposed scheme not only can embed confidential data while maintains the same image quality as the original AMBTC-compressed version from Table 1.

5   Conclusions

In this paper, a novel reversible data hiding scheme for AMBTC-compressed image has been proposed. The secret data were embedded according to the proposed reference matrix RM. Experimental results show that the embedded images had high image quality after embedding and the original, AMBTC-compressed image can be recovered completely after extraction of the secret data.

References

1. Provos, N., Honeyman, P.: Hide and seek: An introduction to steganography. IEEE Security & Privacy 1(3), 32–44 (2003)
2. Diffie, W., Hellman, M.E.: New directions in cryptography. IEEE Transactions on Information Theory 22(6), 644–654 (1976)
3. Hu, Y.C.: High-capacity image hiding scheme based on vector quantization. Pattern Recognition 39(9), 1715–1724 (2006)
4. Yu, Y.H., Chang, C.C., Hu, Y.C.: Hiding secret data in images via predictive coding. Pattern Recognition 38(5), 691–705 (2005)
5. Chang, C.C., Lin, C.Y., Fan, Y.H.: Reversible steganography for btc-compressed images. Fundamenta Informaticae 109(2), 121–134 (2011)
6. Weng, S., Chu, S.C., Cai, N., Zhan, R.: Invariability of mean value based reversible watermarking. Journal of Information Hiding and Multimedia Signal Processing 4(2), 90–98 (2013)
7. Lin, C.C., Tai, W.L., Chang, C.C.: Multilevel reversible data hiding based on histogram modification of difference images. Pattern Recognition 41(12), 3582–3591 (2008)
8. Marin, J., Shih, F.Y.: Reversible data hiding techniques using multiple scanning difference value histogram modification. Journal of Information Hiding and Multimedia Signal Processing 5(3), 461–474 (2014)

9. Alattar, A.M.: Reversible watermark using the difference expansion of a generalized integer transform. IEEE Transactions on Image Processing 13(8), 1147–1156 (2004)
10. Gray, R.M.: Vector quantization. IEEE ASSP Magazine 1(2), 4–29 (1984)
11. Delp, E.J., Mitchell, O.R.: Image compression using block truncation coding. IEEE Transactions on Communications 27(9), 1335–1342 (1979)
12. Lema, M., Mitchell, O.R.: Absolute moment block truncation coding and its application to color images. IEEE Transactions on Communications 32(10), 1148–1157 (1984)

Towards Time-Bound Hierarchical Key Assignment for Secure Data Access Control

Tsu-Yang Wu[1,2], Chengxiang Zhou[1], Chien-Ming Chen[1,2], Eric Ke Wang[1,2], and Jeng-Shyang Pan[1,2]

[1] Shenzhen Graduate School, Harbin Institute of Technology,
Shenzhen, 518055, China
[2] Shenzhen Key Laboratory of Internet Information Collaboration,
Shenzhen, 518055, China
{wutsuyang,hitcms2009,chienming.taiwan,jengshyangpan}@gmail.com,
962982698@qq.com

Abstract. Time-bound hierarchical key assignment (TBHKA) scheme is a cryptographic method to assign encryption keys to a set of security classes in a partially ordered hierarchy. Only the authorized subscriber who holds the corresponding key can access the encrypted resources. In 2005, Yeh proposed a RSA-based TBHKA scheme which is suitable for discrete time period. However, it had been proved insecure against colluding attacks. Up to now, no such TBHKA schemes were proposed. In this paper, we fuse pairing-based cryptography and RSA key construction to propose a secure TBHKA scheme. In particular, our scheme is suitable for discrete time period. The security analysis is demonstrated that our scheme is secure against outsider and insider attacks (including colluding attacks). Finally, the performance analysis and comparisons are given to demonstrate our advantage.

Keywords: Access control, key assignment, bilinear pairings, Cryptography.

1 Introduction

The access control (AC) problem is to deal with users who can access some sensitive resources in a system. According to users' priority, users are organized in a hierarchy formed by several disjoint classes (called security classes). These classes have different limitations on the resources. In other words, some users own more access rights than others. In the real world, the AC problem is applied to several applications such as hospital system, computer system, etc.. For example, in computer system, administrator has the high priority to access all files (including sensitive files), but general users only access some common files. Up to now, several famous hierarchical key assignment schemes [1,3,5,6,10,11,13,14,16] had been published to solve the AC problem and address the data security.

In some situations, time-bound property may be involved in the AC problem such as Pay-TV system. In Pay-TV system, subscriber desires to subscribe

L.S.-L. Wang et al. (Eds.): MISNC 2014, CCIS 473, pp. 437–444, 2014.
© Springer-Verlag Berlin Heidelberg 2014

some channels for some certain time periods such as one week, one month, or one year. Hence, subscribers should be assigned different keys for each time period. If the time period expires, the subscriber should not derive any keys to access subscribed channels. Time-bound hierarchical key assignment scheme is a cryptographic method to assign encryption keys to a set of security classes in a partially ordered hierarchy, where the keys are dependent on the time. Note that if two classes form a relation, the subscriber who is in the higher class can access the resources in the lower class, however not vice versa.

In 2002, Tzeng [20] proposed the first time-bound hierarchical key assignment scheme by using Lucas function. However, Yi and Ye [27] pointed that his scheme suffered from a colluding attack in 2003. In 2004, Chien [9] proposed an efficient time-bound hierarchical key assignment scheme by using two hash values. Unfortunately, his scheme was also suffered from a colluding attack mentioned by Yi [26]. In 2005, Yeh [25] proposed an RSA-based hierarchical key assignment scheme. However, Ateniese et al. [2] pointed that Yeh's scheme [25] is insecure against colluding attack in 2006. Meanwhile, they proposed the unconditionally secure and computationally secure setting for a time-bound hierarchical key assignment scheme with a tamper-resistant device. In the same year, Wang and Laih [21] proposed a time-bound hierarchical scheme by using merging. In 2009, Sui et al. [18] proposed the first time-bound access control scheme for support dynamic access hierarchy. In 2012, Chen et al. [7] proposed a time-bound hierarchical key management scheme without tamper-resistant device. In the same year, Tseng et al. [19] proposed two pairing-based time-bound key management schemes without hierarchy. In their two schemes, one scheme combines Lucas function and is suitable for continuous time period. Another scheme fuses the RSA construction and is suitable for discrete time period. In 2013, Chen et al. [8] proposed the first hierarchical access control scheme in cloud computing. However, their scheme did not consider the time-bound property. Recently, Wu et al. [24] extended Chen et al.'s scheme [7] to propose the first time-bound hierarchical key management scheme in cloud computing.

Up to now, no secure time-bound hierarchical key assignment (TBHKA) scheme which is suitable for discrete time period is proposed. In this paper, we fuse pairing-based cryptography and RSA key construction to propose a secure TBHKA scheme. In particular, our scheme is suitable for discrete time period. The security analysis is demonstrated that our scheme is secure against outsider and insider attacks (including colluding attacks). Finally, the performance analysis and comparisons are given to demonstrate our advantage.

The rest of this paper is organize as follows: In Section 2, we introduce the concept of partially ordered hierarchy, bilinear pairings, and RSA cryptosystem. Our concrete scheme is proposed in Section 3. In Section 4, we demonstrate the security analysis of our scheme. The performance analysis is given in Section 5 and the conclusions are draw in Section 6.

2   Preliminaries

In this section, we brief review the concept of partially ordered hierarchy, bilinear pairings, and the RSA cryptosystem.

2.1   Partially Ordered Hierarchy

Consider a set of resources organized into a number of disjoint classes. A binary relation $\preceq$ partially orders the set of classes $\mathfrak{C}$. The pair $(\mathfrak{C}, \preceq)$ is called a partially ordered hierarchy. For any two classes C_i and C_j in $\mathfrak{C}$, the notation $C_j \preceq C_i$ means that the user in C_i can access the resource in C_j and the opposite is forbidden. It is easy to see that $C_i \preceq C_i$ for any $C_i \in \mathfrak{C}$. The partially ordered hierarchy $(\mathfrak{C}, \preceq)$ can be represented by a directional graph, where each class corresponds to a vertex in the graph and there exists an edge form class C_j to C_i if and only if $C_j \preceq C_i$. For the detailed descriptions about partially ordered hierarchy, readers can refer to [2,17].

2.2   Bilinear Pairings and Its Security Assumptions

Let G_1 and G_2 be two groups with a same large prime order q, where G_1 is an additive cyclic group and G_2 is a multiplicative cyclic group. A bilinear pairing e is a map defined by $e : G_1 \times G_1 \rightarrow G_2$ which satisfies the following three properties:

1. Bilinear: For all $P, Q \in G_1$, $a, b \in \mathbb{Z}_q^*$, we have $e(aP, bQ) = e(P, Q)^{ab}$.
2. Non-degenerate: For all $P \in G_1$, there exists $Q \in G_1$ such that $e(P, Q) = 1_{G_2}$.
3. Computable: For all $P, Q \in G_1$, there exists an efficient algorithm to compute $e(P, Q)$.

The detailed descriptions for bilinear pairings can be referred to [4,22,23].

2.3   Integer Factorization Problem and RSA Cryptosystem

As we all known, given two large prime number p and q to compute $n = p \times q$ is easy. However, given a value n to find p and q is intractable. It is well-known the integer factorization problem.

The security of RSA cryptosystem is based on the difficulty of integer factorization problem. In this cryptosystem, the two large primes p and q are selected firstly and then the two values $n = p \times q$ and $\phi(n) = (p - 1) \cdot (q - 1)$ can be computed. Then, a public value e is selected which satisfies $\gcd(e, \phi(n)) = 1$ and $1 < e < \phi(n)$. According to e, a secret value d can be chosen which satisfies $e \cdot d \equiv 1 \mod \phi(n)$. Note that given two values n and e, an adversary without p and q is unable to compute the secret value d. The detailed descriptions for the integer factorization problem and the RSA Cryptosystem can be referred to [12,15].

3   A Concrete Scheme

In this section, we propose a concrete time-bound hierarchical key assignment scheme for secure data access control. The proposed scheme combines the pairing-based public key system with the RSA cryptographic method and is suitable for subscribers in discrete time intervals. In our scheme, we assume that each user can access some resources in some discrete time interval T_i such as one weak, one month, etc.. These resources are stored in a set of classes $\mathfrak{C}$. Without loss of generality, the maximal system life time is defined as $T = \{1, 2, \ldots, z\}$, ie. $T_i \subset T$ and there are n classes, $\mathfrak{C} = \{C_1, C_2, \ldots, C_n\}$. Note that the n classes form a directional graph with the relation $\preceq$ mentioned in Subsection 2.1. The proposed scheme consists of following four phases: *System setup, User subscribing, Encryption key generation*, and *Decryption key derivation* phases.

System setup phase: Firstly, the system vender (SV) constructs a set of classes $\mathfrak{C}$ and deploys the resources into n classes. In other words, a directional graph $(\mathfrak{C}, \preceq)$ is produced. Then, the SV generates the needed keys and parameters as follows. The SV selects a bilinear pairing $e : G_1 \times G_1 \to G_2$ mentioned in Subsection 2.2. A generator $P \in G_1$ is generated and then a public value $P_{pub} = s \cdot P$ is computed, where $s \in \mathbb{Z}_q^*$ is a secret value kept by the SV. Meanwhile, the system vender selects two prime numbers p_1, q_1 and computes $n = p_1 \times q_1$ and $\phi(n) = (p_1 - 1) \cdot (q_1 - 1)$. Then, the SV determines two RSA key pairs (e_i, d_i) and (g_t, h_t) such that $e_i \cdot d_i \equiv 1 \bmod \phi(n)$ and $g_t \cdot h_t \equiv 1 \bmod \phi(n)$ for $i = 1, 2, \ldots, n$ and $t = 1, 2, \ldots, z$, where d_i and h_t are kept secret. Finally, the SV defines a cryptographic hash function $H : \{0, 1\}^* \to \mathbb{Z}_q^*$ and publishes the public parameters $\{e, G_1, G_2, q, P, P_{pub}, n, e_1, \ldots, e_n, g_1, \ldots, g_s, H\}$.

User subscribing phase: When a user subscribes class C_i to access some resource in some time period $T_i \subset T = \{1, 2, \ldots, z\}$, the system vender computes a key pair $(\alpha = a \prod_{C_k \preceq C_i} d_k P, \beta = e(P, P)^{\prod_{y \in T_i} h_y})$, where $a \in \mathbb{Z}_q^*$ is a secret value kept by the SV. Finally, (α, β) is sent to user via a secure channel.

Encryption key generation phase: For each time $t \in T = \{1, 2, \ldots, z\}$, the system vender computes a encryption key $K_{i,t} = H(k_i \| k_t)$ to protect the resource in class C_i, where

$$k_i = e(\prod_{C_k \preceq C_i} d_k P, P_{pub})^a = e(P, P)^{\prod_{C_k \preceq C_i} sad_k} \text{ and } k_t = e(P, P)^{h_t}.$$

Note that we can use the symmetric encryption algorithm such as AES with the key $K_{i,t}$ to encrypt the resource in class C_i for time period t.

Decryption key derivation phase: For any user who is in class C_i with her/his subscribing time period T_i, she/he can compute the decryption key $K_{j,t} = H(k_j \| k_t)$ of class C_j if and only if $C_j \preceq C_i$ and $t \in T_i$. The key derivation is shown as follows:

$$k_j = e(\alpha, P_{pub})^{\prod_{C_k \preceq C_i, C_k \npreceq C_j} e_k} = e(a \prod_{C_k \preceq C_i} d_k P, s \cdot P)^{\prod_{C_k \preceq C_i, C_k \npreceq C_j} e_k}$$
$$= e(P, P)^{\prod_{C_k \preceq C_j} s a d_k}$$

and

$$k_t = (\beta)^{\prod_{y \in T_i, y \neq t} g_y} = e(P, P)^{h_t}.$$

4 Security Analysis

In this section, we demonstrate the security of our proposed scheme. It is easy to see that the security of our scheme is based on the computation of both k_i and k_t because the encryption key $K_{i,t} = H(k_i || k_t)$. In the following Lemmas 1 and 2, we will demonstrate the security of k_i and k_t, respectively.

Lemma 1. *Under the security of the RSA cryptosystem, the value k_i for $i \in \{1, 2, \ldots, n\}$ of the proposed scheme is secure against outside and inside attacks.*

Proof. Here, the security proof of Lemma 1 is divided into following three parts.

Part 1. *Any outside attacker cannot compute the value k_i.* An outside attacker only knows the public values $e_1, e_2, \ldots, e_n$, and P_{pub}. Since the value $k_i = e(\prod_{C_k \preceq C_i} d_k P, P_{pub})^a$ is generated by the secret values $d_1, d_2, \ldots, d_n$, and a, she/he has no way to know them. In other aspect, the security of d_k relies on the security of the RSA cryptosystem. The pair (e_i, d_i) is a pubic/private key pair and nobody can derive d_i from e_i.

Part 2. *Any legal user with the value k_j still cannot derive the value k_i for the two cases: (1) $C_j \preceq C_i$ and (2) $C_j \npreceq C_i$.* For the case 1, the key point is how to find d_i such that $k_i = (k_j)^{d_i}$. However, it is impossible by the same reason mentioned in Part 1. Similarly, to find d_i such that $k_i = (k_j)^{e_j \cdot d_i}$ is also impossible for the case 2.

Part 3. *The value k_i is secure against colluding attacks.* Without loss of generality, assume that two legal users with the two values k_j and k_l and they want to derive the value k_i for the two cases: (1) $C_l \preceq C_j \preceq C_i$ and (2) $C_l \preceq C_i$ and $C_j \preceq C_i$. For the case 1, to compute k_i they must find d_i or d_j such that $k_i = (k_j)^{d_i} = (k_l)^{d_i \cdot d_j}$. However, it is impossible by the same reason mentioned in Part 1. Similarly, it is also impossible for the case 2.

Lemma 2. *Under the security of the RSA cryptosystem, the value k_t for $i \in \{1, 2, \ldots, s\}$ of the proposed scheme is secure against outside and inside attacks.*

Proof. By the similar approach in Lemma 1, we can prove (a) *any outside attacker cannot compute the value k_t*, (b) *any legal user with the value k_{t_1} still cannot derive the value k_{t_2} for $t_1 \neq t_2$*, and (c) *the value k_t is secure against colluding attacks.*

Based on the above two lemmas, the following theorem demonstrate the proposed scheme is a secure time-bound hierarchical key assignment scheme.

Theorem 1. *Under the security of the RSA cryptosystem and the security of hash function, any outside and inside attackers cannot compute the encryption key $K_{i,t} = H(k_i \| k_t)$.*

Proof. By Lemmas 1 and 2, we have proven that k_i and k_t are secure against outside and inside attacks. If the outside and inside attackers can obtain a value $v = k_i \| k_t$ such that $K_{i,t} = H(v)$, it is a contradiction for the security property "collusion resistance" of the hash function H.

5 Performance Analysis and Comparisons

For convenience to evaluate the performance of our scheme, we define the following notations:

- TG_e: The time of executing a bilinear pairing operation, $e : G_1 \times G_1 \to G_2$.
- TG_{mul}: The time of executing a scalar multiplication operation of point in G_1.
- T_{exp}: The time of executing a modular exponentiation operation.
- T_{mul}: The time of executing a modular multiplication operation.
- T_H: The time of executing a one-way hash function H.
- T_{syme}: The time of executing a symmetric encryption algorithm.
- d: The path length between the subscribing class and its lower level classes.
- l: The number of subscribing time interval.

In the user subscribing phase, $TG_e + TG_{mul} + T_{exp} + (d+1)T_{mul}$ is required to compute (α, β). In the encryption key generation phase, it requires $TG_e + TG_{mul} + 2T_{exp} + dT_{mul} + T_H$ to compute $K_{i,t} = H(k_i \| k_t)$. In the decryption key derivative phase, $TG_e + 2T_{exp} + (d+l-2)T_{mul}$ is required to derive $K_{i,t} = H(k_i \| k_t)$.

Table 1. Comparisons between our scheme and the recent proposed time-bound hierarchical key assignment schemes

	Chen et al.'s scheme [7]	Yeh's scheme [25]	Our scheme
Key construction	Pairing-based	RSA	Pairing-based + RSA
Type of time interval	Continuous	Discrete	Discrete
User subscribing	TG_{mul} $+2T_{exp} + 2T_{mul}$	$2T_{exp}$ $+(d+l-1)T_{mul}$	$TG_e + TG_{mul}$ $+T_{exp} + (d+1)T_{mul}$
Encryption key generation	$TG_e + 3T_{exp}$ $+2T_{mul} + T_{syme}$	$2T_{exp} + dT_{mul}$	$TG_e + TG_{mul}$ $+2T_{exp} + dT_{mul} + T_H$
Decryption key derivative	$TG_e + T_{syme}$	$2T_{exp}$ $+(d+l-2)T_{mul}$	$TG_e + 2T_{exp}$ $+(d+l-2)T_{mul}$
Security	Provably secure	Existing attack [2]	Provably secure

Then, we compare the recent presented time-bound hierarchical key assignment schemes [7,25] in terms of key construction, performance, and security properties. The results are summarized in Table 1. We can see that Yeh's scheme [25] is based on the RSA key construction, Chen et al.'s scheme [7] is based on the pairing-based key construction, and our scheme fuses the pairing-based and the RSA key constructions. In other aspect, Chen et al.'s scheme focuses on continuous time interval. Our scheme and Yeh's scheme are suitable for discrete time interval. Though Yeh's scheme is efficient, it suffered from colluding attack mentioned in [2]. Our scheme and Chen et al.'s scheme are provably secure.

6 Conclusions

In this paper, we have proposed a time-bound hierarchical key assignment scheme. Our scheme fuse pairing-based cryptography and RSA key construction and is suitable for discrete time interval. The security analysis is demonstrated that our scheme is secure against and outsider and insider attacks (including colluding attacks). In the future, we will extend our scheme to the cloud environments.

Acknowledgments. This work is supported by Shenzhen Peacock Project of China (No. KQC201109020055A), Shenzhen Strategic Emerging Industries Program of China (No. ZDSY20120613125016389 and No. JCYJ20120613151032592), and National Natural Science Foundation of China (No. 61100192).

References

1. Akl, S.G., Taylor, P.D.: Cryptographic solution to a problem of access control in a hierarchy. ACM Transactions on Computer Systems (TOCS) 1(3), 239–248 (1983)
2. Ateniese, G., De Santis, A., Ferrara, A.L., Masucci, B.: Provably-secure time-bound hierarchical key assignment schemes. Journal of Cryptology 25(2), 243–270 (2012)
3. Blanton, M., Fazio, N., Frikken, K.B.: Dynamic and efficient key management for access hierarchies. In: Proceedings of the ACM Conference on Computer and Communications Security (2005)
4. Boneh, D., Franklin, M.: Identity-based encryption from the weil pairing. SIAM Journal on Computing 32(3), 586–615 (2003)
5. Chen, C.M., Lin, Y.H., Lin, Y.C., Sun, H.M.: Rcda: recoverable concealed data aggregation for data integrity in wireless sensor networks. IEEE Transactions on Parallel and Distributed Systems 23(4), 727–734 (2012)
6. Chen, C.M., Wang, K.H., Wu, T.Y., Pan, J.S., Sun, H.M.: A scalable transitive human-verifiable authentication protocol for mobile devices. IEEE Transactions on Information Forensics and Security 8(8), 1318–1330 (2013)
7. Chen, C.M., Wu, T.Y., He, B.Z., Sun, H.M.: An efficient time-bound hierarchical key management scheme without tamper-resistant devices. In: 2012 International Conference on Computing, Measurement, Control and Sensor Network (CMCSN). pp. 285–288. IEEE (2012)
8. Chen, Y.-R., Chu, C.-K., Tzeng, W.-G., Zhou, J.: CloudHKA: A cryptographic approach for hierarchical access control in cloud computing. In: Jacobson, M., Locasto, M., Mohassel, P., Safavi-Naini, R. (eds.) ACNS 2013. LNCS, vol. 7954, pp. 37–52. Springer, Heidelberg (2013)

9. Chien, H.Y.: Efficient time-bound hierarchical key assignment scheme. IEEE Transactions on Knowledge and Data Engineering 16(10), 1301–1304 (2004)
10. Jiang, T., Zheng, S., Liu, B.: Key distribution based on hierarchical access control for conditional access system in dtv broadcast. IEEE Transactions on Consumer Electronics 50(1), 225–230 (2004)
11. Kayem, A.V., Martin, P., Akl, S.G.: Heuristics for improving cryptographic key assignment in a hierarchy. In: 21st International Conference on Advanced Information Networking and Applications Workshops, AINAW 2007, vol. 1, pp. 531–536. IEEE (2007)
12. Lenstra, A.K.: Integer factoring. Designs, Codes and Cryptography 19, 101–128 (2000)
13. Lin, C.W., Hong, T.P., Chang, C.C., Wang, S.L.: A greedy-based approach for hiding sensitive itemsets by transaction insertion. Journal of Information Hiding and Multimedia Signal Processing 4(4), 201–227 (2013)
14. Lin, C.W., Hong, T.P., Hsu, H.C.: Reducing side effects of hiding sensitive itemsets in privacy preserving data mining. The Scientific World Journal 2014, Article ID 235837, 12 pages (2014)
15. Menezes, A.J., Van Oorschot, P.C., Vanstone, S.A.: Handbook of applied cryptography. CRC Press (2010)
16. Naor, D., Naor, M., Lotspiech, J.: Revocation and tracing schemes for stateless receivers. In: Kilian, J. (ed.) CRYPTO 2001. LNCS, vol. 2139, pp. 41–62. Springer, Heidelberg (2001)
17. Sandhu, R.S., Samarati, P.: Access control: principle and practice. IEEE Communications Magazine 32(9), 40–48 (1994)
18. Sui, Y., Maino, F., Guo, Y., Wang, K., Zou, X.: An efficient time-bound access control scheme for dynamic access hierarchy. In: 5th International Conference on Mobile Ad-hoc and Sensor Networks, MSN 2009, pp. 279–286. IEEE (2009)
19. Tseng, Y.M., Yu, C.H., Wu, T.Y.: Towards scalable key management for secure multicast communication. Information Technology and Control 41(2), 173–182 (2012)
20. Tzeng, W.G.: A time-bound cryptographic key assignment scheme for access control in a hierarchy. IEEE Transactions on Knowledge and Data Engineering 14(1), 182–188 (2002)
21. Wang, S.Y., Laih, C.S.: Merging: an efficient solution for a time-bound hierarchical key assignment scheme. IEEE Transactions on Dependable and Secure Computing 3(1), 91–100 (2006)
22. Wu, T.Y., Tsai, T.T., Tseng, Y.M.: A revocable id-based signcryption scheme. Journal of Information Hiding and Multimedia Signal Processing 3(3), 240–251 (2012)
23. Wu, T.Y., Tseng, Y.M.: An id-based mutual authentication and key exchange protocol for low-power mobile devices. The Computer Journal 53(7), 1062–1070 (2010)
24. Wu, T.-Y., Zhou, C., Wang, E.K., Pan, J.-S., Chen, C.-M.: Towards time-bound hierarchical key management in cloud computing. In: Pan, J.-S., Snasel, V., Corchado, E.S., Abraham, A., Wang, S.-L. (eds.) Intelligent Data Analysis and Its Applications, Volume I. AISC, vol. 297, pp. 31–38. Springer, Heidelberg (2014)
25. Yeh, J.H.: A secure time-bound hierarchical key assignment scheme based on rsa public key cryptosystem. Information Processing Letters 105(4), 117–120 (2008)
26. Yi, X.: Security of chien's efficient time-bound hierarchical key assignment scheme. IEEE Transactions on Knowledge and Data Engineering 17(9), 1298–1299 (2005)
27. Yi, X., Ye, Y.: Security of tzeng's time-bound key assignment scheme for access control in a hierarchy. IEEE Transactions on Knowledge and Data Engineering 15(4), 1054–1055 (2003)

Research of Automated Assessment of Subjective Tests Based on Domain Ontology

Lu-Xiong Xu[1,2], Na Wang[1], Lin Xu[2,3,*], and Li-Yao Li[1,2]

[1] The School of Mathematics and Computer Science
Fuqing Branch of Fujian Normal University, China
[2] The Institute of Innovative Information Industry
Fuqing Branch of Fujian Normal University, China
[3] The School of Ecomomic, Fujian Normal University, China
xulin@fjnu.edu.cn

Abstract. Automated scoring technology aims to reduce the workload of homework or examinations and to ensure the fairness. Therefore, to study automated scoring technology and its implementation process has great practical meaning. A method of automated assessment of subjective tests based on domain ontology and corpus is proposed to solve the existence problems of automated scoring systems. The conception of domain ontology is introduced and a software engineering domain is built constructed. Some key technologies includes Chinese word segmentation, the TF-IDF algorithm which is utilized to calculate the importance of each keyword in texts and text similarity calculation are described in detail. The method mentioned in the paper has been applied in the automated assessment of short-answer questions of software engineering. Comparison between the results made by the automatic scoring system and the teachers proves reasonableness of the model.

Keywords: Domain Ontology, Chinese Word Segmentation, Automated Assessment, Semantic similarity calculation.

1 Background

The online examination system have been widely used in the actual teaching due to its flexible and efficient features.Examinations are often divided into objective and subjective questions.It uses examinees' answer to comparing with standard answer to achieve marking objective questions,but subjective questions are usally marked by artificial way which is easily affected by marking people's fatigue and mood,other factors like examinees' writing wheather is clearly,so it is difficult to avoid subjectivity. Automated scoring can effectively avoid the subjectivity of manual scoring,and it's more objective ,impartial and efficient.In this respect, domestic and foreign scholars have done a lot of research. Mark Warschauer , etc have developed an automatic essay score (AWE) software, it uses artificial intelligence to evaluate

---

* Corresponding author.

L.S.-L. Wang et al. (Eds.): MISNC 2014, CCIS 473, pp. 445–453, 2014.

papers,and generates feedback[1]. Roth V, etc have developed a job evaluation system based on WEB which is mainly used in advanced mathematics and science courses,the system records the student's behavior change[2]. Jinrong Li, etc have designed and implemented an automatic scoring system which was using C programming language and based on the semantic matching,it gives test score by matching level and operating results between the program and the standard answer[3]. Islam , etc used the generalized latent semantic analysis to develop an automated essay scoring system in Bangladesh[4].Tian Lan Liu takes sample questionsas of NCRE Level Two (VFP) and subjective exam questions of Universities VFP programming as background,and one of kind which is much regular form of the program questions is considered good for the study, on this basis he proposed a machine scoring algorithm combined static semantic understanding with dynamic operation analysis[5]. In order to reduce the workload of homework or examinations, on the basis of previous research,my team proposed an automatic scoring method of assessment of subjective tests based on domain ontology, it builds domain ontology by introducing ontology learning techniques,then extracts relevant concepts contained in the domain ontology and was applied to the automatic evaluation system based on the domain ontology.

2   Overview of Domain Ontology

2.1   Definition of Ontology and Domain Ontology

With the development of computer technology, communication technology and the internet, organizing and managing vast amounts of information is particularly important for every computer user. As a conceptual modeling tool based on semantic and knowledge levels,ontology emerges at this time.Ontology is originated from Philosophy,it's a concept belongs to philosophical categories and generally believed to originated in Ancient Greece which was proposed by Aristotle and Plato. Ontology is a question of primitive things or phenomena of all of the objective world , defined as "the world of objective existence of system description, also called ontological"[6]. Domain ontology is used to describe a specially designated areas of knowledge,it's one kind of Specialized ontology,it gives the physical domain concepts ,relationships, areas activity, characteristics and rules of a formal description in this field [7].

2.2   Construct the Software Engineering Field of Ontology

Software engineering is an important IT industry of courses which has strong practical features, it has played a positive role in promoting stuents' software development capabilities and software project management capabilities. Building software engineering ontology needs the glossary of terms , variety of semantic resources and reusing of existing terminology and ontology relationships, uses the experts' artificial construct in the field as necessary and useful supplement,this can make the building results as accurate and professional as possible.

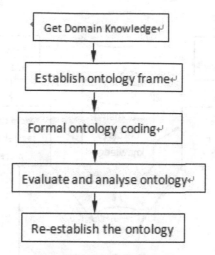

Fig. 1. Domain ontology building process

Ontology build process as shown in figure 1.First, getting the concept of software engineering disciplines is a huge project, so before creating the ontology it must define the coverage areas and objectives of subject. The method used to get ontology information adoptes on the basis of existing knowledge of domain ontology and is completed through the active participation of experts in the field. Second, extracting important terms and concepts, then defining knowledge of software engineering courses and contacts between them for creating their instances, establishing ontology frame. Third, using OWL to describe ontology and formally code for ontology. Fourth, the ontology knowledge of profession and accuracy are given scientific evaluation and analysis. Due to the structural design of onlology is refer to artificial parts, its quality is affected by the designers' extent of knowledge and degree of experience, and the description language of onlology needs not only accuracy , easily to understanding but also to avoid ambiguity, so to build a good ontology for evaluating is very necessary. Finally, we need modify and improve the system. The onlology is not built in a day, will be a long process. This requires a number of experts to participate in the field, through various discussions and exchanges, finding and fixing the problem from discussions, continuously improving the construction of the ontology.

3   The Overall Structure of Subjective Questions Scoring System Based on Domain Ontology

When teachers are marking papers in the manual way, generally give scores by matching score points between examinees' answer and standard answer combined with personal experience. So the key words of examinees' answer became one of the basis for scoring. The overall structure of subjective questions scoring system based on domain ontology as shown in figure 2.

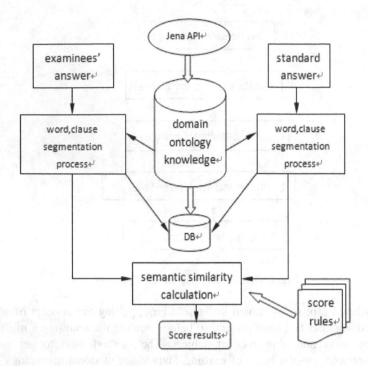

Fig. 2. The overall structure of subjective questions scoring system based on domain ontology

Using computers to mimic human intelligence to realize automatic scoring needs the following steps:

STEP 1: examinees' answer and standard answer must be processed respectively as clause and word . maybe there are characters such as spaces and enters in examinees' answer which are nothing to do with the content of the expression, so it's necessary to delete this part of the useless characters.The signification of process is define the mode of clause,also the clause of examinees' and standard answer which is not relevant should be deleted.Because of the own characteristics of chinese is unlike english word that has a direct seperation effect, so it need automatic word processing, the purpose is to make sure the keywords of examinees' answer and standard answewer can be matched after processed.

STEP 2: The system apply the concept model basedon domain ontology as a tool to calculate semantic similarity between words in text which are written in natural language by examinees and ontology,thus get the similarity of sentences,last we can get the similarity of answer.The calculation of similarity is shown as figure 3.

STEP 3: Jena API is used to build a good domain ontology which is mapping to relational databases, so the ontology database is transformed into MYSQL database.

STEP 4: It gets scoring results through combining semantic similarity calculation and the certain scoring rules.

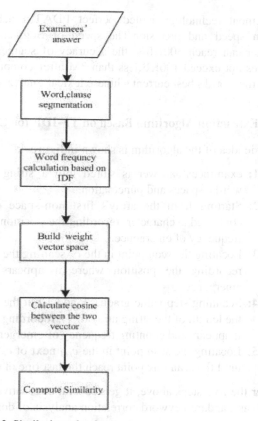

Fig. 3. Similarity calcution process

4 Key Technologies

4.1 Segmentation Technology

Segmentation algorithms include methods based on string matching,understanding and statistics etc. Each method has advantages and disadvantages,in order to achieve better segmentation results, the article uses a method of Chinese word segmentation tool of Chinese Academy of Sciences which is based on string matching to reach a preliminary segmentation.

Whether segmentation system can meets the practical requirements depends on two factors: segmentation accuracy and speed of analysis,they are mutual restraint,so it's difficult to get balance. Institute of Computing Technology of Chinese Academy has developed a Chinese Lexical Analysis System called ICTCLAS which was based on accumulation of years of research,and developed a large-scale knowledge base management technology called perfect PDAT,it achieved a major breakthrough between speed and precision.The speed of Segmentaion in ICTCLAS of single computer can reach 500KB/s, the accuracy of segmentation can reach 98.45%,also API does not exceed 100KB,less than 3M after compression of variety of dictionary data, is the world's best current Chinese lexical analyzer[8].

4.2   Extraction Algorithm Based on TF-IDF for Chinese Keywords

The basic idea of the algorithm is shown in figure 4:

STEP 1: examinees' answer is stored into a string array, and removed function words, spaces and punctuation.

STEP 2: Starting from the array's first non-space character c_i,scaning the array contained c_i character, recording the position where it appears and counting frequency of emergence.

STEP 3: Locating the scan point in the c_i, scaning the array contained $c_i c_{i+1}$ character, recording the position where it appears and counting frequency of emergence.

STEP 4: Scanning step value gradually add 1,but the maximum length can't over of the length of the string array of 1), recording the position of $c_i c_{i+1}...c_k$ where it appears and counting frequency of emergence.

STEP 5: Locating the scan point in the c_{i+1} next of c_i, repeating the step of 3) and 4) until the scanning point reach the last one of the array.

After the five steps above, it gets several alternative concepts in the field. Use the TF-IDF as standard keyword correlation analysis of domain ontology[9]:

$$W_{ij} = TF_{i,j} \times IDF_i \qquad (1)$$

For a particular word tf_i,

$$TF_{i,j} = \frac{tf_{i,j}}{MAX(tf_{i,j})} \qquad (2)$$

$tf_{i,j}$ is the times of the word appeared in the document j, IDF is a common measure of the importance of words. IDF is calculated by taking the logarithm which is generated through total number of document file divided the number of document file that contains the words, can be expressed as:

$$IDF_i = \log\left(\frac{N}{n_i}\right) \qquad (3)$$

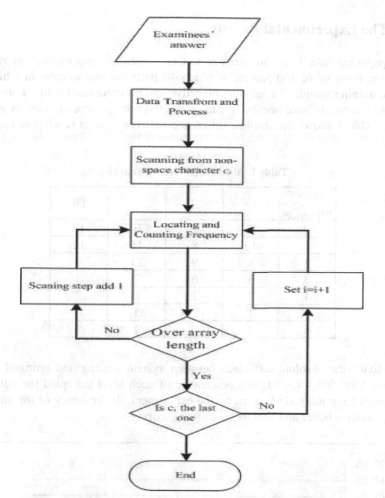

Fig. 4. Extraction algorithm based on TF-IDF for chinese keywords

4.3   Text Similarity Computing

Assuming K is thesaurus, examinees' answer is A, K_i and A_i represent respectively of the weight of the i-th feature items, $1<=i<=n$, n is the dimensions of feature vectors. Similarity $Sim\ (K, A)$ calculated between A and K can be represented by the cosine of the angle between the two vectors, the formula is:

$$sim(K, A) = \frac{\sum_{i=1}^{n} A_i \times K_i}{\sqrt{\left(\sum_{i=1}^{n} A_i^2\right)\left(\sum_{i=1}^{n} K_i^2\right)}} \tag{4}$$

5   The Experimental Results

In the paper,we take four short-answer test of "software engineering" as research object, receive total of 160 papers, of which 40 parts are test sample, of which 120 parts are training sample. To get more objective and accurate results, the short-answer test of all the papers have beenby five teachers, the average score is used in artificial scoring. Table 1 shows the absolute difference between system scoring and artificial scoring.

Table 1. Absolute Difference Statistics

question difference	一	二	三	四
0	11	14	10	13
1	13	8	15	12
2	8	9	11	9
3	5	6	2	3
4	2	1	0	2
5	1	2	1	0
6	0	0	1	1

We divide the absolute difference between system scoring and artificial scoring into three level 0-2,3-4,5-10,get a percentage of each level occupied the number of topic which have marked. As seen by the experiment, the accuracy of the automatic scoring model is better,and is close to artificial scoring.

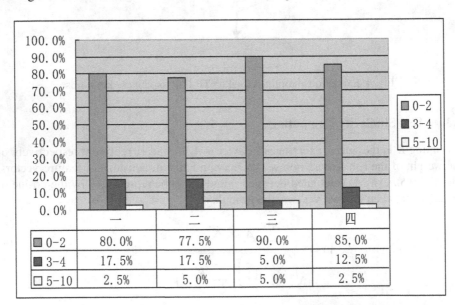

	一	二	三	四
■ 0-2	80. 0%	77. 5%	90. 0%	85. 0%
■ 3-4	17. 5%	17. 5%	5. 0%	12. 5%
□ 5-10	2. 5%	5. 0%	5. 0%	2. 5%

Fig. 5. The percentage of each level of absolute difference

6   Conclusion

In order to make teachers get free from the workload of homework or examinations, the paper proposes an Automatic Scoring System based on domain ontology, further,for the purpose of achieving better segmentation results, we uses a method of Chinese word segmentation tool of Chinese Academy of Sciences which is based on string matching,and get a good balance between accuracy and speed. The system contains some parts such as domain ontology building process,chinese word segmentation,keywords extraction and text similarity calculation which are described in detail. The model can simulate the process of artificial scoring,can give automatically score to short-answer,and the results have a high accuracy rate. On the other hand, the paper is about automatic marking for short-answer of software engineering,the scope of the research is relatively limited. In future work, to improve the accuracy of text similarity calculation,we intends to expand the scope of the research and make further supplement and complement for software engineering ontology which has established.

Acknowledgement. This research was partially supported by the Department of Fujian Provincial Education, under grant JK2012063, by the the School of Mathematics and Computer Science of Fujian Normal University,by the the Institute of Innovative Information Industry of Fujian Normal University,by the The School of Ecomomic of Fujian Normal University.

References

1. Mark, W., Douglas, G.: Automated Writing Assessment in the Classroom. Technology and Literacy 3(1), 22–36 (2008)
2. Roth, V., Ivanchenko, V., Record, N.: Evaluating student response to WeBWorK, a web-based homework delivery and grading system. Computers & Education 50(4), 1462–1482 (2008)
3. Li, J., Pan, W., Zhang, R., Chen, F., Nie, S., He, X.: Design and Implementation of Semantic Matching Based Automatic Scoring System for C Programming Language. In: Zhang, X., Zhong, S., Pan, Z., Wong, K., Yun, R. (eds.) Edutainment 2010. LNCS, vol. 6249, pp. 247–257. Springer, Heidelberg (2010)
4. Monjurul Islam, M., Latiful Hoque, A.S.M.: Automated Bangla essay scoring system: ABESS. In: 2013 International Conference on ICIEV, pp. 1–5 (2013)
5. Liu, T.-L.: The research on automatic scoring algorithm of program topic based on semantic understanding and operational analysis. Hunan Normal University, ChangSha (2013)
6. Maeche, A.: Ontology Learning for the Semantic Web. Kluwer Academic Publishers, Norwell (2002)
7. Zhu, H.-M., Ji, X.-L., Huang, W.-D., et al.: Research on the method of constructing Telecom domain ontology. Modern Information (1) (2008)
8. http://www.ictclas.org/ictclas_feature.html
9. Chen, R.-C., Lee, I.-Y., Lee, Y.-C., et al.: Upgrading domain ontology based on latent semantic analysis and group center similarity calculation. In: IEEE International Conference on SMC, pp. 1495–1500 (2008)

Author Index